AF270438

Requiem for the Santa Cruz

Requiem for the Santa Cruz

An Environmental History of an Arizona River

ROBERT H. WEBB, JULIO L. BETANCOURT,
R. ROY JOHNSON, AND RAYMOND M. TURNER

FOREWORD BY BERNARD L. FONTANA

THE UNIVERSITY OF
ARIZONA PRESS

TUCSON

The University of Arizona Press
© 2014 The Arizona Board of Regents
All rights reserved

www.uapress.arizona.edu

Library of Congress Cataloging-in-Publication Data
Webb, Robert H., author.
Requiem for the Santa Cruz : an environmental history of an Arizona river / Robert H. Webb, Julio L. Betancourt,
R. Roy Johnson, and Raymond M. Turner ; foreword by Bernard L. Fontana.
    pages cm
  Includes bibliographical references and index.
  ISBN 978-0-8165-3072-4 (hardback)
  1. Natural history—Santa Cruz River (Ariz. and Mexico)   2. Stream ecology—Santa Cruz River (Ariz. and
Mexico) 3. Santa Cruz River (Ariz. and Mexico)   I. Title.
  QH104.5.S26W43 2014
  577.6'40979179—dc23
                              2013039493

Publication of this book is made possible in part by funding from the US Geological Survey.

Manufactured in the United States of America on acid-free, archival-quality paper containing a minimum of 30 percent
post-consumer waste and processed chlorine free.

19   18   17   16   15   14   6   5   4   3   2   1

We owe much of what we know about the natural history of southern Arizona to a tightly knit cohort of dedicated field scientists in the middle part of the twentieth century. We dedicate this volume to two such field scientists who very much influenced the content of this book: James Rodney (Rod) Hastings (1923–1974), a University of Arizona climatologist, and Joe T. Marshall (1918–), a University of Arizona and Smithsonian ornithologist. Hastings's early interest in the changing southwestern landscape led to the establishment of the world's largest collection of repeat photographs. His introductory study of changes along the Santa Cruz River was the springboard for our own investigations. Marshall, a student of the Joseph Grinnell system at the University of California, Berkeley, kept meticulous field journals about birds and their habitats, on which we relied heavily to reconstruct changes in bird assemblages along the Santa Cruz River. The final biological work in the Great Mesquite Forest was completed by Marshall and his students immediately prior to its demise due to groundwater depletion after the 1950s. Marshall and Hastings helped us compose this requiem of an aridlands river.

Contents

List of Illustrations viii

Foreword by Bernard L. Fontana xi

Preface and Acknowledgments xiii

1. The Problem of Riverine Change 1

2. Characteristics of a Desert River 9

3. Causes of Arroyo Downcutting 27

4. Perennial Flow and Discontinuous Arroyos, 1691–1872 36

5. Land Use, Climate, and Floods, 1873–1888 52

6. Arroyo Downcutting and Widening, 1889–1915 68

7. Water Development and the Great Mesquite Forest, 1916–1942 92

8. The City and the Arroyo, 1943–1975 112

9. Arroyo Management in the Time of Floods, 1976–1995 130

10. Channel Filling and River Restoration Efforts, 1996–2012 149

11. Summary of the Past and Some Possible Futures 166

Appendixes 183

Notes 237

References 251

Index 273

Illustrations

Figures

1.1. Map of Santa Cruz River drainage basin   5

1.2. Photograph of Tucson ca. 1890   6

2.1. Map of Santa Cruz River in Tucson Basin   10

2.2. Photographs of Santa Cruz River north of 22nd Street in Tucson   11

2.3. Idealized Holocene stratigraphy of Santa Cruz   13

2.4. Seasonal precipitation for Tucson University of Arizona climate stations   15

2.5. Seasonal precipitation anomalies for Upper Sonoran Desert climate stations   16

2.6. Number per year of tropical cyclones, hurricanes, and tropical storms in eastern North Pacific Ocean   18

2.7. Annual peak discharges recorded at Santa Cruz River at Tucson   21

2.8. Hydrograph of monsoon-generated flash flood in Santa Cruz River at Tucson   22

3.1. Schematic diagram showing stages of arroyo development   34

4.1. Historic map of Santa Cruz valley in Tucson– Mission San Xavier area   37

4.2. Upstream view of Acequia de Punta de Agua   41

4.3. Map of northeast portion of San Xavier Indian Reservation in 1882   43

4.4. Maps of San Xavier Indian Reservation in 1888 and 1891   44

4.5. Repeat photographs of Silver Lake   46

5.1. Solomon Warner's house and mill in 1880   55

5.2. Repeat photographs of Santa Cruz River valley from base of Sentinel Peak   56

5.3. Repeat photographs of Santa Cruz River valley from Sentinel Peak   57

5.4. Repeat photographs of Warner's Lake   61

5.5. Map of Sonora and Arizona showing effects of 1887 earthquake   64

5.6. Diagram showing longitudinal profile of intercept ditch   65

5.7. Headcut of Sam Hughes's intercept ditch at St. Mary's Road in 1889   66

5.8. Headcut of Sam Hughes's intercept ditch at St. Mary's Road in 1889   66

6.1. View looking west across Santa Cruz River at St. Mary's Road   70

6.2. Upstream view of Santa Cruz River at St. Mary's Road   72

6.3. Santa Cruz arroyo between Congress Street and St. Mary's Road   73

6.4. Downstream view of confluence of West Branch and Santa Cruz River from Sentinel Peak   74

6.5. Manning Ditch near Sentinel Peak   81

6.6. Plan map of Tucson Farms Company Crosscut and distribution system   82

6.7. Vertical profile of Tucson Farms Company Crosscut   83

6.8. Crosscut under construction, just downstream of the former dam at Silver Lake   84

6.9. Sector of crosscut water distribution system   84

6.10. Map of Greene's Canal and lower Santa Cruz River   85

6.11. Downstream view of Santa Cruz River in flood, at Congress Street Bridge, 1915   86

6.12. Downstream view of Santa Cruz River after 1915 flood, at Congress Street Bridge 87

6.13. Repeat photographs of Santa Cruz River from Rillito Peak 88

7.1. Upstream and downstream views of Santa Cruz River at St. Mary's Road 93

7.2. Repeat photographs of Santa Cruz River from Martinez Hill 96

7.3. Repeat photographs of Santa Cruz River near 22nd Street 98

7.4. Groundwater levels for two wells along Santa Cruz River 100

7.5. Composite aerial photograph from 1936 showing extent of Great Mesquite Forest 101

7.6. Repeat photographs looking south across Santa Cruz River from Martinez Hill 104

7.7. Photograph of second-growth mesquite in Great Mesquite Forest 109

8.1. Census and livestock data for Tucson and Pima County, 1880–2007 113

8.2. Repeat photographs of Congress Street from Powderhouse Hill 114

8.3. Repeat photographs looking east across Santa Cruz River from Sentinel Peak 116

8.4. Repeat photographs looking east-northeast across Santa Cruz River from Sentinel Peak 117

8.5. Repeat photographs looking northeast across Santa Cruz River from Sentinel Peak 118

8.6. Repeat photographs of Santa Cruz River south of Congress Street Bridge 120

8.7. Repeat photographs of Santa Cruz River valley and Tucson from Sentinel Peak 122

8.8. Repeat aerial photographs of Santa Cruz River south of Martinez Hill 124

8.9. Number of floods above base discharge for Santa Cruz River at Tucson 126

9.1. Repeat photographs of Santa Cruz River upstream from A Mountain 134

9.2. Repeat photographs of Santa Cruz River at Silverlake Road 135

9.3. Repeat photographs of Santa Cruz River at Drexel Road ford 137

9.4. Repeat aerial photographs of Santa Cruz River at Congress Street Bridge 138

9.5. Photograph of 1983 flood near its peak discharge at the St. Mary's Road Bridge 141

9.6. Upstream aerial view of active headcut of Greene's Canal arroyo 142

9.7. Photograph of 1993 flood near its peak discharge downstream from Congress Street Bridge 143

9.8. The 100-year flood for Santa Cruz River at Tucson 144

9.9. Water deliveries by Tucson Water, 1899–2011 147

10.1. Repeat photographs of Santa Cruz River at Congress Street Bridge 152

10.2. Repeat photographs of Santa Cruz River at Congress Street 154

10.3. Repeat photographs of Santa Cruz River from Congress Street Bridge 156

10.4. Channel cross section downstream from Congress Street Bridge 158

10.5. Photograph of Santa Cruz River showing concrete sill downstream from Congress Street Bridge 158

10.6. Stage-discharge relations for three time intervals for Santa Cruz River at Tucson 159

10.7. Combined effluent discharge into Santa Cruz River 161

10.8. Repeat photographs of Santa Cruz River at confluence with Rillito River 162

11.1. Climate and annual peak discharge for Santa Cruz River at Tucson 169

11.2. Groundwater levels for well near Santa Cruz River between Valencia and Drexel Roads 173

11.3. Oblique aerial photograph of Santa Cruz River upstream from Martinez Hill 173

11.4. Photographs of Santa Cruz River showing channel-management scenarios 178

Tables

7.1. Cavity-nesting birds of Santa Cruz–Rillito River system 107

9.1. Estimates of 100-year flood on Santa Cruz River at Tucson, made by previous investigators after 1970 132

10.1. Dates of stage-rating curves for Santa Cruz River at Tucson after 1992 160

Appendixes

A. Summer birds reported from Great Mesquite Forest before mid-twentieth century   184

B. Summer birds recorded during late 1800s and early 1900s from Rillito River   190

C. Current status of summer birds along Santa Cruz River   195

D. Birds from Mexico at their northern breeding limits in southern Arizona   199

E. Newly named species and subspecies of birds discovered in Tucson vicinity   201

F. Amphibians and reptiles of Great Mesquite Forest, Santa Cruz River, and Sonoran desertscrub   203

G. Mammals of Great Mesquite Forest and Santa Cruz River   206

H. Special status of species for Great Mesquite Forest and Santa Cruz River   208

I. Ornithologists who conducted studies in Santa Cruz–Rillito River system   219

J. Comparison of birds of Great Mesquite Forest, Rillito River, and Blue Point Cottonwoods   224

K. Common names and Latin equivalents for plants   228

L. Summer birds recorded in nonnative saltcedar stands   229

M. Summer birds historically recorded for Great Mesquite Forest and Santa Cruz and Rillito Rivers compared with those along present-day Santa Cruz River   232

Foreword

Bernard L. Fontana

It was October 1983. My wife and I sat on the front porch of our adobe house and listened to an unaccustomed roar, a noise something like the steady rumble of a freight train. In the twenty-seven years this had been our desert home we'd never heard anything quite like it before.

That sound, coming to us from the east a little more than a mile away, was the Santa Cruz River in flood stage. The water had not come from a late summer thunderstorm. It was the product of many days of slow but steady rain that had fallen throughout all of southern Arizona and much of neighboring northern Sonora in Mexico.

Curiosity led us to drive to the river's edge. There we saw a raging torrent with waves and whitecaps threatening to overflow its banks and to flood the surrounding flatlands. It was, indeed, a torrent that later took out two of the four lanes of an interstate highway bridge, as well as an entire bridge linking the interstate to the San Xavier Indian Reservation.

I cancelled a planned flight to California, afraid that all of the Santa Cruz's bridges might disappear and I would be left stranded on the east side of the river at the airport.

It is likely that over the millennia the drainageway we now call the Santa Cruz River had seen many similar storms. The valley through which the river flows is an alluvial plain, its relatively flat surface made up of sands, clays, and gravels brought down from surrounding ranges of mountains and deposited there in geologic times. Throughout its long history the river has meandered. It has flowed on the surface; it has carved deep fissures into the soils; it has widened; it has narrowed; it has filled. As the reader of this book will learn, these are events that also have taken place in historic times.

Today an overriding question in all of this is to what extent the mammal presently at the top of the food chain, *Homo sapiens*, can be held accountable for recent and future episodes in the river's story. A partial answer lies in the relationship between human beings and the river through time. That is what this book proposes to do. An understanding of our past and present connection to the river ideally provides guidance for what our future actions should be.

And these chapters do more. They offer a model of ways in which people and their environments, whatever those environments may be, can be discerned.

Preface and Acknowledgments

Between about 1862 and 1915, arroyos developed in alluvial valleys of the southwestern United States across a wide variety of hydrological, ecological, and cultural settings. That they developed in the span of a few decades has encouraged the search for a common cause, some phenomenon that was equally widespread and synchronous. As with most environmental changes, whether global or local, efforts to understand arroyo formation have been hindered by the inability to discriminate between natural and cultural factors in the great uncontrolled experiment that is Planet Earth. Furthermore, there are few southwestern streams for which we have even a qualitative understanding of timelines and processes involved in initiation and extension of historic arroyos. The Santa Cruz River through Tucson, a reach commonly cited as a prime example of arroyo cutting in the literature, offers a unique opportunity to chronicle the arroyo legacy, evaluate its causes, and consider its aftermath.

Our history of the Santa Cruz River reconstructs the physical, biological, and cultural circumstances of its entrenchment, widening, and subsequent partial filling. Primary data before 1930 include newspaper accounts, notes and maps of General Land Office surveys, eyewitness accounts, legal depositions, bird surveys, and extensive historical photography, followed by extensive streamflow gaging, groundwater measurement, survey data, and remote sensing. For events after 1930, we used published papers, remote sensing, hydrologic and topographic data, and other sources to document changes. After the late 1950s, our personal experiences with this river and its changes heavily influence our presentation.

R. Roy Johnson, an ornithologist, was a student of Joe Marshall's in the late 1950s, and he spent time with Marshall in what we now refer to as the Great Mesquite Forest counting birds and documenting habitat loss. Also in the 1950s, Ray Turner, a plant ecologist, worked with his close colleague Rod Hastings on a multipronged effort to document large-scale change in the Sonoran Desert using permanent plots and repeat photography. In the late 1970s, Julio Betancourt worked as an archaeologist/historian for the Santa Cruz River-park Plan, a precursor to today's Rio Nuevo downtown redevelopment project; in the course of this project, he gained an appreciation for the history of this river. In the 1980s, this led to a collaboration between Betancourt and Turner that used archival evidence and replication of historical photographs to reconstruct a detailed chronology of when, where, and how this arroyo initiated and evolved in the nineteenth and early twentieth centuries. Finally, in the early 1980s, Robert Webb became interested in the changing flood frequency of the Santa Cruz River, which eventually involved examination of the histories that Betancourt and Turner had assembled.

Our career-long perspectives on western US hydroclimatology, flood frequency, geomorphic processes, and large-scale hydroecological dynamics were very much shaped by the spectacular flood of 1983 on the Santa Cruz River. There is nothing quite like an extreme flood to focus one's attention on a river's behavior. Webb based much of his work on arroyos and flood frequency on the history of the Santa Cruz River, and Betancourt and Webb helped several students, notably John Parker, write theses and dissertations documenting and analyzing recent channel changes and sediment transport. Using an interdisciplinary approach, we approached the issue of late twentieth-century channel change from a climatic perspective as well as trying to determine the role that land-use practices, and particularly the rapidly growing metropolitan area, might have had on the size of this flood. In response to several floods and the need for increased flood control, the Santa Cruz River through Tucson was mostly channelized and stabilized using bank protection, and this began another legacy of change in this river.

Johnson had a long-term interest in the regional avifauna and fond memories of fieldwork in the Great Mesquite Forest before its final demise. Although many before us acknowledged the decline of wetlands and the bosque along the Santa Cruz, the research was poorly documented. Here we use more than a century of observations and collections to chronicle wide-ranging impacts on birds and other vertebrates.

Several previous books have addressed some of the questions discussed here in detail. Ronald Cooke and Richard Reeves (1976) viewed the arroyo of the Santa Cruz River as one of the most important to illustrate the myriad of factors that contributed to channel downcutting in the region. In *The Birds of Arizona*, Allan Phillips, Joe Marshall, and Gale Monson (1964) highlighted the Great Mesquite Forest and its downfall. Michael Logan (2002) discussed the human history of the Santa Cruz River basin and drew conclusions as to the magnitude and rates of change based on assumptions of how land uses have changed the watershed. Douglas Kupel (2003) reviewed the history of groundwater development of the Santa Cruz Basin within a legal framework, emphasizing pump technology and its development to explain the timeline of fall in the alluvial aquifer underlying Tucson. Finally, Ken Lamberton (2011) discussed changes in the Santa Cruz River and their effect on the current human inhabitants of the watershed, using a first-person account of a hike down its channel as a metaphor for change.

In this book, we address how the arroyo downcutting and widening of the nineteenth and early twentieth century was followed by a prolonged period of little or no change during a persistent drought. In the period between arroyo formation and large-scale groundwater mining, the large bosque south of Martinez Hill reached its zenith. The late twentieth century was marked by excessive groundwater extraction, large floods, and extensive bank protection installed to stabilize the channel and reduce the damage in urbanized areas during such floods. The Santa Cruz River of yesteryear is barely recognizable, inspiring nostalgia and even hopes of turning back the clock. Today, we face the following conundrum: do we manage ephemeral rivers through urban areas for flood control, or do we attempt to restore them to some previous state of naturalness? To provide a long-term perspective on management of this aridlands river, we explore the channel-change legacy, the efficacy of attempts to stabilize the incised channel and keep it from widening, and the nascent attempts at river restoration.

Acknowledgments

This book grew from several converging publications, including a dissertation completed by Julio L. Betancourt in 1990, a monograph by Robert H. Webb and Betancourt on flood frequency in the Santa Cruz River published in 1992, extensive information collected by Roy Johnson on biotic change, several master's theses and doctoral dissertations, and a wide-ranging interest in landscape change documented using repeat photography by Raymond M. Turner and colleagues, which are epitomized by *The Changing Mile Revisited,* published in 2003. Three of us (RHW, JLB, and RMT) worked on this book while employed by the National Research Program, Water Resources Division (now called Water Mission Area), US Geological Survey, which also provided a subsidy to help the University of Arizona Press produce this volume.

Our accounting of temporal change along the Santa Cruz benefited from interactions with many friends and colleagues. A great deal of our time was spent examining the myriad of records preserved in the archives of the Arizona Historical Society and Special Collections at the University of Arizona Library. We especially thank the late Susan Peters, the late Margaret Bret-Harte, Joan Metzger, Heather Hatch, Roger Myers, and Louis Hieb for pointing the way to key documents and photographs and for occasionally bending the rules to satisfy our needs.

Several colleagues and friends shared our interest in the Santa Cruz River, the topic of arroyos, and the importance of historical information: Byron Aldridge, Vic Baker, the late Don Bufkin, Tony Burgess, Russell Davis, Doug Duncan, Alan Ferg, Bunny Fontana, Al Gardner, Kathy Groschupf, Polly Hayes, Vance Haynes, Richard Hereford, Katie Hirschboeck, David Hyndman, Ricky Karl, the late Keith Katzer, Ken Kingsley, Pete Kresan, Doug Kupel, the late Paul Martin, John T. C. Parker, Tom Peterson, the late Charlie Polzer, Dick Reeves, Brian Reich, Richard Roberts, Martin Rose, Cecil Schwalbe, Tom Sheridan, the late Leland Sonnichsen, Larry Stevens, and Mike Waters. Numerous people assisted with questions concerning the Great Mesquite Forest, including Dave Brown, Julia Fonseca, Joe Marshall, Amadeo Rea, and Philip Rosen. Evan Canfield and Fernando Molina provided data on Tucson water use. Charles Sternberg drafted and helped design many of the maps and diagrams of the early manuscript, which have been digitally updated and modified in this presentation; Jeanne DiLeo helped prepare some digital illustrations. Diane E. Boyer prepared the digital photographic matches and helped with editing of the manuscript. Richard Hereford and Evan Canfield reviewed the initial book manuscript, and Bunny Fontana graciously provided additional comments, along with his much-appreciated foreword.

1

The Problem of
Riverine Change

Between 1862 and 1915, most watercourses in the southwestern United States downcut to create arroyos in reaches where flow formerly was at or near the overall valley surface. The Spanish word *arroyo* generally refers to an ephemeral stream, but late nineteenth-century geomorphologists in the United States applied this word to a channel—ephemeral or perennial—that has downcut below its historic floodplain. In a matter of decades, and sometimes during a single flood, streams that flowed unincised in shallow, meandering channels became entrenched between vertical walls or gullies several meters below valley surfaces. Accelerated erosion resulted in destruction of farm- and grazing lands, obliteration of irrigation systems and other waterworks, lowering of local groundwater levels, drastic changes in riparian ecosystems, and reduction in reservoir volume owing to increased sedimentation rates. Some settlements in the region were abandoned in response to this change, and nearly all of those built along watercourses were negatively impacted.

The economic consequences of arroyo downcutting were devastating. At the end of the nineteenth century, arroyo downcutting was perceived as anomalous and undesirable, a change that brought economic ruin to what had been or might have become productive land. Initiation of arroyos received little national attention, however, partly because it reinforced "the forbidding image of an American Sahara"[1] and threatened economic development and aspirations of statehood. Few heeded John Wesley Powell's caution that only a fraction of the West was irrigable,[2] and arroyo downcutting did little to discourage the tenacious American dream of making that area west of the 98th Meridian, the Great American Desert, bloom. Arroyos were merely a temporary setback that challenged settlers to develop new ways of tapping water resources. For sci-

entists, channel downcutting raised the specter of other environmental disasters, natural and man-made alike. For engineers, however, the accelerated gullying posed a formidable challenge. If they could determine the causes of the ongoing arroyo cutting, something possibly could be done to minimize or reverse the damage. Could the study of arroyo formation lead to a better understanding of geomorphic process and landscape evolution? These issues and questions stimulated a vigorous scientific debate on the causes, roles, and consequences of gullying that still rages more than a century later.

At the beginning of the twentieth century, the immediate incentive for studying arroyos was to design and implement erosion-control measures to reduce further agricultural losses and stabilize water supplies. Few cared about floodplain management or ecosystem stability when starvation was a reality. Because the disciplines of geomorphology and hydrology were still in their infancy, several decades passed after the initial phase of arroyo downcutting before the origin and dynamics of ephemeral channels attracted scientific attention. Exciting new theories, including Powell's concept of base level, G. K. Gilbert's theory of the graded stream, and W. M. Davis's geographic cycle, were born from nineteenth-century explorations in the West, and these concepts were little more than a decade old when arroyos downcut. Few geologists were on hand to witness or report on accelerated erosion before the turn of the century: Davis was one exception,[3] and Herbert Gregory was another.[4] Beginning at Chaco Canyon in New Mexico,[5] scientists relied on historical information, now referred to as "anecdotal evidence," including the recollections of ranchers or farmers, photographs, and newspaper accounts. Arroyos made historians out of earth scientists, and the history of channel and landscape change dominated the hypotheses of geomorphologists.[6]

At the same time as arroyo cutting, biologists became increasingly attracted to the diverse flora, fauna, and habitats of southern Arizona, especially those spanning desertscrub, grassland, and forests and linked by a network of intermittent streams. In particular, ornithologists were fascinated with the mixture of neotropical and temperate waterbirds and the resident species that seasonally or annually used the ecosystems of the arid environment. Tucson was a mecca for naturalists, lay observers, and scientists alike, all of whom were drawn to the Rillito and Santa Cruz Rivers for their bird populations. Although southern Arizona remains an ornithological hotspot, primarily for other river reaches and certain canyons that attract certain types of birds—especially "Mexican species"—the Santa Cruz River in the Tucson Basin was the place that once held the attention of birdwatchers. The earliest observers took specimens and made notes that are preserved in museums, government reports, special collections at libraries, and other archives.

The history of landscape change became paramount to early researchers. Archives, long the exclusive domain of historians and social scientists, are essential in reconstructing landscapes and climates of the past where conventional or standard measurements are unavailable. In the Southwest, the historical record would seem indispensable for arroyo studies, more so as critical eyewitnesses passed away during the first half of the twentieth century. Even so, few studies fully exploit archival information in their analyses of historical channel change.[7] Insufficient documentation of change that occurred in still-remote places may explain this oversight in remote places but certainly not where towns and cities were established along the banks of changing rivers. There, the evidence for channel change is chronicled, sometimes with surprising detail, in newspapers, letters, diaries, and other documents. These long-forgotten information sources seldom are consulted when today's land managers discuss important concepts, such as flood control and restoration of riparian habitat.

In the twenty-first century, arroyos remain a problem for much different reasons than when they were initiated. Irrigation water mostly comes from long-distance canals and deep wells, so the issue of diminished surface-water supplies resulting from nineteenth-century arroyo downcutting became irrelevant long ago. The entrenched channel contains floodwaters, and as long as those waters stay in the channel and the banks are stabilized, development can safely abut the arroyo banks. Despite scientific evidence to the contrary, the downcutting of arroyos is still associated with certain kinds of land use, particularly livestock grazing. If we abandon these practices, so think some environmentalists and land managers, perhaps the damage could be undone and the habitat restored. But restored to what? It is here, in an aridlands river that winds its way through an increasingly urbanized area, that the twin agendas of flood control and environmental quality clash.

This study, which focuses on the Santa Cruz River in southern Arizona (fig. 1.1), grew out of appreciation for the vast archives that accrued as Tucson evolved from a mud-walled Spanish village (fig. 1.2) to a modern American metropolis.[8] Historically, the Santa Cruz watershed is important in southern Arizona for settlement, ranching, and economic development of Native Americans (especially the Tohono O'odham). In the late 1700s, Spaniards joined the Tohono O'odham, once called the Papago, and the related Sobaipuri living near or along the river, making this watershed the site of the first European colonization in what is now Arizona and one of the earliest in North America. The descendants of these early Spanish colonists became citizens of Mexico, and then they became US citizens following the Gadsden Purchase in 1854.[9] Settlers from other parts of the United States and Mexico continued to arrive in the late nineteenth century, and they became increasingly dependent on surface water in the river for irrigation and domestic supplies.[10] Because of its proximity and utility, the settlers observed and recorded a long history of channel and floodplain change for the Santa Cruz River as they struggled to develop and maintain their water supply.[11]

The headwaters of the Santa Cruz River are in the San Rafael Valley north of the Arizona-Mexico border. The river flows south into northern Sonora and curves northwest back into the United States and through Tucson (fig. 1.1) to join the Gila River south of Phoenix. With the exception of seasonal flooding, there is no historical evidence that the Santa Cruz River had continuous flow from its headwaters to its terminus at the Gila River; instead, local reaches of perennial flow, particularly the flow that attracted prehistoric and historic settlement near Tucson, punctuated a stream that was ephemeral along most of its length.[12] Like the San Pedro, its twin to the east, the Santa Cruz River drains an extensive area south of the Gila River and some area in Sonora, Mexico, and both rivers join the Gila River upstream from the confluence with the Salt River.

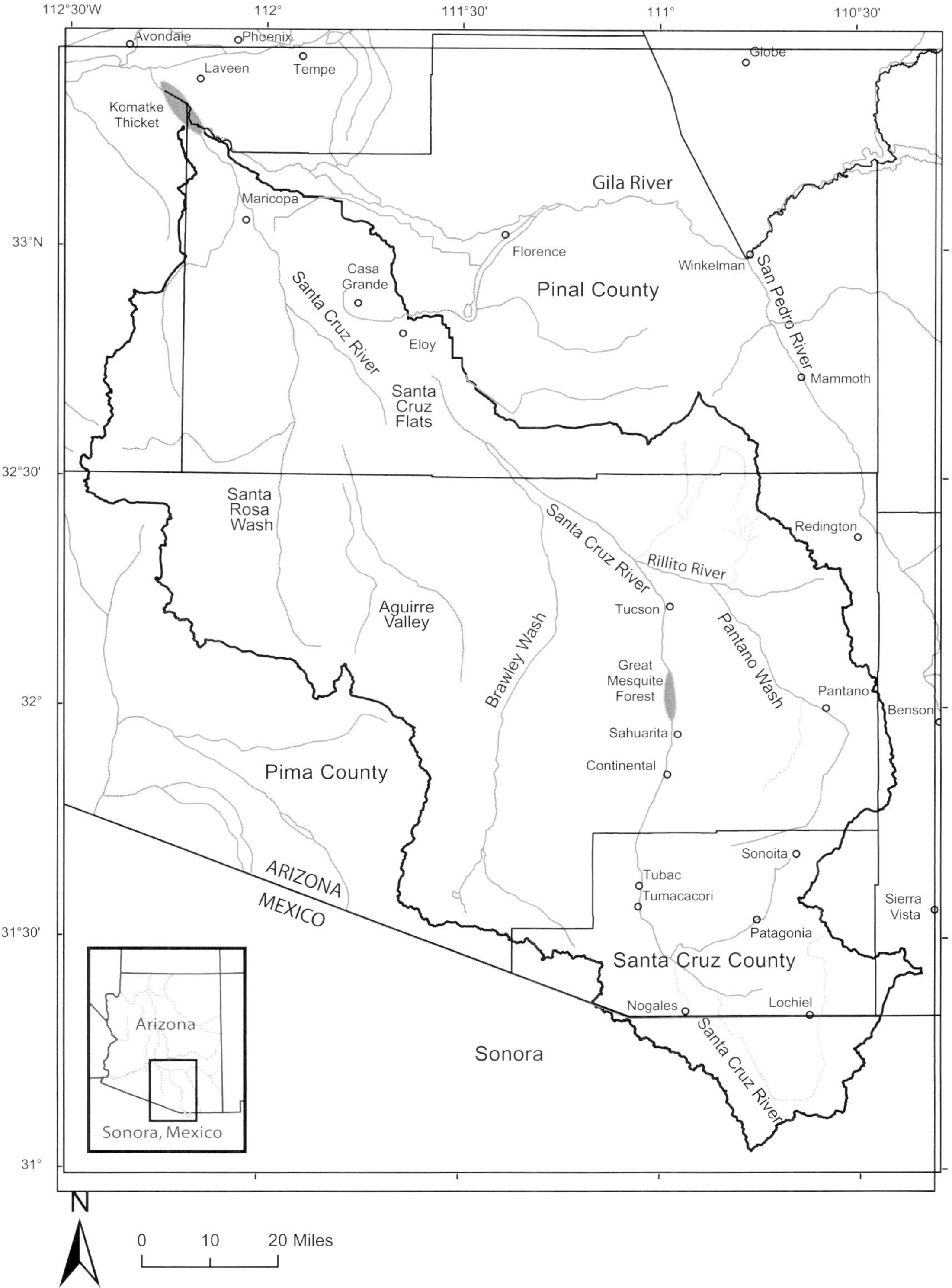

Figure 1.1.
Map of the Santa Cruz River drainage basin in south-central Arizona and Sonora, Mexico.

Figure 1.2.
(ca. 1890) Henry Buehman, a long-time Tucson photographer, took this northwest-looking view of Tucson and the Santa Cruz Valley from the Pima County courthouse. The Plaza de Armas (now Presidio Park) is in the foreground, and agricultural fields appear beyond the town. The channel of the Santa Cruz River is marked by the line of trees beyond the agricultural fields in the right midground but before the desert vegetation in the background. (Buehman, 45079, courtesy of the Arizona Historical Society.)

The Santa Cruz River has long been used as an example by some protagonists in the arroyo-cutting debate.[13] However, a full accounting of how and when the channel changed does not exist. Here, historical data are developed to reconstruct both the physical and the cultural circumstances of arroyo formation along the Santa Cruz and to evaluate potential causes, principally those having to do with land use and climate. The stream's history is chronicled from Spanish colonial times through the early twenty-first century to determine what pre-arroyo conditions were like, to show how and when the arroyo developed relative to land use and runoff events, and how the arroyo legacy affects—even stymies—floodplain management in the modern, urban setting.

Regional synchroneity of arroyo downcutting is one of the key concepts that pervade the scientific literature on the geomorphology of alluvial channels. As the term typically is used, the geologic interpretation of regional synchroneity is where the same type of change occurs simultaneously (and leaves geologic evidence) within the uncertainty constraint of various age-dating techniques, such as radiocarbon dating. In the historic era, we assume that downcutting is synchronous when it happened at approximately the same time throughout a region. Again there is a caveat of uncertainty in timing because of spatial heterogeneity of major floods, compounded by the fog of history in a sparsely settled landscape. Both definitions could imply concurrent shifts in regional climate and (or) deleterious land uses that exerted their impact all at the same time. However, asynchroneity in arroyo downcutting would suggest that differences in the physical characteristics and the local history of each drainage basin are more important influences. The standard for synchroneity used in both concepts is the most recent downcutting episode, which happened regionally over a period of five decades,[14] placing the time frame of "synchroneity" within a geologic-time framework and outside the common definition of the word.

Regional synchroneity is more generally claimed for prehistoric arroyo downcutting episodes,[15] but tests of synchroneity are complicated. Although evidence of paleoarroyos is commonly exposed in the banks of modern arroyos, large uncertainties are associated with dating the erosion, since the stratigraphic evidence of the erosional event itself is usually missing. In the stratigraphic record, erosion cannot be directly dated because there are no deposits; instead, downcutting of paleoarroyos is dated within an interval defined as the youngest obtained age of deposition prior to downcutting and the oldest obtained age of fill in the paleoarroyo. The resolution is usually no better than a few hundred years, although this interval may in fact define the time needed to complete a single cut-and-fill cycle. Nevertheless, there is little to indicate that the synchroneity in paleoarroyos is comparable to the suddenness of more recent arroyo formation over such a large region. Despite this obvious limitation, knowledge of the details of historic arroyo development informs how channels may have behaved at other times in the Holocene (approximately the past 11,700 years)[16] or how they may behave in the future.

By necessity, we have taken an unorthodox historical approach to what is essentially a geomorphic problem, relying on written observations and photographs as primary evidence of channel change in the nineteenth and early twentieth centuries, followed by analyses of ground and aerial photography to document channel changes into the early twenty-first century. Widening of the historic arroyo during late twentieth-century floods removed some of the stratigraphy indicative of pre-entrenchment conditions;[17] most of the other evidence now lies beneath pavement, soil cement, housing developments, or businesses. The marriage of written accounts—some of which are vague at best and potentially misleading—with photographs provides an objective account of what actually occurred during the critical period of downcutting and widening.

A few qualifications are perhaps warranted in the use of historical sources. Firsthand accounts, obtained years after the fact, may be wildly inaccurate or spot-on.[18] Secondhand accounts, newspaper coverage, reminiscences, correspondence, and legal depositions must be evaluated for personal, economic, or political motives that may taint the accuracy of observations. For example, to drum up support for establishing a mission in the Tucson Basin, the Jesuit missionary Eusebio Kino likened the irrigation potential along the Santa Cruz to that of Mexico City, a hyperbole that casts suspicion on other observations by Kino.[19]

Other sources of historical data are the cadastral surveys commissioned by the US General Land Office (GLO).[20] The GLO surveys generally can be trusted, although a few exceptions are notable. During Henry Atkinson's tenure (1876–1884) as surveyor general of New Mexico, false applications for township surveys and manufactured field notes were common. In his notes, one surveyor described dense mesquite growing on coppice dunes in what was actually a barren, gypsiferous playa; this playa is the sediment source for the White Sands dune field in New Mexico.[21] Fictitious surveys and gross errors during the Atkinson administration also rendered the GLO surveys worthless for study of channel entrenchment in certain areas, such as Chaco Canyon, New Mexico.[22] Those types of problems have not been documented for southern Arizona, however.

Another challenge in interpreting historical data is the reliability of negative evidence. Nineteenth-century surveyors faithfully recorded channel widths but seldom recorded channel depths. Along the San Pedro River in southern Arizona, the journals of itinerants between 1849 and 1884 describe discontinuous arroyos, with perpendicular banks ten to twenty feet deep;[23] these observations generally can be trusted, since those itinerants had to pass wagons through those channels. In 1873, however, GLO surveyor Theodore White did not record channel depths along these reaches.[24] As a result, the lack of mention of channel depths in GLO surveys cannot be taken as evidence for unincised floodplains without corroborating information from other sources.

Many historical sources, particularly newspapers, tend to focus on extreme and rare events of economic consequence, such as floods, while ignoring more commonplace details that could have been equally important. The detail of such reports is understandably proportional to proximity to the nearest settlement or to the amount of damage to waterworks and farmland; where erosion occurred on uninhabited rangelands, it seldom made the newspapers. This uneven coverage may give the false impression that some river reaches were impacted more than others. Some of the most important information comes from the notes and journals of ranchers and travelers, who viewed the landscape more in terms of their immediate needs.

Old photographs add clarity to the written word, as will be evident in this book. Photographs can be thought of as temporal-change benchmarks, as anyone who has reoccupied the original camera station and documented changes can appreciate.[25] Finding old photographs is relatively easy in any southwestern town, where archival efforts are fueled by public nostalgia for the past and the inevitable historical society. In Tucson, a concerted and sustained effort at repeat photography not only has resulted in identification and preservation of numerous historical photographs but also has added a legacy of frequent replication of those photographs, creating a time series of visual change.[26] Like written accounts, historical photographs and their repeats have limitations, including missing or incorrect dates, misidentification of the location, and the limited field of view that is photographed. Some of these limitations can be overcome by careful historical research about photographers, their travels, and their interests; other problems eliminate the usefulness of some tantalizing views.

In the Southwest, environmental historians tend to generalize about so-called presettlement or even pre-Columbian conditions—what were rangelands and floodplains like before the onslaught of European settlers? Observations made over decades, if not centuries, are conveniently lumped with the underlying assumption that presettlement landscapes were relatively stable until disrupted by settlement. At the risk of redundancy, we took a different tack to reconstruct pre-arroyo conditions along the Santa Cruz River. Repeated observations of the same phenomena, whether they are unincised floodplains or groundwater discharge along the streambed, were duly noted to illustrate long-term stability (or instability). Equal attention was given to the possibility that perennial reaches may have elongated or shortened, or that discontinuous arroyos developed during the two centuries prior to accelerated erosion.

We quote directly from many original sources, paraphrasing repetitious views of the same event or phenomenon.[27] Though some readers may find quotations cumbersome, we want our historical sources to speak for themselves, allowing the reader to judge the accuracy and objectivity of the observations. Many of the accounts we reproduce constitute primary data, and the original prose may be fresher—and in some cases more colorful—than ours. The narrative, which attempts to weave observations about the Santa Cruz River into the cultural and historical context of the times, is organized chronologically; the periods for each of these chapters were defined on the basis of either channel or climate history and not arbitrarily. Wherever possible, great care was taken to record the reactions of Tucsonans to the river's metamorphosis, particularly when an opinion was rendered as to the cause for change.

In considering the role of climate in channel change, we are indeed fortunate to be writing at a time when knowledge about the complex link between global and regional climates is unfolding. The El Niño event of 1982–1983, the most catastrophic occurrence of this phenomenon in the recorded history of our region, heightened scientific and public awareness worldwide.[28] The 1982–1983 El Niño and the floods it spawned captured the imagination of those embroiled in the arroyo debate, many who now recognize the coincidence between the warm events in the tropical Pacific Ocean and the floods that produced significant channel changes during the past century.[29] Following the late twentieth-century advances, climatic research has broadened and branched into the twin—and compatible—areas of global climate-system modeling and development of new approaches and indices used to describe past climatic fluctuations.

Finally, no matter what climate and climatic change may have contributed, historic arroyos coincided with intensified land use as the Southwest was settled and livestock were introduced onto rangelands. The earliest observers of arroyo downcutting pointed the finger at livestock,[30] and this association continues to be used today to justify reducing or removing livestock on public rangelands. Hence, the question of what caused arroyos to downcut into their floodplains duplicates the dilemma now facing most environmental scientists: can climatic and land-use influences be disentangled when addressing the causes for ecological changes in the southwestern United States? In considering the causes for change, be they global, local, rangeland, or channel, we first must determine whether natural and cultural factors can be separated using historical evidence combined with geomorphic theory.

2

Characteristics of a Desert River

The Santa Cruz River gained its name in a manner that has been called convoluted.[1] By most accounts, the name is attributed to Father Eusebio Kino, who visited the region in the late seventeenth century and established Mission San Xavier south of present-day Tucson (see chapter 4). Kino originally called this watercourse the Río de Santa María, after a settlement of Native American Sobaipuri who lived along its upper reaches. In 1691, Kino used the name Santa Cruz in reference to one or more Sobaipuri villages on the San Pedro River to the east of the Santa Cruz watershed. These villages shifted locations, combined, separated, and recombined over the following century, owing to battles with other Native American groups, and the name Santa Cruz persisted as part of the village name(s). Hostilities drove the Sobaipuri westward, until they and a Spanish garrison moved into the abandoned settlement of Santa María Soamca in what today is called the San Rafael Valley. The settlement was renamed Santa Cruz, and the river through the valley became the Santa Cruz River, probably in 1787.

This river drains 8,581 square miles of southern Arizona and northern Sonora (see fig. 1.1). The headwaters of the watershed are in oak woodlands above 5,200 feet on the east slope of the Patagonia Mountains, the south slope of the Canelo Hills, and the west slope of the Huachuca Mountains. Its terminus is at the Gila River south of Phoenix in desertscrub vegetation dominated by saltbushes. Through much of its mostly northwesterly course, it traverses landscapes of the Sonoran Desert in south-central Arizona and is joined by numerous tributaries from the east and west. At its mouth near Laveen, the entire Santa Cruz River basin has an average-basin elevation and annual precipitation of 3,060 feet and 13.0 inches, respectively.[2]

From its headwaters, this river has a shallow channel with perennial or intermittent (occasionally dry) flow and courses south through the desert grasslands of the San Rafael Valley, draining 82 square miles of southern Arizona before passing into Mexico at a point 2 miles east of Lochiel, Arizona.[3] In Sonora, much of the perennial flow is captured by wells and infiltration galleries for agricultural and municipal use. Much of this reach sustains groves of native riparian trees, notably, cottonwood and black (Goodding) willow.[4] The river makes a 35-mile loop past the Sonoran settlement of Santa Cruz and before reentering Arizona 6 miles east of Nogales (fig. 2.1). At this point, the average annual flow is 3.9 ft^3/s from 533 square miles of drainage area, of which 348 square miles are in Mexico.[5]

As it flows north from its border crossing to its confluence with the Gila River (figs. 1.1, 2.1), the Santa Cruz River is joined by several notable tributaries flowing from the higher mountains on the east side of the drainage basin. Sonoita Creek drains the west slopes of the Canelo Hills, passing between the Patagonia and Santa Rita Mountains before joining the river north of Nogales. The river is perennial to Tubac, as it was historically, but since the late 1960s this flow mainly is effluent discharged from the Nogales International Wastewater Treatment Plant downstream from Nogales (fig. 2.1);[6] flow was increased to 23 ft^3/s following plant capacity expansion in June 2009. From Tubac to Amado, infiltration into the sandy streambed occurs at a rapid rate,[7] and the stream is normally dry at the north end of this reach, as it was historically. In winter, because of decreased water consumption by riparian vegetation upstream during this season, the channel has intermittent flow to just south of Continental.

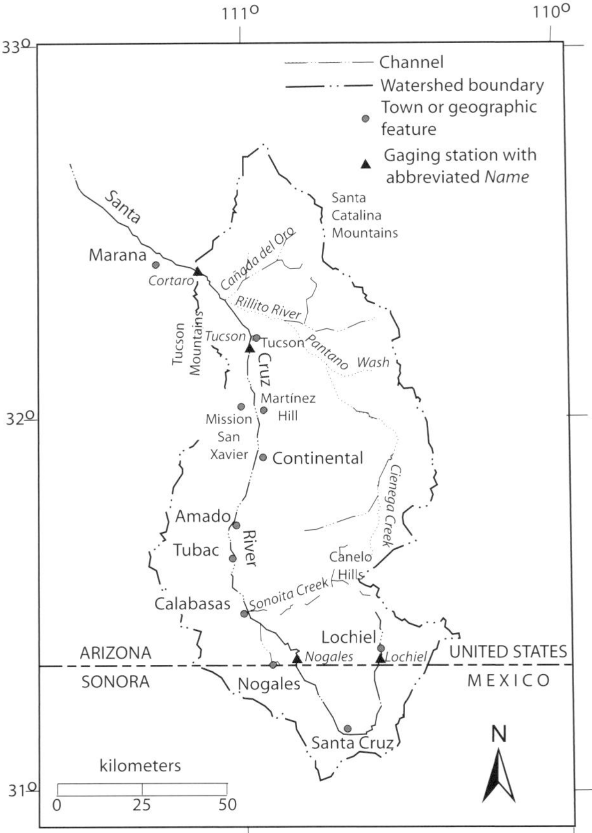

Figure 2.1.
Map of the Santa Cruz River in the Tucson Basin.

At Continental, a deep, continuous arroyo enters the Tucson Basin (fig. 2.1), a northward-trending, structural depression of about 100 square miles. The Santa Cruz River is ephemeral through the San Xavier Indian Reservation, with vertical banks up to 30 feet high and 300 feet apart where the river meanders around the base of Martinez Hill (fig. 2.1). To the north, even the largest floods are confined within the channelized reach that has partial to continuous banks of soil cement, an artificial bank protection that inhibits flood damage to the heavily urbanized floodplain (fig. 2.2). Historically, the channel was in a different position on the west side of the valley, and that largely abandoned channel, known as the West Branch, has its own history of change, as we discuss in later chapters. The Santa Cruz River flows 128 miles from its headwaters to Tucson, falling 20.1 feet per mile over that distance. The drainage basin at the Tucson gaging station is 2,222 square miles with mean-basin elevation and annual precipitation of 4,050 feet and 16.9 inches, respectively.

Northwest of Tucson, where the Rillito River and Cañada del Oro join the Santa Cruz from the east, the channel gradually becomes shallower and wider.[8] Treated wastewater is discharged into the river upstream from the confluence with the Rillito River (Roger Road Water Reclamation Facility) and immediately downstream of the Cañada del Oro confluence (Ina Road Water Reclamation Facility),[9] contributing to a diurnally varying streamflow of about 20–70 ft^3/s at Cortaro Road in the absence of runoff from upstream.[10] A narrow band of riparian vegetation, a mixture of native and nonnative species, lines the channelized river in this reach. The most common species are Athel tamarisk and tree tobacco, both nonnative species that can be invasive; the presence of black willow, Frémont cottonwood, burrobrush, and other native species varies, depending on depth to groundwater.[11]

North of Tucson, the Santa Cruz River flows another 97 miles before it joins the Gila River south of Phoenix. At the downstream end of the Tucson Basin, north of the northernmost ridges of the Tucson Mountains and the town of Marana (figs. 1.1, 2.1), floodwaters spread out onto a broad and deep alluvial plain, where, prior to human modifications, deposition was uninterrupted for centuries, if not millennia, without a well-defined channel. This featureless plain, typified by the Santa Cruz Flats near Eloy, is interrupted only by the deep arroyo emanating from Greene's Canal, a ditch that was built in about 1910 (see chapter 6) and became an active headcut during the high runoff in 1915. North of Santa Cruz Flats, the Santa Cruz River could be considered to be a tributary of Santa Rosa Wash, which has a much larger channel at the confluence. The Santa Cruz River ends its 241-mile course and 4,300-foot drop south of Phoenix near the small villages of Komatke and Laveen at an elevation of 1,017 feet.

Little if any sediment entrained upstream of Marana makes it through the Santa Cruz Flats to the Gila River except during rare, large floods. Indeed, most maps do not show a channel crossing this nearly featureless plain. Most of the time, the lower Santa Cruz valley functions as a closed basin, with all water and sediment from the Tucson Basin trapped on the alluvial plain downstream from Marana. One could speculate that the sediment yield of the Santa Cruz River contributes to that of the Gila River only when streamflow is sustained during extremely large floods, which occurred in 1891, 1915, 1983, and 1993.[12]

Geologic History of the Tucson Basin Reach

The Tucson Basin was formed by uplift of mountain blocks and downthrow of the intervening landscape

Figure 2.2.
The Santa Cruz River north of 22nd Street in Tucson. **A.** (16 April 1903) This upstream view shows the channel of the Santa Cruz River just downstream of the present-day 22nd Street Bridge. (D. Griffiths, courtesy of the National Archives.) **B.** (27 July 2001) Soil cement lines both banks of this ephemeral channel, which supports native and nonnative riparian vegetation. (D. Oldershaw, Stake 2483, courtesy of the Desert Laboratory Collection.)

during the Early Tertiary (about 55 million years ago), giving the region its distinctive Basin and Range physiographic characteristics.[13] To the north and northeast, the Santa Catalina and Rincon Mountains are metamorphic core complexes dominated by a core of granitoid rocks, mostly gneiss of Proterozoic age, with a halo of metamorphic rocks, some bearing economic quantities of copper and other metals.[14] To the southeast and southwest, the Santa Rita and Sierrita Mountains are also metamorphic core complexes, but they have granitic cores of younger age, mostly Early Tertiary. To the west, the Tucson Mountains are mostly highly fractured volcanic rocks with some scattered and fractured sedimentary lithologies mixed in, particularly limestones.[15]

Additional changes to the Tucson Basin resulted from extensional faulting around 17 million years ago that down-dropped the center of the valley.[16] The total depth of the Tucson Basin, documented in one deep well,[17] is 12,000 feet of sedimentary rock overlying granitoid rocks. The Pantano Formation occurs at depths of 6,170–8,260 feet and represents the initial period of erosion of the surrounding mountain ranges, particularly the Santa Catalina and Rincon Mountains.[18] Sediment eroded from these highlands moved across alluvial fans and into the depression that formed over much of the landscape between the mountains, including the area of present-day Tucson. Surficial exposures of these alluvial fans remain, notably, northwest of Tucson. Westward tilting of the valley fill during the Miocene accounts for the dipping beds of the Pantano Formation. The result of either climatic change or faulting, the original fill was eroded several times, leaving a series of terraced surfaces sloping down to the present alluvial valley upstream from Tubac.[19]

The upper 6,170 feet of the Tucson Basin is unconsolidated fill deposited from local sources that constitute the major groundwater aquifer of the Tucson Basin. Most of the sediments constituting this fill are fluvial deposits of braided streams that cross alluvial fans into the basin. The lithology of deposits intercepted by the well indicates a shifting source from the various sides of the basin, which probably reflects the history of uplift and erosion. Several lacustrine units also are present, indicating that either persistent lakes or ephemeral playas once were present in the Tucson Basin.[20] Several outcrops of this fill on the basin margins have been named, including the Tinaja beds, an informal designation for extensive deposits to the southwest, and the Fort Lowell Formation, which underlies much of the surface of the Tucson Basin.[21]

At an unknown time during the middle Pleistocene, about a million years ago, the northeast side of the closed basin was breached by the ancestral Santa Cruz River.[22] Nine geomorphic surfaces have been identified in the vicinity of Tubac, indicating that climatically and (or) tectonically driven cut-and-fill sequences occurred from the mid-Tertiary to the early Holocene.[23] These geomorphic surfaces are ancient alluvial fans emanating from the eastern side of the drainage basin, and these mostly disappear where the river passes from Santa Cruz County into Pima County (figs. 1.1, 2.1),[24] suggesting that the terraces may have graded to a much lower base level in the Pleistocene, probably to one of the levels intercepted in the deep well near the middle of the Tucson Basin.

Perhaps the most comprehensive mapping of geomorphic surfaces in the Tucson Basin was conducted along the Cañada del Oro in what is now the town of Oro Valley.[25] Sixteen geomorphic surfaces in total were mapped in three basic time intervals that encompass the main geomorphic surface, known as the Cordonnes surface. This surface, of Pleistocene age, consists of coalescing alluvial fans emanating from the western Santa Catalina Mountains and occupies the largest surface area between the Santa Cruz River and the Santa Catalina Mountains. Downstream from Marana (see fig. 1.1), Holocene alluvium, delivered from the combined yield of the Santa Cruz River and its tributaries, the Rillito River and Cañada del Oro, has buried the youngest Pleistocene surface, known locally as the Jaynes terrace.[26] The depth of this unconsolidated fill ranges from 100 feet to 130 feet along the central axis of the valley.

Holocene Development of Arroyos

Little is known about the channel form of the Santa Cruz River through the Tucson Basin at the end of the last Ice Age and the beginning of the Holocene approximately 12,000 years ago. Before 9,000 years ago (calendric age, not radiocarbon age), the Santa Cruz was a braided stream flowing across bottomlands about 20–30 feet below the land surface adjacent to the present-day river.[27] The extent of Holocene filling is manifested by the present-day land surface adjacent to the river, and Holocene fill covers a large area of the Tucson Basin.[28] Valley aggradation, associated with both valley-wide flooding and the formation of marshes upstream of the San Xavier Mission and Tucson, occurred throughout most of the Holocene and was punctuated by brief but extensive episodes of arroyo downcutting and degradation circa 9,000–6,400, 4,500, 2,000, 1,000, and 500 years ago (fig. 2.3).

About 9,000 years ago, the channel was approximately 30 feet below what is now abandoned floodplain, and groundwater discharged at the surface of the valley near San Xavier and Tucson.[29] Discharge of groundwater in a vaguely defined and shallow channel formed *cienegas*, wetlands characterized by standing water ringed with aquatic and riparian vegetation. The early Holocene alluvium was removed by downcutting and channel widening between 9,000 and 6,400 years ago,[30] during the middle Holocene or so-called Altithermal,[31] which thus left little depositional evidence.[32] Deposition of a new floodplain began before 6,400 years ago, and alternating braided streams, high water tables, and arroyo downcutting occurred at least five times afterward, including initiation of the late nineteenth-century

arroyo.[33] Cienega development, reflecting high groundwater levels, characterized the period between 4,500 and 2,500 years ago.

The Santa Cruz River generally follows the regional framework of downcutting and filling over the past 4,000 years.[34] Two paleoarroyos of comparable width and depth to the modern arroyo and following a similar course were incised into the floodplain around 2,000 and 500 years ago (fig. 2.3). The paleoarroyo of 500 years ago was about 560 feet wide and 18 feet deep in the San Xavier reach. Its downstream extent is undetermined, but this paleochannel extended through Tucson,[35] and it represents an episode analogous to downcutting of the historical arroyo. The paleoarroyo filled rapidly.[36] Between AD 1020 and 1160, two tributary arroyos on the western side of the valley were discontinuous and became shallower as they approached the floodplain of the Santa Cruz River.[37] At the time that these arroyos

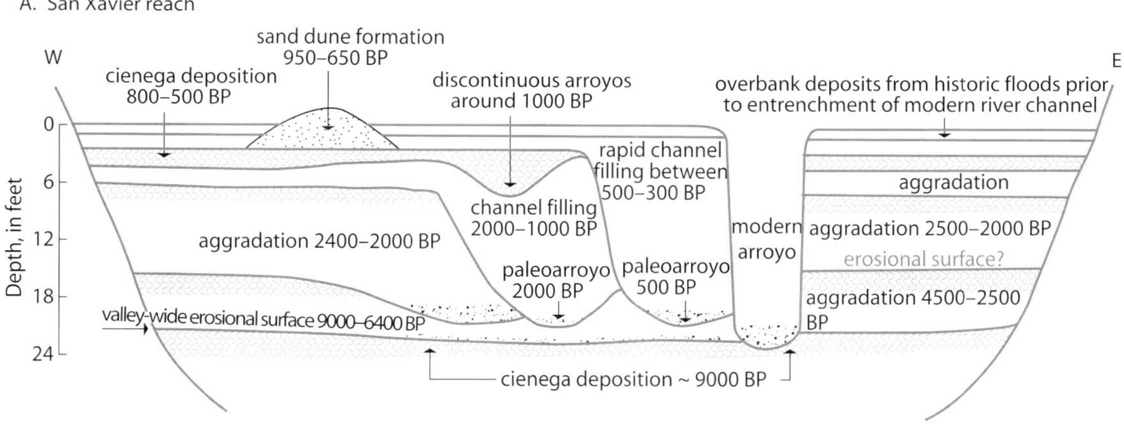

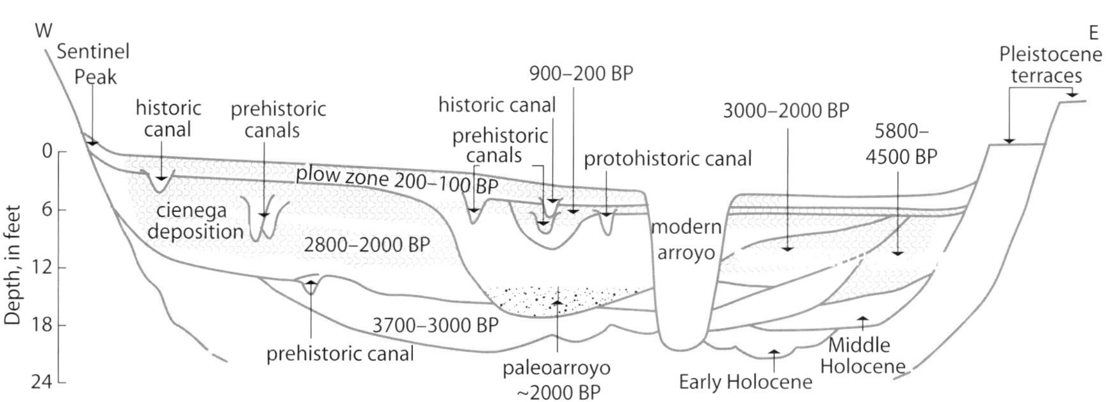

Figure 2.3.
Idealized Holocene stratigraphy of the Santa Cruz River (not to scale; modified from Waters 1988 and Mabry 2006b: fig. 20.3). The stratigraphic dates shown are calendric ages. **A.** San Xavier reach. **B.** Tucson reach.

were present, sand transported downwind from their channels accumulated to form low dunes that now outcrop near the historical source of the Spring Branch or Agua de la Misión (see chapter 4).[38]

Extensive analyses of the past 2,000 years of alluvial stratigraphy in the Tucson Basin (fig. 2.3) have led to a rich story of floodplain aggradation and channel downcutting and how groundwater levels have responded.[39] Correlation of stratigraphy in various reaches suggests that cienegas developed near the end of deposition of stratigraphic packages;[40] in other words, groundwater rise follows alluviation, and groundwater drop follows arroyo downcutting. This interpretation differs from other explanations of arroyo change, which hypothesize that arroyo downcutting may be related to, or even caused by, death or reduction of riparian vegetation caused by a drop in groundwater levels.[41] In this book, the history of arroyo downcutting in the Santa Cruz River is used to test this hypothesis.

Cultural Impacts of Paleoarroyo Downcutting

Human occupation of southeastern Arizona began around 11,500 years ago with the Clovis culture, which is best preserved in the San Pedro River valley east of Tucson.[42] There is good reason to assume that these peoples, and several groups that followed, would have also occupied the Santa Cruz River valley in the Tucson Basin, given later aboriginal uses. However, channel erosion evidently has removed all the geologic and cultural evidence of early people in the vicinity of Tucson.[43] The earliest evidence of human occupation in the vicinity of Tucson is in the late Archaic period, 4,500 to 2,000 years ago.[44]

Beginning about 3,500 years ago, Early Agricultural/San Pedro phase people living in the Tucson Basin manipulated the river by creating canals to divert water to agricultural fields.[45] These canals are the oldest known in North America north of central Mexico;[46] they predate the better-known extensive Hohokam canal system throughout the Salt River valley near Phoenix,[47] as well as at Snaketown on the Gila River south of Phoenix.[48] The past 2,500 years represent vertical aggradation of some twenty-three feet, punctuated by short periods of arroyo downcutting when channels incised into a narrow trench but eroded only a small amount of the adjacent floodplain. Particularly following the most recent episode of arroyo downcutting—approximately 1,000 years ago—Hohokam culture underwent a major

upheaval, at least in part in response to destruction of their canal network.[49] The canals did not turn into arroyos; instead, channel downcutting lowered available water below the elevation of the diversion points. Much of the canal geometry was preserved, suggesting that not all human manipulation of floodplains resulted in channel downcutting.

There was a close correspondence between the intensity of Hohokam agriculture, including shifting settlement patterns in the Martinez Hill area (fig. 2.1), and floodplain stability.[50] The peak of Hohokam activity corresponded with periods of net aggradation and cienega development; for example, aggradation occurred during the Rillito and early Rincon phases between 850 and 1,150 years ago. The location of several village sites shifted during the middle Rincon phase (900–1,000 years ago) to the fans of newly formed discontinuous arroyos. As these paleoarroyos filled and cienegas developed during the Tanque Verde phase (700–850 years ago), the number of villages increased, particularly in the eastern sector of the floodplain, which remained unincised. The paleoarroyos that developed about 500 years ago may account for abrupt abandonment of the area. As this paleoarroyo filled between 500 and 300 years ago, prehistoric farmers known as the Sobaipuri occupied the Tucson Basin. When Kino first visited San Xavier and Tucson in the 1690s (see chapter 4), their population in the Santa Cruz River valley was greater than at any other location in southern and central Arizona but still far less than the peak Hohokam population in the ninth through the twelfth centuries.

Hydroclimatology of Southern Arizona

Climate and Climate Variability

The basin-wide precipitation for the Santa Cruz River basin of greater than 10 inches per year makes it a semiarid watershed by common definition. Two long-term precipitation records, at the University of Arizona and Tucson International Airport, indicate that Tucson's precipitation is 11.35 and 11.56 inches, respectively, which is higher than the 9.84-inch average annual precipitation that generally defines the upper limit of desert climate. Precipitation at the University of Arizona has been measured from 1868 to the present (fig. 2.4),[51] and the long-term averages are 38 percent precipitation in winter (November–March) and 49 percent in sum-

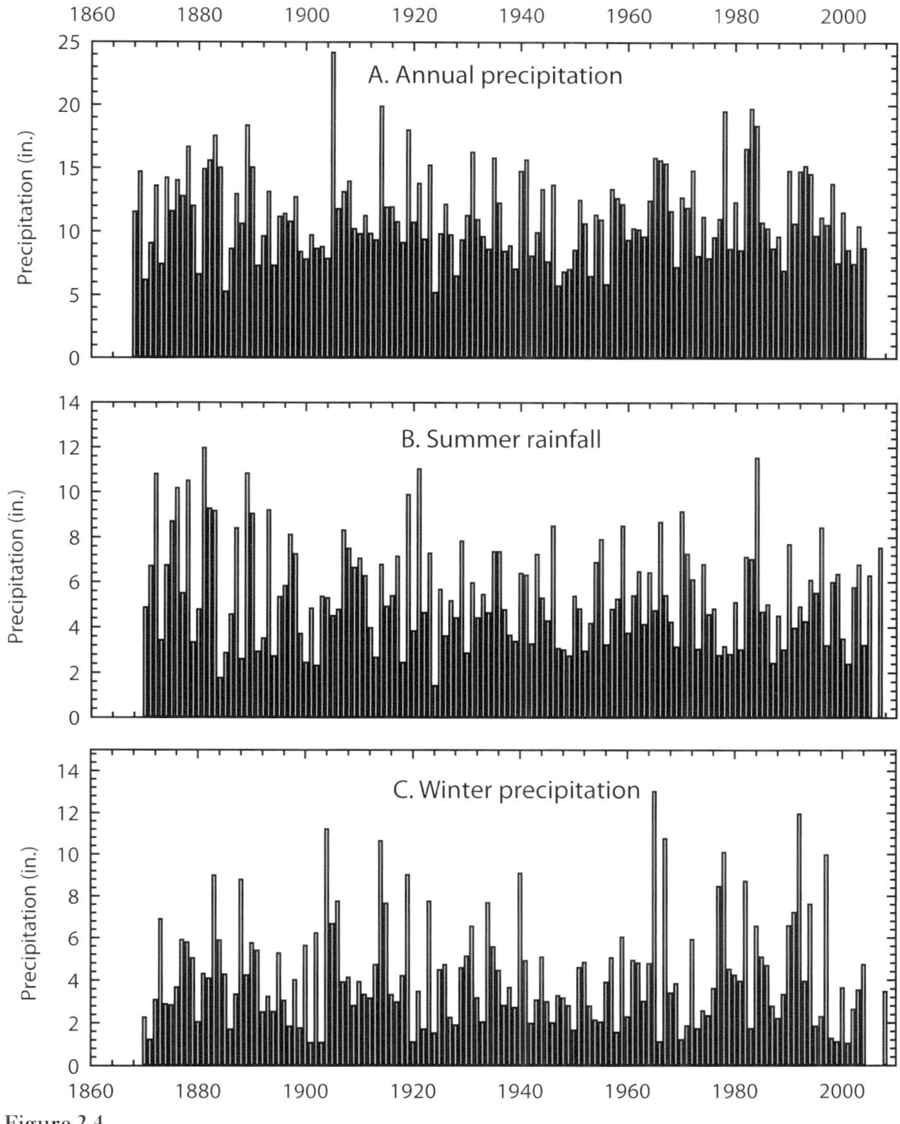

Figure 2.4.
Seasonal precipitation for the climate stations collectively known as Tucson University of Arizona. The period of record is 1868–2009; missing data are common for 2005–2009. **A.** Annual precipitation (January–December). **B.** Summer rainfall (July–September). **C.** Winter precipitation (previous November–present March).

mer (July–September). Although Tucson's high temperatures for June and July average nearly 100°F, the basin also experiences seventeen days of frost per year with an average minimum temperature of 38.6°F in January.[52]

The climate of southeastern Arizona has varied historically.[53] One metric of climatic variability is the interdecadal variation in precipitation, calculated as a standardized anomaly index with the mean and variance removed to create a dimensionless time series (fig. 2.5). For long-term climate stations in the Upper Sonoran Desert,[54] these data indicate that periods of more-or-less-stable precipitation occurred at various time periods from the beginning of the record in 1868 to the present (fig. 2.5). This type of time series has been used to define and name historical periods of climate that have affected the southwestern United States following settlement.[55]

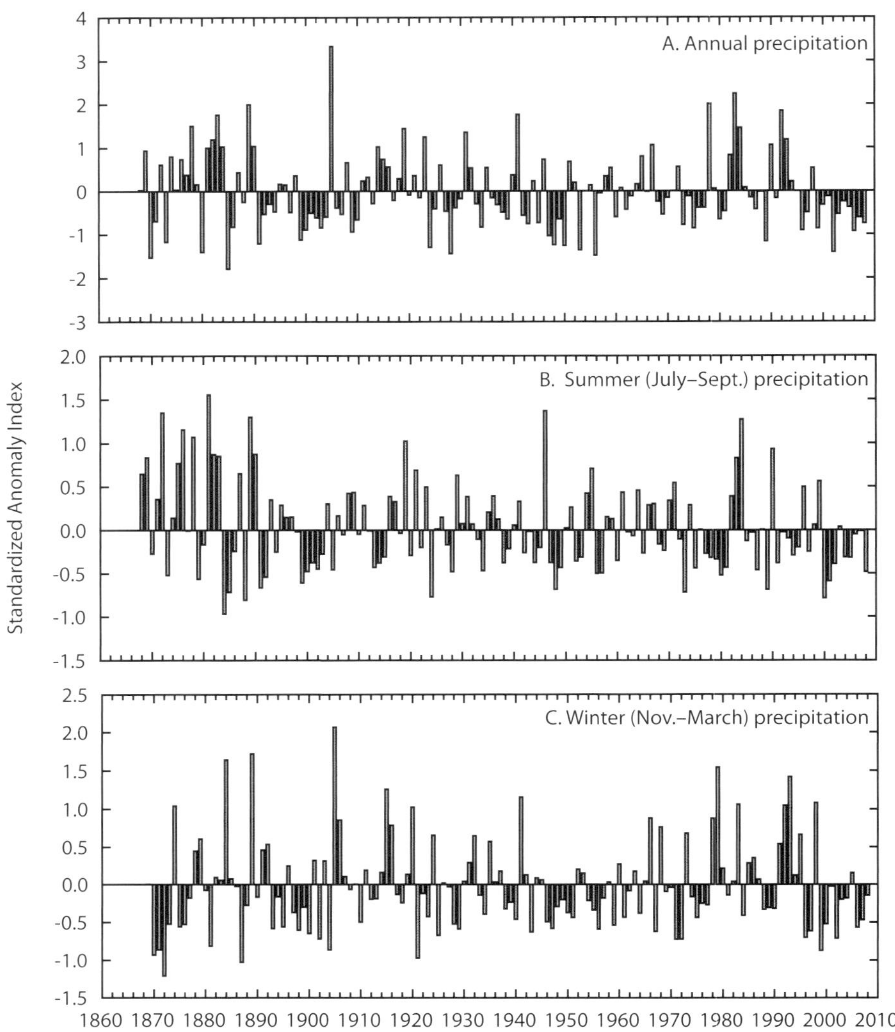

Figure 2.5.
Seasonal precipitation anomalies at a group of climate stations collectively known as the Upper Sonoran Desert (revised from Turner et al. 2003). The number of stations reporting varies among years, with a maximum number of thirteen in 1953. Only the Tucson University of Arizona gage reports from 1868 through 1892. **A.** Annual precipitation (January–December). **B.** Summer precipitation (July–September). **C.** Winter precipitation (previous November–present March).

The late nineteenth-century pluvial, which is poorly known but likely extended from about 1860 to 1891, was a time of increased summer precipitation but also large winter storms, particularly in 1861–1862 and 1884. The early twentieth-century drought, the most severe in the region's history, started and stopped at different times across the region, but it extended from the summer of 1891 to the fall of 1904 in southern Arizona. The early twentieth-century pluvial,[56] which caused overestimation of Colorado River flow,[57] extended from the winter of 1904–1905 to about 1920 in southern Arizona, al-

though some researchers extend this period to 1940 or 1942 in southern Utah and northern Arizona.[58] The midcentury drought, sometimes referred to as the 1950s drought, extended from the mid-1940s to the early 1960s and possibly as late as 1976. From 1977 through 1995, an extended period of increased fall and winter precipitation caused substantial flooding in the region, setting record floods on most rivers.[59] From 1996 to 2004, and arguably through 2012, the early twenty-first century drought has prevailed, punctuated in 1997–1998 and 2003–2004 by El Niño events that had little effect on increasing precipitation above normal. The early twenty-first century drought differs from other historic droughts in that summer precipitation mostly has remained about normal or has increased, as occurred in 2006 (fig. 2.4B).[60]

Hydroclimatology

Hydroclimatic research in the Southwest links flood-producing storm types to large-scale circulation patterns in the atmosphere.[61] Here, we reduce these to four principal types and associated upper-atmospheric circulation patterns that are interrelated. These storm types generally fall within the well-defined seasons of spring to arid foresummer (April–June), summer to early fall (July–September), fall (September–October), and winter (November–March). In large watersheds such as the Santa Cruz River basin, floods often occur under a special set of climatic conditions that combine general circulation over North America and sea-surface temperatures in the Pacific Ocean.[62] Thus, floods may integrate climatic information that might be difficult to detect in more-direct measurements of the climate system.[63]

Summer thunderstorms of typically local extent are part of the seasonal circulation system that has variously been called the Arizona, Mexican, and North American monsoon.[64] Monsoonal precipitation is highly variable spatially but consistent from year to year at a given station;[65] monsoonal floods on the Santa Cruz River exhibit similar characteristics and do not appear to change with time.[66] Tropical storms, cutoff lows, and winter frontal activity associated with heavy flooding result from unique atmospheric conditions, linked to high sea-surface temperatures in the northeastern Pacific Ocean and broadscale circulation anomalies. These storms appear to be the source of most significant sources of precipitation for large floods in the southwestern United States. Precipitation from one or more tropical storms may contribute more than a third or even most of the summer precipitation at a given station in Arizona.[67] Tropical storms that are known to have tracked inland usually produced significant rainfall somewhere in the Southwest,[68] in many cases causing large floods.[69]

Frontal Systems
(Fall, Winter, and Spring)

Winter storms in southern Arizona originate from two related mechanisms that in turn are related to large-scale general circulation of the atmosphere over the Northern Hemisphere. Low-pressure frontal systems embedded in the westerlies can track over Arizona, most commonly in the winter months of December through February. The storm track moves southward in conjunction with seasonal expansion of a low-pressure cell, called the Aleutian Low, that occurs in the North Pacific Ocean. During dry winters, storms track north of a high-pressure ridge off the California coast and move into the Pacific Northwest. In wet winters, this ridge is displaced westward and a low-pressure trough develops over the western United States, pushing storms southward into Arizona. When a subtropical jet bearing moisture from the Central Pacific Ocean, known colorfully as "the Pineapple Express," joins the low-pressure trough, rainfall may be high in Arizona. An example of a frontal system that caused a flood on the Santa Cruz River is the storm of 17–18 December 1978.[70] The rainfall during this storm ranged from 2.8 inches to 9.8 inches in central Arizona and caused widespread flooding.[71]

Cutoff Low-Pressure Systems
(Fall and Spring)

When a high-pressure ridge in the Pacific is well developed, low-pressure systems can detach from the pattern of general circulation and form cutoff low-pressure systems, also known as cutoff lows. These are essentially low-pressure eddies that move between tropical and extratropical circulation systems, and occasionally these eddies can become stationary for many days off the Pacific Ocean coast of California. Cutoff lows that affect Arizona typically form between 30°N and 45°N latitude and 105°W and 125°W longitude and have maxima in spring and fall.[72] Cutoff lows typically intensify off the California coast before moving inland into Arizona, where they can produce substantial rainfall.[73] In fall, cutoff lows may stall over warm tropical waters and steer dissipating tropical cyclones inland, creating conditions for the idealized probable-maximum precipitation in Arizona.[74]

Dissipating Tropical Cyclones
(Summer and Fall)

Occasionally in late summer and early fall, tropical cyclones, which include hurricanes and tropical storms, move northward or northeastward, causing widespread and intense rainfall in the southwestern United States.[75] In Mexico, these storms are included in a group of storms generally called *chubascos* (literally, squalls) or El Cordonazo de San Francisco (the Lash of St. Francis), the latter in reference to their occurrence around the

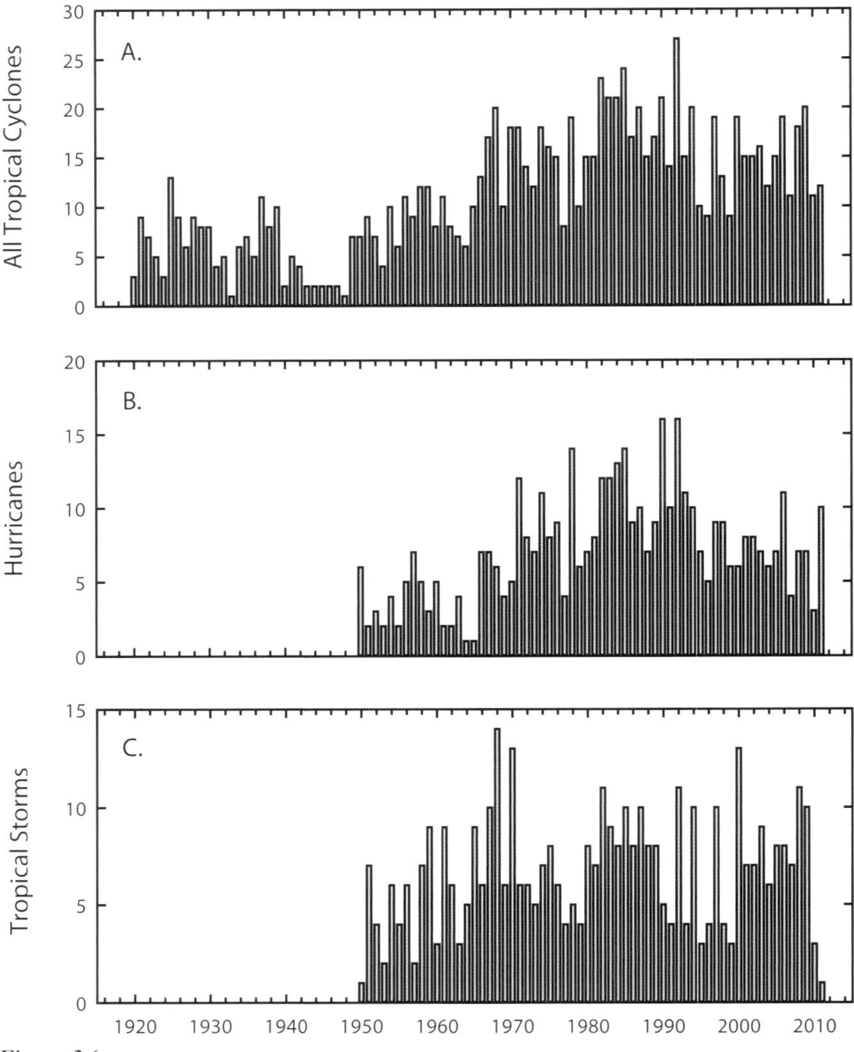

Figure 2.6.
The number per year of (**A**) all tropical cyclones, (**B**) hurricanes, and (**C**) tropical storms in the
eastern North Pacific Ocean from 1921 through 2008. Before satellite coverage was attained in 1965,
many tropical cyclones were not recorded, and tropical storms and hurricanes were differentiated
only after 1950 (1921–1946 data from García et al. ca. 1976).

feast day of St. Francis on October 4. On average, 15.9 tropical cyclones are generated each year in the eastern North Pacific Ocean (fig. 2.6),[76] including an average of 7.3 tropical storms and 8.4 hurricanes (1965–2011). July and August have the greatest number of tropical cyclones, with 3.4 and 3.5 cyclones per month, respectively.[77] The main area of cyclone generation is in the eastern North Pacific Ocean off the west coast of Mexico between 10°N and 15°N latitude and between 95°W and 100°W longitude; most tropical cyclones originate more than 185 miles south of Cabo San Lucas, the southernmost point in Baja California.[78] Some storms originating in the Atlantic Ocean or Gulf of Mexico cross Central America and become Pacific Ocean storms.

Several studies have grouped the types of movements of tropical cyclones in the eastern North Pacific Ocean into three to five categories.[79] Most of these tracking categories do not have significant effects on southern Arizona, because they track westward toward Hawaii. The most common ones that do affect this region track (recurve) west-northwestward from their points of origin, intensifying into tropical storms or hurricanes and either making landfall in Mexico or dissipating over the ocean. Some tropical cyclones recurve toward the north and east, steered by either southerly winds ahead of a low-pressure trough centered over the Pacific Northwest, a weak trough between two subtropical high-pressure cells, or circulation associated with a cutoff low-pressure system.[80] These cyclones dissipate over the ocean and their moisture is advected into Mexico and the United States, particularly southern Arizona, causing intense precipitation and regional flooding. Precipitation from dissipating tropical cyclones can range from several tenths of inches to more than 11.8 inches in two to four days.[81]

Recurving cyclones that have affected southern Arizona were generated most frequently in September and October (72 percent) compared with July and August (27 percent).[82] Between 1965 and 1984, an average of 1.4 tropical cyclones per year caused precipitation in the southwestern United States. Tropical Storm Octave in late September and early October 1983 is an example of the interaction between a tropical cyclone and a cutoff low that caused record flooding in the Santa Cruz River.[83] More commonly, dissipating tropical cyclones contribute isolated to widespread showers embedded within the overall monsoon season.

The difference in seasonality of cutoff low-pressure systems and tropical cyclones explains the greater incidence of recurvature during fall.[84] Although generation of tropical cyclones is at a maximum in July and August, cutoff low-pressure systems have a maximum incidence in October. The greater incidence of recurvature in the fall also is associated with the weakening and southern migration of the Pacific subtropical high and the more frequent appearance of midlatitude troughs at lower latitudes.[85] These two phenomena can behave synergistically, because dissipating tropical cyclones may contribute moisture to early fall extratropical cyclones from the North Pacific, while the extratropical cyclones provide extra lifting to enhance precipitation.

Monsoonal Storms (Summer)

The summer rainy season in Arizona begins near the end of June and early July, when subtropical high-pressure cells shift rapidly northward and induce advection of moist tropical air into Arizona.[86] Monsoonal storms typically are isolated or complex groups of thunderstorms that have a duration of less than several hours.[87] Analyses of broad-scale patterns in precipitable water, water-vapor flux, low-level winds, and regional precipitation suggest that much of the moisture originates from both the Pacific Ocean and the Gulf of California,[88] although the Gulf of Mexico may be the largest source for day-to-day summer precipitation in the Southwest. These storms tend to have weak atmospheric steering systems and mostly are not associated with broad-scale patterns of general circulation. Floods caused by monsoonal storms have occurred in every year of record for the Santa Cruz River.[89]

The largest storms during the summer monsoon typically are mesoscale-convective complexes, consisting of extremely large or coordinated thunderstorms.[90] These storms tend to be associated with larger-scale circulation features, prompting one researcher to refer to them as "monsoonal frontal systems."[91] The second-largest flood on the Santa Cruz River downstream from the Rillito River was caused by mesoscale-convective thunderstorms in July 2006. Generally, floods generated by monsoonal storms are more numerous and extreme on small tributaries of the Santa Cruz River instead of on the mainstem, which has more extreme floods during winter storms and dissipating tropical cyclones.

El Niño and La Niña

The relative importance of flood-producing storm types appears to vary through time.[92] Like drought years, heavy rainfall events tend to cluster in time, suggesting that they are symptomatic of persistent anomalies in atmospheric circulation. The atmosphere generally shifts between two different states of large-scale motion, the stable one dominated by zonal flow, and the unstable one by meridional circulation. Shifts between these two states have been linked to decadal differences in global temperature trends and regional climate.[93] Several large-scale atmospheric-oceanic states have been associated with change in atmospheric circulation, none more strongly than the El Niño–Southern Oscillation (ENSO) phenomenon.[94]

ENSO exhibits two polar states termed El Niño and La Niña.[95] In Southern Hemisphere summers, a southward-flowing current brings warm waters to the normally cold coast of Peru and Ecuador, signaling the end of the fishing season. Because it occurs around Christmastime, local fishermen named this current El Niño (The Christ Child). Climatologists and oceanographers now reserve the term for an amplification of this seasonal warming, which occurs at intervals of two to ten years and lasts for a year or more, crippling the local fishing industry, producing torrential rains in the Peruvian coastal desert, and affecting weather worldwide.[96]

Conversely, La Niña conditions produce colder oceanic waters off the western coasts of North and South America and typically produce drought conditions.

The link between the ENSO phenomenon and anomalous weather has been made not only in the tropics but also well beyond, in subhumid and temperate regions worldwide. For example, during the strong ENSO episode of 1982–1983, links were established between heavy rains and flooding in coastal areas of northern Peru and southern Ecuador; severe drought in northeastern Brazil, much of Africa, Australia, Indonesia, and India; a relatively hurricane-free season in the tropical Atlantic; and a wet, stormy winter over California, the southwestern United States, and the Gulf states. The long-term stability of these apparent spatial coordinations, termed *teleconnections*,[97] is undetermined and, according to some authors, questionable.[98]

Although instrumental records of ENSO, determined primarily by sea-level pressure differences between Darwin (in Australia) and Tahiti, have been collected for more than a century, other data have been used to reconstruct ENSO conditions for far longer.[99] This long-term perspective on ENSO sheds some light on conditions in the twentieth century that may help to explain events on rivers in this region, particularly the Santa Cruz River and its tributaries. One reconstruction estimated El Niño and La Niña occurrences since AD 1525.[100] This record indicates that an estimated 43 percent of extreme and 28 percent of protracted ENSO events (both El Niño and La Niña events) occurred in the twentieth century, and 30 percent of these occurred after 1940.[101] This work suggests that extreme events that affect the Santa Cruz River—increasing the magnitude of both floods and droughts—have been more severe historically, particularly in the latter half of the twentieth century.

The ENSO phenomenon helps to explain why major floods may cluster in time in the southwestern United States. Some researchers have observed a correlation between El Niño activity in the tropical Pacific during the previous summer and heavy precipitation in the fall,[102] as well as in the following winter and spring, in the southwestern United States.[103] It should not be surprising to find some correspondence between the list of El Niño years and the roster of major floods in the southwestern United States over the past century.[104] One of the more intriguing features of ENSO effects in the Southwest is the changing frequency of events over time and its correspondence to annual precipitation patterns. The period of droughts between 1930 and 1960 is characterized by three strong El Niño events (1932,

1940–1941, and 1957–1958), which produced high precipitation and some flooding in the region. A much higher frequency of ENSO episodes prior to 1930 and after 1960 coincides with relatively wet periods with numerous large floods.

Hydrologic Data Collection

Between its headwaters and terminus, numerous streamflow gaging stations document flow and flood frequency in the Santa Cruz River.[105] These gaging stations were established for different purposes at different times, and they document everything from flow into and out of Mexico (gaging stations at Lochiel and near Nogales, respectively; see fig. 1.1) and wastewater effluent flowing from Pima County into Pinal County (gaging stations at Cortaro and Trico-Marana Roads in Marana). All of these gaging stations document flow and floods over a water year, defined as 1 October to 30 September, coinciding with the federal budget year. This artificially separates the fall runoff season, which creates problems with the assumption of interannual independence in annual floods.[106] This is particularly important in regard to the flood of record, in October 1983, which is officially recorded in the 1984 water year because the peak discharge occurred on 2 October. The hydroclimatic water year for southern Arizona, defined as 1 November to 31 October, helps to correct the problem of the split in the tropical cyclone season; the 1983 flood then is properly placed in the 1983 hydroclimatic water year.

Two long-term gaging stations have been maintained on the Santa Cruz River in Pima County. The gaging record for the Santa Cruz River at Tucson is the longest but is discontinuous because of a complicated station history. Although the first gaging station was installed in 1905,[107] the continuous gaging record accepted by most hydrologists began in 1915. The station was discontinued in 1981 and was reestablished in 1986,[108] although annual peak discharges were measured or estimated for the missing years.[109] Peaks above a base discharge of 1,700 ft^3/s (the partial-duration series) were measured for 1930–1981; peaks above base discharge are not known for July and August 1984 or for water year 1985. The gaging station named the Santa Cruz River at Cortaro, Arizona, is at the Cortaro Road Bridge north of Tucson and has records for 1939–1947, 1950–1984, and 1990 to the present.[110] The drainage area is 3,544 square miles above this gaging station, which includes the Rillito River and Cañada del Oro drainages.

Runoff in the Santa Cruz River at the Tucson gaging station occurs mainly from December through February and July through October with high variability in monthly streamflow.[111] The annual flow volume for the Santa Cruz River at Tucson (upstream from effluent discharge points) is 22.5 ft^3/s, averaged over the period of 1915–1981; the Rillito River, the major tributary in the Tucson Basin, has an annual average flow volume of 14 ft^3/s for the period of 1914–1975 at a now-discontinued gaging station that was at the 1st Avenue Bridge.[112] During certain periods of time, the Rillito River has had a higher annual flow volume and once was considered to be a potential source of irrigation water for the Tucson Basin.[113] Between about 1913 and 1934, the Rillito River produced 21,500 acre-feet of runoff annually compared with 17,600 acre-feet for the Santa Cruz River.[114] Because it drains both the Santa Catalina and Rincon Mountains, storms impacting these mountain ranges, such as occurred in July 2006, could greatly increase runoff in the Rillito River compared with the Santa Cruz River, but over time, the Santa Cruz River clearly is the master watercourse.

Flood Frequency at Tucson

The gaging station just downstream of the Congress Street Bridge in Tucson is named the Santa Cruz River at Tucson, Arizona, and its record of floods is one of the longest in the state of Arizona (fig. 2.7). Through much of the twentieth and early twenty-first centuries, the channel at this location typically has been dry, with flow occurring only thirty to sixty days per year. The average size of annual flood at Congress Street is

7,060 ft^3/s, and in many years, the runoff is local with little or no contribution from most of the upstream watershed. The flood of record, which occurred in October 1983, had a peak discharge of 52,700 ft^3/s, an amount that is one-third higher than the second-highest peak discharge of 37,400 ft^3/s in January 1993. The time series of annual flood series suggests that the seasonality of flooding has shifted over the ninety-five years of gaging record at this site (fig. 2.7), and past research has shown that the hydroclimatology of floods at this gaging station has changed through time as well.[115]

The Wall of Water

One of the greatest myths concerning flash flooding in southern Arizona, and throughout the region at large, is the question of how fast the river rises during a flood. The story is still told, particularly by those who drive into a wash, fooled by the depth of a crossing, that the flood rose with "a wall of water." While there is no question that flow rises quickly at the start of a summer flood and that flash floods pose potentially fatal hazards to motorists, they rarely have front flood waves higher than a foot or so, and that front generally consists of some combination of foam, debris, trash, and water, particularly early in the runoff season. The "wall of water" generally is a phenomenon that occurs in narrow, steep channels instead of a meandering arroyo such as the Santa Cruz River across the Tucson Basin.

An example of a summer flood hydrograph appears in figure 2.8. This flood, which is part of the partial-duration series because its peak discharge was 4,470 ft^3/s, began with a dry channel. Within a half-hour period,

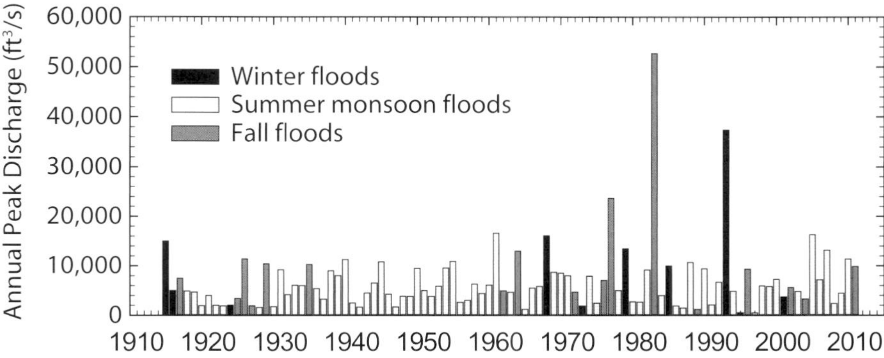

Figure 2.7.
Annual peak discharges for the Santa Cruz River at Tucson, Arizona. The gaging station recording these floods has been either on or just downstream from the Congress Street Bridge, and the record between 1981 and 1988 was estimated using indirect discharge techniques (for the 1983 flood) or other records (data from Webb and Betancourt 1992).

the stage, which roughly translates to flow depth at the gaging station, rises about two feet during the rising limb of the hydrograph. Several hours later, the flow rises 3.75 feet in 1.75 hours (fig. 2.8A). In 11.25 hours, the flow decreases from its peak discharge to a nonflowing channel during the flow recession. While changes in water level and discharge of this flood are rapid, they do not represent anything approaching a "wall of water" during either the rising or the recessional limb of the hydrograph.

There is no "typical" hydrograph for the Santa Cruz River. Winter floods generally rise gradually in response to long-distance runoff from the headwaters to Tucson and may lead to weeks of sustained flow at Congress Street.[116] Summer floods rise and fall rapidly, as discussed above, but the shape of the hydrograph depends on the distance to the source of runoff; if the floods are generated in Mexico, the flood will rise more gradually than if the flood is generated in Tucson or just upstream.

Floods generated during dissipating tropical cyclones can either be flashy or have gradual rises, or both.[117] How hydrograph shape affects erosion is uncertain; most observers relate flood erosion to the duration of the hydrograph, not to its shape.[118]

Muddy Water

Flow in the Santa Cruz River seldom is clear. Summer floods tend to have higher sediment concentrations than winter floods, but very few sediment-transport data have been collected from this river and only during summer runoff.[119] During 1988 and 1989, summer storms produced floods that passed by the Congress Street gaging stations, and crews collected sediment-transport samples during those floods. For discharges of 400–9,580 ft³/s, sediment concentrations ranged from 3,100 to 64,400 parts per million (ppm) of suspended sediment.[120] Sediment concentrations were higher on

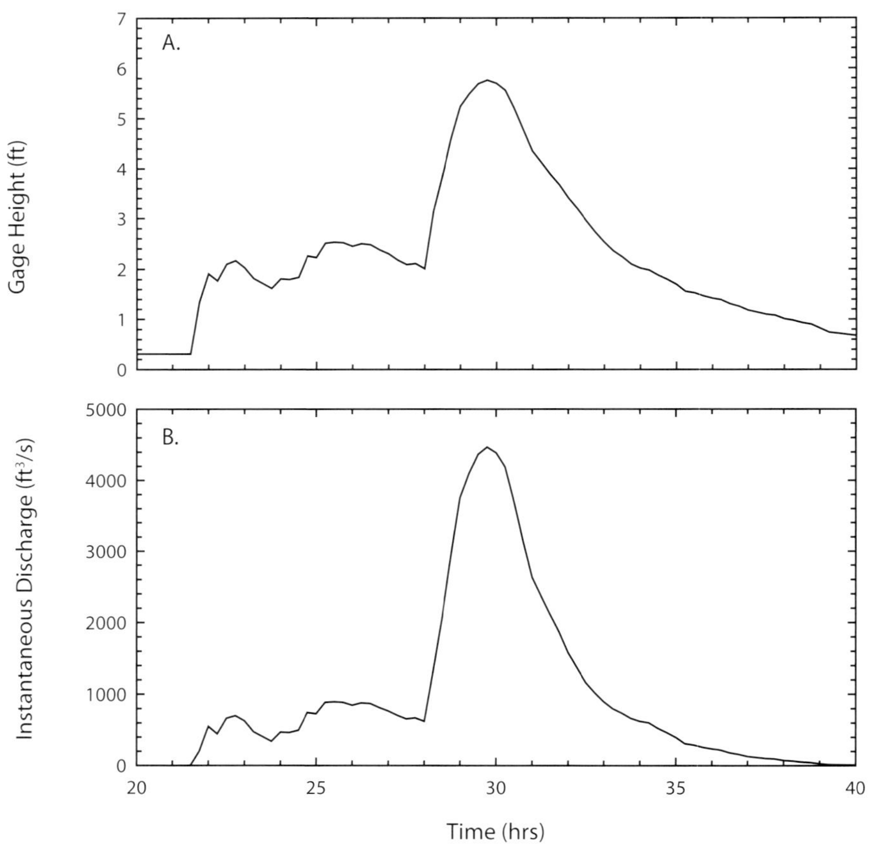

Figure 2.8.
Gage height (**A**) and instantaneous discharge (**B**) in the hydrograph of a monsoon-generated flash flood in the Santa Cruz River at Tucson, 3 July 2009.

the rising limb of the hydrograph than on the recessional limb, and these concentrations are not unusually high. For comparison, concentrations exceeding 400,000 ppm have been measured for the Paria River at Lee's Ferry, Arizona.[121]

Where sediment transport has been measured in both winter and summer, the differences in sediment concentrations are striking.[122] Winter flows on the Salt River in central Arizona were less than about 600 ppm in 1991–1992, partly because the river is regulated with multiple dams upstream. In contrast, summer floods in 1992 on the Hassayampa River yielded sediment concentrations of about 13,000 to 130,000 ppm. As for the Santa Cruz River, sediment concentrations were highest before the peak discharge was reached. This indicates that most sediment entrainment occurs on the rising limb of the hydrograph and that deposition occurs on the falling limb. Many times on the Santa Cruz River, local scour followed by filling back to the original bed surface has been reported, which agrees with the general results of sediment-transport data collection.

Riparian Ecosystems and Channel Change

The history of riparian vegetation growing along the Santa Cruz River is one of extreme change, change that is related to the types of vegetation and their relation to groundwater. Three general types of woody vegetation have grown in the vicinity of Tucson. Vegetation dependent on shallow groundwater for survival, known as *obligate riparian species*, have been present along the reach from Martinez Hill to below the confluence with Rillito River throughout recorded history, although their locations have shifted considerably. The most common plants in this category are Frémont cottonwood and willow trees, although arrowweed, seep willow, and coyote willow have also been reported at various times.

Riparian species that can use—but do not require—shallow groundwater, known as *facultative riparian species*, include mesquite and blue paloverde trees,[123] as well as burrobrush, a shrub that commonly grows on floodplains of the Santa Cruz and Rillito Rivers.[124] These species are common in desert settings but grow larger and denser when additional water is available along channels. Finally, desert plants, also known as *xeroriparian* or *xerophytic species*, also can grow in or adjacent to channels and include many of the most common species of the Sonoran Desert, including creosotebush.

Desert broom, in particular, is common in upland vegetation settings, typically in disturbed sites, as well as along ephemeral washes.

Several types of riparian ecosystems occur in southern Arizona.[125] Riparian woodlands are rare in the Sonoran Desert and traditionally are referred to by the Spanish name of *bosque*.[126] True wetlands, or cienegas (from the Spanish word *ciénaga*), still occur in the region and sustain what is known as obligate riparian species, or those trees and shrubs that require perennial water at or close to the ground surface. One term once used to describe this ecosystem is *hydroriparian*.[127] The best examples of obligate riparian species in southern Arizona are Frémont cottonwood, black (or Goodding) willow, coyote willow, sycamore, and netleaf hackberry, among other species. In contrast, facultative riparian species—and velvet mesquite may be the best example in southern Arizona—grow in or near riparian settings but can also grow in dry upland environments. Mixtures of obligate and facultative riparian species occur in what have been termed semiriparian ecosystems, and riparian zones occupied primarily by facultative riparian species have been termed pseudoriparian ecosystems.[128] Another term once used to describe the semiriparian ecosystems, especially along intermittent streams, is *mesoriparian*. The term *xeroriparian* is now well accepted to describe the ecosystem developed along ephemeral washes, replacing *desert riparian*, an ambiguous term used in the past to describe ecosystems along intermittent and perennial desert streams, as well as those along ephemeral washes.

Riparian areas provide numerous benefits for rivers, water resources, and the regional ecology. Bosques and other areas of dense riparian vegetation create a flow impediment that spreads water over a greater surface area, decreasing channel erosion and the sediment load carried downstream, as well as allowing water to recharge into the ground. The dense network of roots helps to diminish channel erosion by increasing sediment strength. Riparian areas provide substantial habitat to everything from insects to predators, and that habitat is unique in the desert Southwest. The beneficial functions of bosques in specific and riparian areas in general have been termed *ecosystem services*,[129] reflecting a need to show a societal benefit of the presence of certain ecosystems.

Riparian areas along southwestern rivers have among the highest densities of birds for the United States.[130] Cottonwood-willow gallery forests and mesquite bosques are of special interest to ecologists,[131] and they are critically important to birds.[132] In spring and fall, these corridors are important routes for birds migrating from the American Tropics, through the Southwest, and into the northern and eastern parts of North America. Birds that move seasonally between Central and North America, or even farther in their annual pilgrimages, depend on patches of riparian vegetation for food and water to fuel their travels, as "stopover habitat" for resting during their long flights, or even as places to raise their young.[133]

Birds are drawn to riparian ecosystems for their abundant food supply, especially insects and plants.[134] Riparian areas in the arid lowlands of the desert Southwest typically contain five to ten times the number of both individuals and bird species compared to that of the surrounding uplands.[135] Phenology (timing of recurring events in the annual life cycle of plants and animals) is of critical importance to neotropical migrants, whose arrival may be timed to coincide with insect abundance associated with flowering of mesquite and other trees and shrubs.[136] The mixed canopy height and structure in mesquite bosques and adjacent cottonwood-willow stands are critical for bird diversity. Neotropical migrants also use vegetation along ephemeral washes;[137] continuous bands of riparian trees are not as important to migrating bird populations because migrating birds use both continuous gallery forests and isolated patches along their flight path.[138]

Long-Term Change in Riparian Ecosystems

There is little question that extensive stands of obligate riparian vegetation were lost along the Santa Cruz River upstream from Tucson, primarily in response to groundwater overdraft.[139] Misconceptions abound concerning how much riparian vegetation the Santa Cruz River in the Tucson Basin sustained—and where it grew—before the arroyo downcut and groundwater was mined. Fanciful descriptions of extensive cottonwood-willow forests extending through Tucson aside, the reality was that riparian vegetation was patchy along the river's course, not surprisingly congregating in reaches of perennial flow, such as in the Great Mesquite Forest and the reach in the vicinity of Sentinel Peak (A Mountain). In other reaches (for example, downstream from the current Grant Road Bridge), there is little evidence except for references to cottonwood trees in the 1860s of any significant and continuous reaches of riparian vegetation other than the ubiquitous mesquite and paloverde. An exception is the reach where the Santa

Cruz River is joined by the Rillito River and Cañada del Oro, where a grove of cottonwoods and perhaps even a small mesquite bosque were present in 1937.[140]

Historical observations of southern Arizona, beginning in the mid-nineteenth century, document the presence of regionally important riparian ecosystems.[141] Riparian ecosystems in southern Arizona are seriously threatened by water resources development,[142] particularly groundwater extraction ("mining") at levels beyond natural recharge rates.[143] Little is known historically about the occurrence or distribution of riparian ecosystems regionally, although specific studies and accounts document specific riparian areas that have changed significantly.[144] What is generally known is that cienegas, or riparian marshes, occurred along many watercourses in southern Arizona, particularly along the Rillito and Santa Cruz Rivers. As the accounts in this book show, those riparian zones typically were of limited extent, related to high groundwater levels, and separated by intermittent or ephemeral reaches.

We know little about Sonoran Desert riparian bird assemblages before European settlement, because few studies or surveys were conducted before water development eliminated many of the largest riparian areas in the mid- to late 1900s. The Santa Cruz River is one of the few exceptions due to its strategic location and biodiversity. The Santa Cruz is situated at the juncture between tropical deciduous forests of western Mexico and the southwestern US deserts, and it connects several biomes, including conifer forest in the *Sky Islands* (isolated high mountain ranges in southeastern Arizona and northwestern Mexico), oak woodland, desert grassland, and Sonoran desertscrub. The few studies in riparian woodlands and forests of the Arizona lowlands during the late 1800s and early 1900s are of special interest, particularly for the Santa Cruz River, one of the most well-documented losses of riparian ecosystems in the southwestern United States.[145]

Because of the lack of records before the late 1800s, the biota, including avifauna, of southeastern Arizona at the time of early Euro-American settlement will never be known. The watercourses of this region were altered to an unknown extent by Spanish settlers beginning with the arrival of Eusebio Kino in the late seventeenth century. Within the later groups of Anglo explorers, military men, and settlers, there was a general absence of both ornithological and botanical description of the area. As early as the 1870s and 1880s, water projects that impacted the riparian vegetation had begun along the Santa Cruz River, and the first ornithological records from this region were made at this time (see chapters 5 and 7).

The Regional Significance of Riparian Ecosystems in Southern Arizona

The Santa Cruz River, and the San Pedro River to the east, formed a south–north riverine fluvial/riparian corridor from Mexico to the Mogollon Rim of northern Arizona.[146] Historically, this corridor was important not only to migrating birds but also for fish and aquatic amphibians, reptiles, and mammals—such as beaver and rare mammals (for example, ocelot and jaguar).[147] Its importance as a migration corridor was recognized at the start of the twentieth century:

> The river valley runs practically due north and south, and, presenting an abundance of food, water and shelter in a comparatively restricted area, with a barren, practically desert, country on all sides, it forms a natural highway, along which the majority of the birds passing through the region would naturally travel.[148]

The Santa Cruz and San Pedro Rivers both flow north into the Gila River, which courses westerly to its confluence with the Salt River. From there, animals could migrate eastward and upstream along the Salt River to the Verde River, then northward to Sycamore Creek, which drains the southern side of the Mogollon Rim, or Colorado Plateau. Even though there was neither a continuous permanent stream along some of this corridor nor a solid stand of woody riparian vegetation, birds and many other animals were able to cross the spaces of miles or tens of miles of desert landscape between water sources and riparian vegetation by either flying, swimming during flooding, or moving at night.

In spring, migrating birds moved northward along the north–south riverine routes from the Mexican border into central Arizona. They continued northward for approximately 350 river miles through the arid lowlands to the forested Colorado Plateau. A reverse route could be followed by migrants in the fall. The riparian forests and woodlands provided water, food, and "stopover habitat" for transients during their long journey. During the breeding season, lowland neotropical breeding species were able to expand their ranges northward along the Santa Cruz River into habitats not unlike those farther south without having to fly over high mountain ranges. In winter, during times of severe weather, birds could drift downhill from adjacent mountains into river valleys, escaping the cold and finding food and shelter in the valley lowlands. These factors explain the large number of species and individual birds that once occurred in the Tucson Basin and other riparian ecosystems of the Santa Cruz–Rillito River system.

The Great Mesquite Forest (see chapter 7), which historically extended along the Santa Cruz River south of Martinez Hill (figs. 1.1, 2.1), was one of only two mesquite bosques in North America that are named in the scientific literature, and both were on the Santa Cruz River. The second bosque, known as the Komatke Thicket,[149] was near the confluence of the Santa Cruz and Gila Rivers on what is now the Gila River Indian Reservation. The local Pima Indians named it the New York Thicket because it was so crowded with trees and animals. This bosque was approximately three miles wide from north to south and eight miles long from east to west. The south end was a cienega, about one mile wide by two miles long, and was "covered with a heavy growth of bamboo [probably common reed]."[150] Unfortunately, the avifauna and other faunal and floral features of Komatke Thicket were never surveyed beyond studies of breeding White-winged Dove.[151]

Even less is known about the animals that inhabited riparian areas in southern Arizona, except possibly for birds. The Santa Cruz River and one of its major tributaries, the Rillito River, have one of the longest ornithological records for Arizona.[152] As we discuss elsewhere in this book, Tucson became the ornithological capital of the Southwest during the late 1800s and early 1900s. The Santa Cruz and Rillito Rivers locally supported lush riparian vegetation that served as ideal nesting habitat for riparian birds, including some of those breeding "Mexican" birds; attracted waterbirds, shorebirds, and long-legged wading birds; and served as migration routes and stopover habitat for a large number of species during spring and fall.[153]

The Santa Cruz and San Pedro Rivers, the two major tributaries of the Gila River in southeastern Arizona, have several similar physical and spatial characteristics,[154] but they differ biogeographically, and this influences their respective avifaunas. Additionally, while most watercourses in Arizona flow from east to west or north to south, the Santa Cruz and San Pedro Rivers are notable exceptions because they flow from south to north, which encourages birds to follow these watercourses during their north–south migrations. Both rivers flow into Arizona from Sonora, draining significant highland area south of the US-Mexico border. The two are roughly parallel to one another, approximately forty to fifty miles apart along most of their length. The San Pedro River flows through the Chihuahuan biogeographic region at a higher elevation along most of its length. The Santa Cruz River is the only major river in southeastern Arizona that flows through the lower elevation, known as the Sonoran biogeographic region.[155] These characteristics influence the biota, including the avifauna, of the two rivers, and numerous species from extreme northwestern lowland Mexico occur in the Santa Cruz drainage but not in the San Pedro drainage. The diversity of the avifauna of southeastern Arizona is further influenced by a third biogeographic region, the Madrean highlands, in which several species of "Mexican" montane birds nest.[156]

Portrait of a Semiarid Watershed

Our brief review of the Santa Cruz River watershed shows a river that mostly drains a semiarid watershed upstream from Tucson. Farther downstream, the channel of this watercourse abruptly disappears into flats and drainage ditches in its downstream reaches. This "river" little resembles the common definition of this word, as perennial, intermittent, and ephemeral reaches play hopscotch downstream from the headwaters; through most of the watershed, the channel is dry sand for most of the year. In the twenty-first century, this pattern of flow is largely controlled by land-use practices, particularly small flow diversions for agriculture and groundwater pumping for domestic and agricultural usage. From its earliest recorded history, the channel also switched from perennial to ephemeral flow, although the locations of both have changed significantly.

In the Tucson Basin, this river changed from a small, mostly perennial channel that sustained small agricultural and pastoral developments in the eighteenth and nineteenth centuries to a deeply incised channel capable of conveying extremely large floodflows (such as those in 1983 and 1993) across an urbanized floodplain. The details of how this change came about are provided in a recitation of that history in the next few chapters, culminating in nearly a century, from 1915 to 2012, during which detailed hydrologic data were collected. In particular, the growth of the small presidio of Tucson into a major metropolitan area is intertwined with the history of floods and channel change in the Santa Cruz River.

3

Causes of Arroyo Downcutting

No other research topic in fluvial geomorphology in the southwestern United States has generated as much attention as the cause for arroyo downcutting at the end of the nineteenth and beginning of the twentieth centuries. Initially, the reasons were pragmatic. The economic damage in the region called for governmental action to repair water-supply systems, potentially by reducing livestock numbers on watersheds or regulating floodplains to minimize changes affecting surface-water flow. Eventually, the motivation became scientific, as geologists trained to seek a climatic cause for landscape sculpting sought to pin arroyo downcutting on climatic change or, more accurately, variation. A tension developed between environmental concerns and economic development, because livestock reductions on watersheds, viewed as beneficial to reducing channel erosion, ran head-on into the viability of pastoral economies in the region. Each of these needs drove scientists and other observers to propose and defend conclusions as to the cause of arroyo downcutting. The resulting array of proposed causes over a century of published research has forced some to adopt a "many possible causes" argument,[1] reinvigorating the old concept of *equifinality*,[2] in which a landform could result from different causes under different circumstances.

Reports of the disparate causes for arroyo downcutting strongly reflect professional interests and advocacy bias.[3] That arroyo downcutting mostly occurred during a period of overgrazing and drought is indisputable, as the history of the Santa Cruz River demonstrates. The societal cost of overgrazing and drought was revealed in the late nineteenth-century drought, during which more than half of the livestock died on southern Arizona rangelands in 1893 and in Utah in 1896.[4] Range managers and ecologists have been quick to suggest that decimation of plant cover by livestock caused watersheds to deteriorate, leading to increased runoff, sediment production, and ultimately arroyo downcutting. Climatol-ogists naturally sought a climatic explanation, knowing full well that climate is ever-changing, particularly when one looks at the long instrumental records from Europe.[5] Geologists, who reconstruct earth-surface events from the stratigraphic record, sought a climatic interpretation for changes, and they initially were fixated on drought;[6] later, some geomorphologists tried to explicitly account for the half century of temporal variability by invoking several new concepts, including intrinsic responses and geomorphic thresholds.[7] Some geomorphologists and hydrologists, accustomed to evaluating channel change in terms of watershed processes, viewed arroyos as a product of changing runoff and the resulting changes in sediment transport in desert landscapes, whether caused by land-use practices or by climatic fluctuations.[8]

Despite sustained research continuing into the twenty-first century, scientists from various disciplines have failed to achieve a consensus cause on why arroyos downcut at the end of the nineteenth and beginning of the twentieth centuries. Although the ongoing debate continues to spur research, it can confuse and frustrate nonscientists. In 1969, a judge ruling on a controversy over the geomorphic consequences of logging in California redwood forests offered this ironic opinion: "While numerous expert witnesses in the field of geology, forestry, engineering, and biology were presented, their conclusions and the opinions they derived from them are hopelessly irreconcilable in such critical questions as how much and how far particles will be moved by any given flow of surface water. They were able to agree only that sediment will not be transported upstream."[9]

In the arroyo controversy, there are multiple indictments, a veritable army of expert witnesses, insufficient evidence, and no real verdict. The moral of this story, certainly the one that is acknowledged by now, is that historic arroyos are a far better subject for study than for debate.

Hypotheses of the Cause of Arroyo Downcutting

Explanations for arroyo downcutting fall into five general categories:[10] (1) livestock grazing and deleterious changes to soils and vegetation at a watershed scale, (2) direct and indirect manipulation of channels and floodplains by human activities and water-resources development, (3) climatic change or fluctuations, (4) the occurrence of extraordinary floods that may or may not be related to regional climate, and (5) intrinsic geomorphic factors of flow and sediment transport in arid-region rivers. We emphasize that flowing water, and particularly floods, causes erosion and sediment transport, and that any cause proposed for arroyo downcutting must explain a change in the amount of runoff and its erosive potential in channels.

Livestock Grazing

The most long-lived cause, invoked by scores of researchers,[11] involves the role of livestock grazing in modifying watershed characteristics that decrease soil infiltration, thereby increasing runoff and sediment production. That livestock grazing affects soils, rangeland vegetation, and certain species—notably perennial grasses—is undeniable; its influence at a watershed scale is where things get a little hazy. For example, the increases in sediment production from heavily grazed and denuded hillslopes should have yielded deposition along mainstem channels, not channel erosion. Livestock also have major impacts on riparian areas, particularly those dominated by grasses and herbaceous vegetation. Livestock trails in riparian areas, where they are parallel to the channel, may concentrate flow, focusing erosion in the denuded, compacted paths.

Perhaps the most important point is that different types of livestock do not have the same impacts on soils and vegetation. Furthermore, the distinction between livestock and other native herbivores may, in some cases, be nothing more than a change from a native to an introduced species with little if any change in degree of impact. One could not easily determine, for example, whether or not replacement of bison by cattle impacts runoff processes at a watershed scale; it might come down to stocking rates and seasonality of watershed usage. When livestock are introduced to desert environments that previously supported few if any large native herbivores, their impact can be very significant.

Because cattle generally were believed to have the largest impact, because of their large body mass, prefer-ence for grass, and sheer numbers on the landscape, their influence is emphasized in this discussion.[12] Cattle are grazers and generally prefer grass; when stressed by drought or when forage is depleted by overgrazing, they will browse shrubs and cacti. That cows locally depleted desert grasslands is undeniable,[13] even though certain grasslands in southeastern Arizona have proved resilient, and other issues, such as fire suppression, might have had larger effects on initiating a conversion from grassland to savanna.[14] Cattle compact soils, thereby decreasing infiltration rates and increasing runoff at lower rainfall intensities. As stocking rates increase, runoff and sediment yield increase from rangelands,[15] primarily because decreased ground cover allows less aboveground water storage, less protection of the soil surface from raindrop impacts and raindrop soil detachment, and lower infiltration rates.

There are three key objections to grazing as a regional cause for historic arroyos. First, epicycles of downcutting and filling, revealed by alluvial terraces, erosional unconformities, and buried paleochannels,[16] occurred many times before the introduction of cattle and after extinction of megaherbivores in North America (bracketed between 15,600 and 11,500 cal yr BP, based on last occurrences for fifteen species).[17] Following the old axiom that "the past is the key to the present," paleoarroyos and historic arroyos should have the same cause. This reasoning causes many earth scientists to discount grazing as the sole cause of historical arroyo downcutting.[18] Although geologic evidence establishes that arroyos downcut prehistorically, and presumably in response to factors other than land-use practices, it does not necessarily exclude grazing as a significant contribution to the most recent episode of erosion. If past vegetation changes induced by climate account for development of paleoarroyos, a case could be made that overgrazing led to rapid and pervasive deterioration of grasslands, contributing to a climate-induced initiation of erosion.

A second objection to the livestock-caused-downcutting hypothesis has to do with grazing history relative to dates of arroyo initiation. Arroyos downcut at approximately the same time in Sonora, southern Arizona, and western New Mexico, even though extensive stock raising began two centuries earlier in Sonora and Chihuahua.[19] One complication is that range conditions two centuries earlier may have been more favorable to sustainable grazing at high stocking rates. For the upper Rio Grande, high livestock numbers in the period of 1788–1848 resulted in little or no gullying.[20] Other researchers counter that this early grazing and trampling had a lagged effect on the landscape, with the final blow dealt by large flocks and herds of the 1870s and 1880s.[21] How the well-documented impacts could have had lagged effects is not discussed and seems farfetched, given the fact that rangelands recover from grazing over short- and long-term scales. How impacts would in and of themselves become cumulative is not obvious unless rangeland type conversion—for example, from grassland to savannah—occurred prior to downcutting. We know from historical photography and analyses that these ecosystem changes occurred in the twentieth century, after downcutting had already happened.[22]

The third problem revolves around the tenuous link between known livestock impacts and the initiation of arroyo downcutting. Heavy grazing generates erosion on hillslopes, and that sediment then is transported onto floodplains; this should result in aggradation, not erosion, of floodplains. Relevant hypotheses have gone untested; for example, sediment contributed from grazed hillsides may steepen transverse gradients across valley floors,[23] creating "critical valley oversteepening"[24] that could favor the types of intrinsic geomorphic processes invoked to explain arroyo downcutting. Modern studies of rangeland hydrology relative to grazing pressure have been far from conclusive,[25] and livestock grazing as the sole cause for arroyo downcutting remains tenuous at best. Although livestock grazing alone may not have caused channels to incise their floodplains, it may well have aided and abetted erosion, and therefore the downcutting of modern arroyos may have been enhanced by livestock.

Other Adverse Land-Use Practices

Other land-use practices with adverse hydrologic impacts could have contributed to erosion of channel floodplains. Removal or decrease of vegetation in channels and on floodplains decreases flow resistance, thereby increasing the velocity of runoff. Irrigation canals and diversions likewise have lower flow resistance than natural channels, allowing floods to greatly increase in flow velocity. Roads and stock trails directed perpendicular to slopes or parallel to channels cause runoff to concentrate, increasing flow velocity and erosion potential.[26] Anecdotal evidence throughout the Southwest links the presence of roads and irrigation canals with the initial downcutting sites for arroyos. In many cases, headcut migration followed the path of an abandoned wagon road, a ditch, or a railroad grade. Other activities linked to channel erosion include placer mining, deforestation of uplands and floodplains, extermination of beaver populations, draining of natural marshes or cienegas, and fire suppression's encouragement of shrublands over grasslands.[27]

Land-use practices can concentrate flows, decrease hydraulic roughness, and increase peak discharges of flood hydrographs having the same total flow volume. As with the effect of livestock grazing, evidence for previous epicycles of erosion figures prominently in the counterargument against adverse land-use practices being the sole cause for arroyo downcutting. Several scientists argue that widespread erosion during AD 1100–1400, dated mostly by archaeological evidence, was unrelated to man's activities, whether overgrazing or the artificial concentration of flows,[28] even though large aboriginal populations used floodplains. Ironically, prehistoric farmers during this period, be they the Ancestral Puebloans on the Colorado Plateau, the Hohokam of central Arizona, the Sobaipuri of the upper San Pedro and Santa Cruz rivers, or the Tohono O'odham of the Sonoran Desert, may have outnumbered the rural population of the Southwest in the late nineteenth century. These prehistoric farmers harnessed streamflow to grow crops in ways not radically different from European practices, including extensive canal irrigation on the Santa Cruz River and other floodplains in central Arizona.

Prehistoric human impact has been discounted as a cause for twelfth- through fifteenth-century arroyo-cutting, possibly because of the still-romanticized concept of the Noble Savage, mythical beings capable of sustainable living with no ecological impact.[29] Scientists have not been immune from such sentiments, as reflected by statements such as "since prehistoric Indians lacked livestock, and since it is commonly believed that they did not despoil nature, the origin of the fossil trenches [arroyos] cannot readily be attributed to humans."[30] This assumption has repeatedly been challenged, including the following argument:

> It is rather interesting to note that Doctor [Kirk] Bryan has used the evidence of human occupation found in ancient buried channels to strengthen his theory of an arid period causing erosion. . . . With equal force, the available evidence can be used to support a belief that the ancient channels may have been caused by accelerated erosion directly related to human occupation. . . . It seems to be the general conclusion that these ancient peoples practiced some method of flood irrigation, diverting water from ephemeral streams to irrigate their crops. The effect of this practice would be that only the infrequent, high

discharges would be allowed to pass down the valley. That lower portion of the valley, deprived of its plant sustaining low flows, would be subjected to a much greater erosion hazard than would have been the case naturally. . . . It would seem to be more remarkable if erosion did not occur with human occupation than that it did.[31]

Although the grazing hypothesis appears to be thwarted as a primary cause, owing to its association only with historic downcutting, the hypothesis that Native Americans manipulated floodplains and inadvertently caused arroyo downcutting, at least in the most recent prehistoric episode, remains at least a possibility that cannot be easily discounted. It would seem even more remarkable if prehistoric farmers understood ephemeral-stream processes well enough to avoid causing channel erosion as a side effect. Although this argument suggests humans could have been involved, at least partially, in two episodes of arroyo downcutting, what about other episodes of downcutting that occurred during the past four thousand years, when human use of floodplains was less intensive?

Drought and Arroyo Downcutting

Many geomorphologists explain the synchronous downcutting of arroyos throughout a region using climatic variation as the principal cause. There are time-honored concepts in this interpretation that have been used to explain geologic evidence millions of years old. A climatic explanation spans the panoply of geomorphic and stratigraphic details that somehow result in geologically simultaneous changes over a large area repeatedly through geologic time. Although the climatic magic wand is frequently used, how climate affects the alternating phases of erosion and deposition is unclear, but geologists apply the extremes—either persistent drought or wet periods—to induce fluvial change. The explanations became highly nuanced as more scientists addressed the question.

Both erosion and deposition have been linked to cyclical drought, in some cases using the same evidence. One should find this counterintuitive, given that water by definition is scarce during droughts, runoff transports sediment, and high runoff would be required to downcut arroyos. Extreme floods have occurred during historic droughts, and some drought-related watershed changes have been implicated in changing runoff and sediment production that caused arroyo downcutting. Underlying climatic interpretations of arroyo erosion

and deposition is the assumption that vegetative cover is the most important factor controlling watershed processes and that vegetation cover is sensitive enough to periodic drought to alter watershed response to extreme precipitation.

Initial scientific observations of channels in the region viewed stream gradients as the primary control of the balance between transport and deposition of sediment, with the volume and character of the sediment strongly adjusted to climate.[32] Aridity would steepen stream gradients and produce aggradation, while a shift to humid conditions would reduce the gradient and lead to entrenchment. This rationale has been applied to the problem of alluvial terraces and arroyo-cutting in the Southwest. Beginning with Huntington,[33] proponents held that the loss of vegetative cover during droughts promotes rapid removal of soil on hillslopes, overloading streams and bringing on channel filling and valley-wide alluviation. A shift toward more humid conditions would have the opposite effect, inducing channelization and downcutting of base level. Increased vegetation cover on hillslopes would inhibit sediment production, resulting in runoff with lower sediment concentrations that was able to entrain additional sediment from the channel bed and banks. This association of entrenchment with wetter conditions was a minority opinion until the 1960s,[34] even though historic arroyos are known to have been entrenched by large floods that occurred in a sequence of wet years.[35]

Bryan was the first to champion drought as the cause for arroyo downcutting,[36] despite the fact that Huntington served on Bryan's dissertation committee. Bryan reasoned that prolonged drought would deplete vegetation cover on hillslopes, reduce infiltration, and increase storm runoff. Greater discharge along valley bottoms would initiate gullying between reaches with discontinuous arroyos, integrating channels by headcut migration. Bryan hypothesized that arroyo downcutting was imminent when cattle were introduced in the Southwest, invoking the image of the trigger pull of livestock grazing on the gun loaded by climate.[37] Ironically, it was Huntington who first used the trigger-pull analogy in reference to impending change,[38] but the phrase is now attributed to his student Bryan.

Invoking drought to cause watershed changes profound enough to significantly affect flow and sediment transport has significant problems, not the least of which is that dry soils infiltrate far more moisture than wet soils. The association of drought with historic arroyos prompted other scientists to invoke different mechanisms, mostly involving channel and floodplain changes. Some argue that groundwater levels decline during drought to the point of decimating riparian vegetation, thereby decreasing channel roughness and increasing the erodibility of fine-grained sediment on floodplains.[39] This hypothesis has a credible process basis, but there is no direct evidence that groundwater levels decreased prior to historic downcutting; indeed, most downcutting was initiated well before the early twentieth-century drought. One could argue the exact opposite: downcutting during a wet period lowers groundwater levels in alluvial aquifers that were effluent into channels, draining water from channel floodplains and presumably contributing to erosion and sediment transport. This point well illustrates the common conundrum in the arroyo debate: a provocative geomorphic process, having clear relation to channel change, could be caused by two different climatic regimes or could not be related to climate at all.

Bryan's hypothesis was embraced by many other geomorphologists, particularly those working closely with archaeologists, notably, his student John Hack.[40] Their enthusiasm for the drought-causes-erosion hypothesis arose not from analysis of historic arroyos but from the coincidence of prehistoric erosion during a hot-dry Altithermal (5500–2000 BC)[41] throughout the West,[42] as well as the "Great Drought" (AD 1266–1299) on the Colorado Plateau.[43] This influence is readily apparent in stratigraphic summaries for the Colorado Plateau, where three depositional layers—the Jeddito, Tsegi, and Naha Formations—are separated by eolian deposits interpreted to represent droughts followed by channel erosion.[44] Other researchers found further support for the drought-causes-erosion hypothesis from an evaluation of radiocarbon dates in Holocene alluvium.[45]

The amount of geologic time during which arroyos dominated the Holocene landscape is short compared to that of valley-wide alluviation or deposition between low terraces. The issue of why arroyos fill has received scant attention, but this process clearly is important to the overall question of downcutting. Some have argued that arroyos fill during phases of cooler and wetter climates.[46] Others point to the increase in riparian vegetation, aided both by wetter climates (particularly germination-enhancing winter floods) and by the draining of alluvial groundwater, as the major contributor to increased floodplain roughness and deposition.[47] This argument becomes extremely important with regard to arroyo processes on the Santa Cruz River.

Changes in Precipitation Intensity

During the first half of the twentieth century, several geomorphologists and climatologists recognized the need to quantify climatic variability and its possible effects on geomorphic processes. Some have claimed that no significant trends were obtainable in climatic records from the Southwest.[48] Others analyzed rainfall data in New Mexico and found trends.[49] In the critical period between 1850 and 1880, just before arroyos downcut historically in northern New Mexico, no trend is detected in annual rainfall, or at least what we know of annual rainfall from the sparse network of climate stations. Throughout the region, annual rainfall accumulates in two distinct seasons—winter and summer—and precipitation occurs at different spatial scales and different intensities during these seasons.

Higher daily rainfall totals, inferred to reflect rainfall intensities, occurred at various times from the late nineteenth through twentieth centuries.[50] Luna Leopold, an eminent twentieth-century geomorphologist, proposed a partitioning of daily rainfall into light rains (daily rainfall of less than 0.5 inch) and heavy rains (daily rainfall of more than 1 inch) as a way to evaluate climatic influences on arroyo downcutting. He believed that light rains favored plant productivity, so a change to fewer light rains would have negative effects on watersheds. Fewer light rains and more frequent heavy rains would result in greater runoff. Leopold identified changes in light versus heavy rains, which he thought caused the erosional episode of the late 1800s, and he hypothesized that an intensified summer monsoon was the mechanism for this climatic influence.[51]

Later researchers had different perspectives on climatic fluctuations or change. Palynological (pollen) evidence suggests a wet Altithermal in southeastern Arizona,[52] and this suggests that increased summer rainfall accounts for both mid-Holocene and historic arroyos. In southern Arizona, some researchers report no significant trends in annual or seasonal precipitation from 1868 through 1966,[53] although heavy rains increased and light rains decreased in the summers of the 1870s and 1880s. Climate records from southern Utah and northern Arizona revealed increases in summer precipitation from the start of most climate records until around 1940 to 1942.[54] Similar trends in rainfall intensities were also noted in central California, where arroyo downcutting is associated with periods of above-normal daily and annual rainfall (1875–1895 and 1935–1945).[55] If daily rainfall totals are assumed to reflect instantaneous intensities and early climate records are presumed to be accurate and compatible with later measurements, it remains unproven that an artificial definition of light versus heavy rains has meaning in terms of vegetation productivity, and it is equally unclear whether any possible influence of precipitation variability on plant cover affects alluvial processes. Given the wide range of physiological and demographic responses of southwestern species to precipitation, seasonal timing of rainfall may be far more important than seasonal intensities. Winter annuals benefited from the shift in climate at the start of the twenty-first century.[56] At the Santa Rita Experimental Range south of Tucson, precipitation during a relatively brief period in summer accounts for most of the interannual variability in grass biomass.[57] In the warm season, a series of light rains might not be as effective as heavy rains in getting soil moisture down to the rooting zones of the different plant species occupying watershed uplands.

The change-in-rainfall-intensity hypothesis is compelling because it focuses on a changing water source for runoff, which is integral to our understanding of watershed behavior. That its influence extends through vegetation cover is inconclusive because of the extreme range of ecological characteristics of watersheds that experienced arroyo downcutting: shrub- and tree-dominated ecosystems have different hydrological characteristics than grasslands. Rainfall-intensity data are available for only the past hundred years, and we currently have no way to address whether intensity changes occurred at other times when arroyos did not downcut.

The drought and rainfall-intensity hypotheses are both overly reliant on the secondary effect of climate on vegetation as a way to control the balance between erosion and deposition instead of the primary effect of increased precipitation directly leading to increased runoff. What appears to be an undue emphasis on vegetation fluctuations pervades the geomorphic explanations of regional arroyo downcutting, especially in research conducted prior to the 1980s. The real issue is whether or not high rainfall intensities recognized for the late nineteenth century produced unusually large floods, irrespective of changes in plant cover. This question could be resolved with modern watershed models, which take vegetation cover into account along with other important variables, including soil infiltration rates, slopes, aspect, and channel configurations. However, this requires that we precisely know watershed conditions at the time of arroyo downcutting. The only information available is anecdotal information from historical accounts.

Intrinsic Geomorphic Factors

Arroyo downcutting and filling have long been recognized as integral parts of the natural processes by which sediment is transported in alluvial systems.[58] Both field and experimental studies show that headcuts develop in floodplains and erosion is focused when and where sediment stored in fluvial systems achieves a critical threshold slope and thus becomes unstable above a discharge or discharge range.[59] This process has been termed critical valley oversteepening; it generally occurs where tributaries join the mainstem or, in the case of wide valleys, where a channel-fan sequence locally creates oversteepened slopes that encourage channel incision. A key point in this model is that intrinsic geomorphic processes lead to local channel incision, or the formation of a headcut, when and where a threshold discharge is exceeded. That threshold discharge is not necessarily related to climate or even land-use practices but instead could be a natural consequence of sediment transport and deposition in an arid or semiarid climate regime.

The intrinsic-geomorphic-response hypothesis holds that isolated headcuts in a watershed periodically coalesce into a continuous arroyo solely from sediment-transport processes. Because intrinsic geomorphic processes vary, not only within a single drainage but also from one drainage to the next (owing to local conditions of geology, flow, and sediment transport), short-term synchroneity of events on a regional scale should be the exception, not the rule. This becomes a major criticism of alluvial-climatic interpretations based on regional correlations of erosional and depositional episodes. If erosion and deposition are random in time and dependent upon local watershed conditions and not climate, then regional correlations of alluvial cut-and-fill stratigraphy are meaningless without independent dating. The concept of intrinsic geomorphic processes as applied to the cause of arroyo downcutting goes straight to the heart of the synchroneity question: given the nearly fifty-year spread in dates of initial downcutting, could arroyos be randomly downcutting and filling on the landscape?

The concept of intrinsic geomorphic thresholds is essential in the explanation of discontinuous ephemeral streams and places the processes that control such streams into a framework of flow and sediment transport. Broadened to the general question of arroyo downcutting and filling, however, it poses a serious challenge to undated correlative schemes in alluvial stratigraphy unless larger controls, such as extreme floods related to regional climatic fluctuations, are involved. This hypothesis alone does not explain how discontinuous arroyos coalesced into continuous arroyos historically, nor does it explain why, at the onset of the twenty-first century, none of those arroyos has completely filled. Some researchers, therefore, resort to land-use practices to explain the conversion from discontinuous to continuous arroyo.[60]

Large Floods of Regional Extent

One clear consensus in the arroyo controversy is that arroyo downcutting was initiated during floods.[61] The earliest observers noted that large floods in and of themselves could have been the principal cause of channel erosion;[62] the question was whether those floods were random in time, related to climate fluctuation or change, or enhanced by land-use practices.[63] Some researchers have noted that the floods associated with arroyo downcutting and widening were extreme in size when compared with typical annual floods,[64] and that clustering of extreme events was associated with periods when climate fluctuated to higher rainfall and more regional storms.[65] The extreme-flood hypothesis stems from recognition that channel change occurs when the discharge exceeds some erosional threshold and that extreme floods are the mechanism for enabling threshold exceedance.

The occurrence of extreme floods alone is not sufficient to explain synchronous arroyo development across a broad region if these floods represent random events in space and time, or if downcutting initiated during only extremely large regional floods. However, large floods over a large region tend to cluster in time because large-scale atmospheric circulation conducive to unusual rainfall events normally persists for years to decades.[66] If the occurrence of extreme regional floods can be linked to regional climatic processes that fluctuate in time, then the extreme-flood hypothesis becomes a viable cause for arroyo downcutting. Other watershed changes caused by land-use practices or persistent drought or wet periods could exacerbate the amount of erosion but not be its cause. For example, numerous roads and canals reportedly downcut into local arroyos, but if all roads and canals were destined to become arroyos, then why do Hohokam canals that transported irrigation water long distances persist on the floodplains of the Santa Cruz, Salt, and Gila Rivers in southern Arizona?

Historically, most channel erosion in the southwestern United States occurred during large floods that clustered in relatively wet periods. Most streamflow gaging stations were installed decades after arroyos had downcut, making comparison of these floods with later events difficult. That some historic floods were extraordinary in geologic time, especially during the past one thousand to two thousand years, has been determined for numerous drainages in the southwestern United States where paleoflood records can be developed.[67] One compelling but disputed analysis suggests that the occurrence of large prehistoric floods clustered in time periods associated with arroyo downcutting.[68] Further research may be required to determine whether this hypothesis indeed is testable, which is questionable given the inherent uncertainties of dating paleoflood occurrence, dating arroyo downcutting, and comparing the two chronologies.[69]

The climatic setting for large floods in the Southwest has been extensively studied, inspired not by the arroyo debate but by the economic consequences of disastrous flooding in the past few decades and its implications for floodplain management. Although some of the increase in flood damages is related to greater construction in floodplains,[70] floods appear to have fluctuated in magnitude and frequency for some watercourses, notably, the Santa Cruz River.[71] Are watershed changes and channel modifications due to progressive channelization, poor land-use management, increasing urbanization of watersheds, and channelization of floodplains, any of which might cause moderate rainfall to produce higher flood peaks? Or are there periodic shifts in general atmospheric conditions that fluctuate between periods of extreme floods versus more ordinary events?[72]

Proponents of extreme floods as a cause for arroyo downcutting argue that a higher frequency of El Niño events heightened the probability for major floods and regional stream degradation.[73] Conversely, fewer El Niño events between 1930 and 1960 resulted in fewer major floods; during this period, many of the arroyos tended to fill. A climatic cause for large floods and thus regional arroyo downcutting remains a viable hypothesis, and its relation to other factors, such as watershed- or floodplain-degrading land-use practices, remains as a topic for more research.

Arroyo Filling

Only a few researchers have discussed what occurs after arroyos downcut to a new base level.[74] Richard Hereford, who has documented historical changes in arroyo systems for more than three decades,[75] proposed

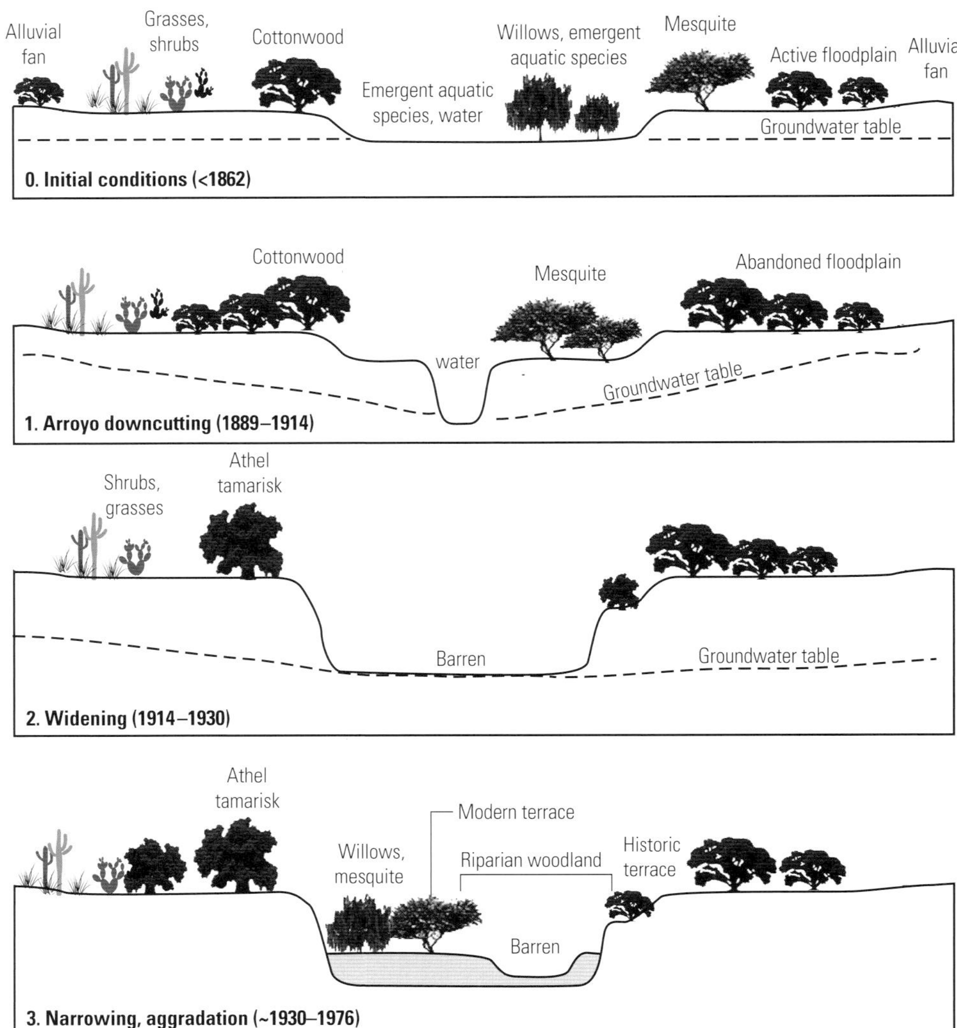

Figure 3.1.
Schematic diagram showing the generalized stages of arroyo development from settlement to the present in southern Arizona (adapted from Webb and Hereford 2010). Many notable exceptions have occurred to this generalized model, particularly with respect to timing of changes and types of riparian vegetation present. The dashed line shows the water level in the alluvial aquifer for perennial streams, and in this general example, water levels remain high, reflecting sustainable groundwater withdrawal (if any).

a series of changes in arroyo morphology based on climatic variability gleaned from analyses of historical records of the southern Colorado Plateau.[76] Subsequent work generalized the sequence of arroyo downcutting, widening, and partial filling (fig. 3.1).[77] Following the initial downcutting, channels widened considerably until about 1940 in most arroyo systems. After 1940, during sustained drought and low flows, floodplains aggraded within the arroyo, providing habitat for woody riparian vegetation.[78] The riparian vegetation traps sediment, creating a positive feedback mechanism that enhances floodplain aggradation. For many arroyos on the Colorado Plateau, this process continues into the twenty-first century unabated for most rivers, with some notable exceptions.[79]

In southern Arizona, channel-change processes are complicated because of a resurgence of large floods from 1977 to 1995.[80] The largest floods in gaging records, some of which date to near the beginning of the twentieth century, occurred during this period.[81] For many rivers in the southern half of Arizona, including the Santa Cruz River, channel widening renewed and some low floodplains were either eroded away or abandoned. Beginning with the early twenty-first-century drought, which began in 1996, low floodplains formed again in this region, and riparian vegetation became established on the deposits. Currently, as shown in chapter 10, channels appear to be aggrading with low floodplain deposition in the Tucson Basin, aided by the establishment of riparian vegetation.[82]

Historical Arroyo Downcutting: The Search for a Cause

Arroyos downcut historically over a wide variety of hydrological, ecological, and cultural settings. They developed over five decades, a short enough period to encourage geomorphologists and others to search for a common cause of some phenomenon that was equally widespread and synchronous. Despite the objections of geomorphologists, it could be argued that development of coalescent arroyos, synchronized within decades and affecting most regional watersheds, happened only once in the past 11,000 years, but this requires a strict view of the uncertainties in dating of erosional episodes. Uncertainties of a few to several centuries cast doubt on regional correlations claimed for the erosional episodes of the middle Holocene and between AD 1100 and 1400.[83] In southern Arizona, where the alluvial stratigraphy of several watersheds has been well documented, periods of degradation and aggradation tend to be out of phase in number, character, and timing from one valley to the next.[84]

Few would dispute that arroyos were a natural part of the Holocene landscape, and even fewer would question the role of floods in initiating downcutting. The circumstances of downcutting varies across the Southwest, suggesting a unique origin or, at the very least, a unique set of circumstances intrinsic to each watershed. Some authors seem content to recognize equifinality,[85] giving up the notion of a unique and singular regional explanation other than the relation between floods and channel erosion. Some forty years after they originally were made, these reflections still hold: "In spite of prolonged interest, some careful work, and an extensive literature, our understanding of gullies in the American West lacks the tantalizing clarity of the landforms themselves."[86]

Part of the problem is that good histories of the details of how and when particular arroyos downcut is anecdotal, incomplete, and inaccurate. In addition, the historical record is scant, incomplete, and occasionally misleading. For only a few southwestern streams do we have even a qualitative understanding of timelines and processes involved in the initiation of coalescent, mainstem arroyos.[87] The Santa Cruz River offers perhaps our best opportunity to chronicle the full development of an arroyo in a reach through the metropolitan area of Tucson. It also provides a test of the general emerging framework of how arroyos might fill and the driving forces behind arroyo filling, particularly in light of intensive land-use and floodplain management.

4

Perennial Flow and Discontinuous Arroyos, 1691–1872

Although humans have been present in the Santa Cruz River basin for more than thirteen thousand years, recorded history began with the arrival of Spanish explorers and missionaries in the late seventeenth century.[1] Although the Spaniards were primarily interested in land, gold, and saving souls, they paid attention to natural resources with the intent of eventual development of their northern lands. The water supply in the vicinity of what is now called Martinez Hill (fig. 4.1) naturally attracted settlement and development, and springs in the wide, open valley downstream meant irrigation for agriculture. Mexico gained its independence from Spain but then sold most of the Santa Cruz watershed to the United States as part of the Gadsden Purchase. Settlers streamed in, slowed at times by animosity from Native Americans, and new observers took stock of the resources available in this dry region. Although there was not much of a river flowing through the Tucson Basin, there was a lot of water available to sustain crops, people, and riparian habitat. For nearly two centuries, these competing resources coexisted, just long enough for us to get a glimpse into what was once present in and along the Santa Cruz River.

The River During Spanish Colonial Days

At the time of the first Spanish explorations, the Tucson Basin sustained a Sobaipuri population of unknown size, although an estimated six thousand Native Americans lived in what is now southern Arizona and northern Mexico.[2] About two thousand Sobaipuri, a Piman tribe, lived along the San Pedro River,[3] east of the Santa Cruz, but they had one settlement at the place later to become San Xavier del Bac,[4] where Kino would build a mission that remains to this day (fig. 4.1). The Sobaipuri, who once lived in small villages in the region, were forced into larger villages as a result of Apache warfare and Spanish aggregation in 1761,[5] and this group eventually merged into the Tohono O'odham Nation.

Members of Kino's parties were the first Europeans to explore the San Pedro and Santa Cruz rivers from near their headwaters to the Gila River. On several occasions in the 1690s, Kino and his men visited Sobaipuri settlements between San Xavier and the Rillito River. Kino first visited Tumacácori in 1691,[6] and he reached the Casa Grande ruins near present-day Coolidge in 1694,[7] presumably passing by the Sobaipuri settlements on the Santa Cruz River near Martinez Hill. In 1697, Kino and Captain Juan Mateo Manje traveled south along the river, reaching a sizeable settlement near Tucson on 23 November:

> [A]fter going six leagues [ca. fifteen miles], we came to the settlement of San Agustín de Oiaur where we were lodged in a big house they had built for us and big enough for all. . . . Here the river runs a full flow of water, though the horses forded it without difficulty. There are good pasture and agricultural lands with a canal for irrigation. . . . We counted 800 souls in 186 houses. . . . On the 26th, . . . we continued south over the plains, passing along the river bed which submerges here.[8]

A total of 3,500 cattle, 2,500 sheep, and 1,200 horses reportedly were at San Xavier. Kino described its

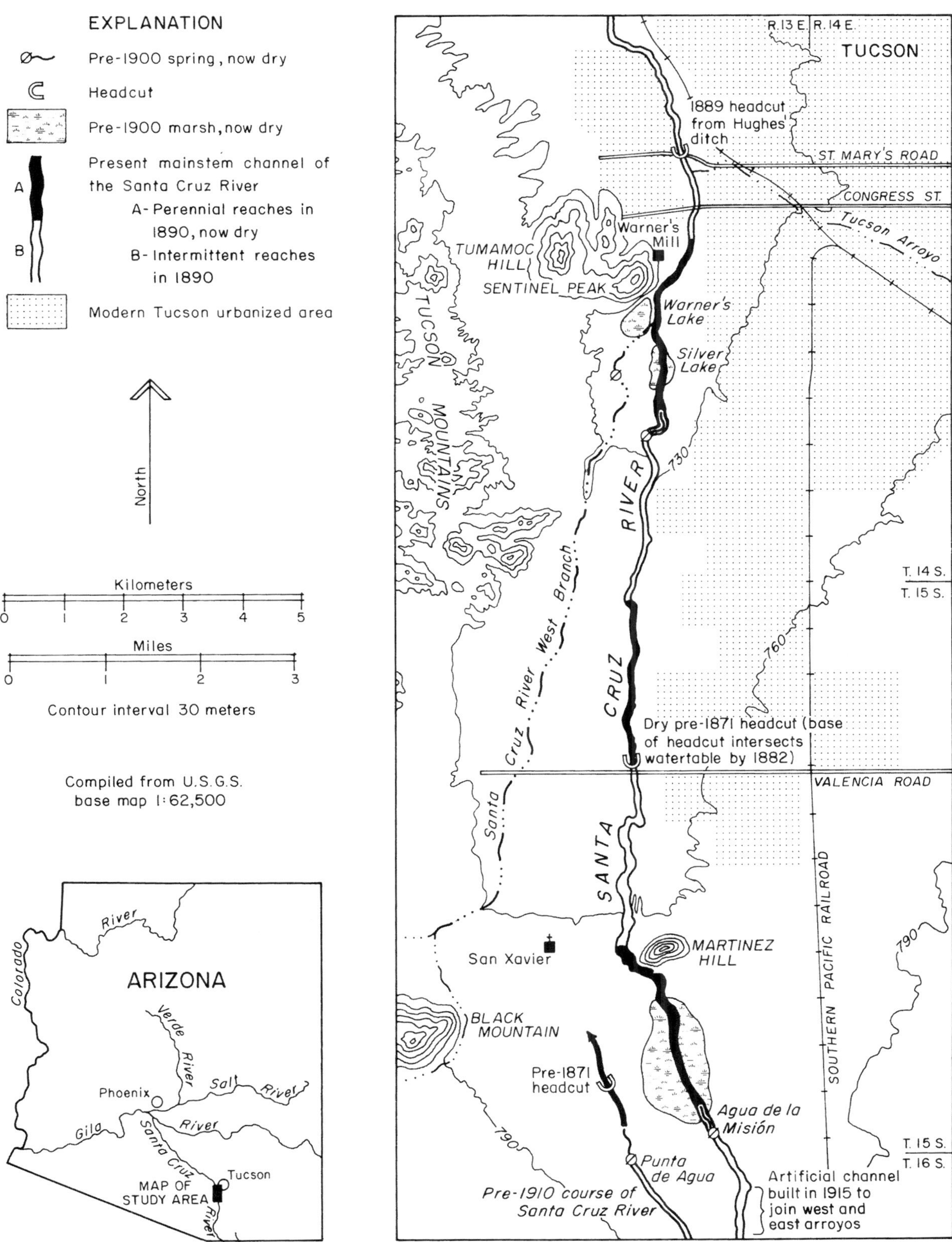

Figure 4.1.

Historic map of the Santa Cruz valley in the Tucson–Mission San Xavier area, summarizing some of the principal features that played a role in arroyo development (from Betancourt and Turner 1988). The Great Mesquite Forest is the area between Martinez Hill and Agua de la Misión.

agricultural potential on 29 October 1699: "The fields and lands for sowing were so extensive and supplied with so many irrigation ditches running along the ground that the father visitor said they were sufficient for another city like Mexico [Mexico City]."[9] That fields were irrigated from ditches indicates that the channel was shallow and its perennial flow could be diverted across the floodplain without pumping or carrying water. On 11 April 1701, Kino and Manje traveled south from San Xavier "over plains and meadows covered with pasture."[10]

Spanish explorations into southern Arizona continued through the eighteenth century. By 1752, the Spanish had a presidio at Tubac. The settlement at San Xavier was vacated by Jesuits in 1767, and Franciscans replaced them nine months later, in 1768. To outdo their Jesuit predecessors, Franciscans pushed to explore an overland route to California, setting up a line of missions along the way, starting with one at San Xavier. Kino, a Jesuit, had no hand in building missions in what would become Arizona. Mission San Xavier del Bac was started in 1783 and finished in 1797, long after Kino's death in 1711.[11]

On 24 October 1775, Pedro Font, a Franciscan with the Juan Bautista de Anza expedition, described the initial days of the journey north from Tubac:

> We set out from La Canoa at two in the afternoon, and at five halted at Punta de los Llanos, having traveled 3 leagues [8 miles] to the north-northwest. At the campsite and in the plains which follow there is grass, but no water. . . . This [San Xavier] is a large pueblo of Sobaypuri [*sic*] Pima Indians. Once it was very large, but now it is much depleted by the hostilities of the Apaches, and more especially because of its waters, which are very injurious, for they are very turgid and salty, so much indeed that a Jesuit father showed by experiment that a bottle distilled by alembic left two ounces of salt and sediment.[12]

Punta de los Llanos, or Point of the Plains, is a place-name that refers to the opening up of the valley north of Continental (see fig. 2.1). Consistent with later reports, there was no perennial flow in the nineteen-mile reach between the Canoa Ranch, south of Continental, and the springs south of Mission San Xavier in 1775. The de Anza expedition continued north from present-day Tucson and found no water until they reached the Gila River.[13]

By the end of the eighteenth century, the Spanish colony in the New World was suffering from financial difficulties in the mother country. A series of evalua-tions by the Real Consulado, an official government tribunal, focused on the status of marginal outposts such as Tucson and Tubac. From Tubac, second ensign Manuel de León gave the following portrayal in his August 1804 response to the tribunal's questionnaire:

> Only in the rainy seasons does it [the Santa Cruz River] enjoy a steady flow. During the rest of the year, it sinks into the sand in many places. Another, which we call the Sonoita River, takes its name from the abandoned Pima mission of the same name. It flows steadily for the first fifteen miles of its westward course, but sinks beneath the sand seven to eight miles before joining the Santa Cruz. This confluence provides water for Tumacácori and Tubac and collects in the marsh lands around San Xavier in great abundance.[14]

This mention of marshlands is the first reference to the area upstream of present-day Martinez Hill and San Xavier that became known as the Great Mesquite Forest.

At the same time, presidio captain José de Zuñiga provided a similar account from Tucson:

> The rivers of the region include the Santa Catalina [Rillito River], five miles from the presidio, which arises from a hot spring [Agua Caliente] and enjoys a steady flow for ten miles in a northwesterly direction, but only in the rainy seasons. It is 33 feet wide near its headwaters. Our major river, however, is the Santa Maria Suamca [Santa Cruz River]. . . . When rainfall is only average or below, it flows above ground to a point some five miles north of Tubac and goes underground all the way to San Xavier del Bac. Only during years of exceptionally heavy rainfall does it water the flat land between Tubac and San Xavier.[15]

The Mexican Period

The early years of the nineteenth century brought prosperity to southern Arizona, the product of a concerted military offensive and subsequent treaties that kept the Apaches content with gifts and rations. Some of the more peaceful groups settled near presidios such as Tucson. As the economy began to thrive under the Apache peace treaty, a number of ranchers scrambled to occupy the choice locations in the Santa Cruz valley. On the eve of Mexican Independence in 1821, Tomás and Ignacio Ortíz of Tubac petitioned for four *sitios* (a sitio is about 1,750 acres) in the vicinity of La Canoa.

A survey party including Manuel de León, the commanding officer at Tucson, measured the property in July. Ignacio Elías Gonzales, commander of the Tubac garrison and Tomás Ortíz's father-in-law, gave the following account of the San Ignacio de la Canoa Land Grant:

> [I]t is a place that contains ample level land through which runs the River of this military post, although without water due to the many sandy places that impede its current half a league to the north [of Tubac]. Only during the rainy seasons, when it receives water from its tributaries does the river flow. Its vast extent is covered by shrubs such as mesquite, acacia, . . .[16] palo verde, saguaro, and very few cottonwoods and willows, it has pasture in all its circumference although not in great abundances, and also sacaton grass along the floodplain and I consider it of some utility for the raising of cattle and horses . . . [after] putting upon it a well[,] . . . which may be done at all times by digging a short distance.[17]

The Mexican War and the Forty-Niners

In 1846, war broke out between the United States and Mexico, and one result of the American victory was that most of the Santa Cruz River valley eventually became American territory. Colonel Stephen W. Kearny and his Army of the West were dispatched to occupy the weakly garrisoned borderlands of Mexico. Some five hundred Mormon youths were recruited at Council Bluffs, Iowa, and mustered into a special unit. The Mormon Battalion under Captain Phillip St. George Cooke was organized not as a combat unit but as a supply train to blaze a wagon trail from the Midwest to California. After leaving the Rio Grande, Cooke crossed the Continental Divide to the upper San Pedro River and traveled west to Tucson, camping just north of the town on 17 December 1846. Cooke described the vegetation along the valley floor on a brief reconnaissance from Tucson to San Xavier:

> The thicket soon became a dense forest of mesquite two feet in diameter. After marching four or five miles, we came to water; and while waiting some time for the footmen to come up, I for the first time spoke freely to the officers and asked their opinion on the prudence of continuing farther in the dense cover which we had found and which the guide stated became worse all the way to the pueblo.[18]

Therefore, Phillip Cooke of the Mormon Battalion became the first to notice the mesquite forest and its extremely large trees present along the Santa Cruz River south of Tucson.

Upon departing Tucson for the Gila, Cooke described the road north to the Point of the Mountains (near Rillito):

> To my surprise, I found water seven miles from town [Nine Mile Water Hole] and plenty of it, instead of an insufficiency for miles reported by Weaver, whom I sent yesterday to examine (he took a different path). . . . The next three miles down the dry creek of Tucson were excessively difficult, with deep sand and other obstacles. Then our beautiful level prairie road was much obstructed by mesquite.[19]

Nine Mile Water Hole was just upstream from the confluence of the Santa Cruz and Rillito Rivers (see fig. 2.1). After the war was over, a column of US Army dragoons visited the Santa Cruz River en route from Monterrey to Los Angeles in September 1848. Lieutenant Cave J. Couts kept a diary and wrote:

> The river, or more properly, branch or creek, disappears in its sandy bottom a little below Ft. Tubac and probably does not rise again[;] its course is northeast, and probably turns to the San Pedro, that or Gila, as it was left to our right. The whole country between the mountains, and from Tubac to Tucson, is remarkably sandy and requires very strong streams to run any distance. Cannot find the Santa Cruz River in any map, reason for thinking it does not rise again.[20]

Each page of Couts's diary is accompanied by a hand-drawn sketch showing the day's route. The river's flow is shown to disappear just below the ford near La Canoa. Approaching San Xavier, he notes an increase in the size and density of mesquite and is forced to amend his earlier conclusions about the river's flow:

> Rio is called San Xavier, though the same as Santa Cruz, which disappears near Ft. Tubac and rises in a spring above Xavier del Bac from whence [it] is called San Xavier. . . . Marched from Ft. de Tucion [sic] about 8 on the morning of 27th. The Church, or Mission as it was at one time, stands some 1/2 mile from the town, on the other side of the branch of San Xavier [then the mainstem of the Santa Cruz]. The town itself is called San Augustine, this mission Tucion [sic]. About here, is where the branch

disappears into the sandy desert which we have passed since leaving. The bed of it can be traced very little farther.[21]

Gold Rush and the Argonauts

Two months after Couts's visit, in December 1848, the Apaches razed Tubac, forcing evacuation of all the inhabitants. Tomás Ortíz's home burned to the ground, destroying the original title papers to the Canoa property. Most of the settlers moved north to the villages of San Xavier and Tucson. That same month, President James K. Polk's annual message to Congress verified newspaper stories about the gold strike in California. Wagon trains of gold seekers, known as *argonauts* in an oblique reference to Greek mythology, passed through Tucson headed west, following Cooke's road and other established trails across southern Arizona. John E. Durivage, a correspondent with the *New Orleans Daily Picayune*, was one of the argonauts:

> We camped eight miles from the last rancho [Tubac] having traveled twenty-five miles during the day. Just below this point [the] river sinks into the sand and appears again only at intervals for many miles. Here [San Xavier] the river is crossed for the last time for fifteen leagues, although the cottonwoods marking its course are frequently in sight. . . . It [Tucson] is eight miles from San Xavier and a miserable old place garrisoned by about one hundred men. Flour and a small quantity of corn were all that could be procured. The Santa Cruz River flows within half a mile of the town and then takes a southerly bend [actually the bend is to the northwest]. Near the town are the remains of an old mission [San Agustín], the gardens of which are well stocked with fruit. The whole valley is exceedingly fertile.[22]

On 29 May 1849, Asa Bennett Clarke, who was traveling with Durivage, describes what probably was the stream emanating from the spring at Punta de Agua just south of San Xavier: "Coming to a grassy meadow, where judging from the nature of the ground, as well as we could in the darkness, that there must be water not far off, we camped. Several men went out in different directions and soon found a small creek with high banks."[23] Clarke's observation, combined with others a few years before and de León's observations from 1804, suggests that this part of the Santa Cruz River consisted of a marshland dominated by perennial grasses and a high groundwater table and surrounded by a dense mesquite forest.

In October 1849, another argonaut, Lorenzo D. Aldrich, depicts the valley between San Xavier and Tucson as a barren plain, which could be viewed as contradictory to repeated references to a mesquite thicket connecting the two settlements: "Driving over a barren plain some five miles in extent, we encamped for the night by the side of a running stream about one mile from the town of Tucson."[24] However, the dense mesquite was south of San Xavier, which is more than five miles south of Tucson.

One of the more detailed journals was kept by H. M. T. Powell, who followed the Santa Cruz River the same month as Aldrich. Powell describes the valley north of Tubac and at Tucson:

> [Nine miles north of Tubac,] we crossed the river to left bank. . . . [T]hree or four hundred yards below where we crossed the river sinks into the sand, and where it rises again we do not know. It sinks in the bend northeast of the point of the double peak mountains. . . . The road from San Xavier to camp, 1 mile short of Tucson, was very level, running throughout mesquite, etc. We encamped in a grassy bottom, much covered with saline efflorescence. The river has divided to a mere brook, the grassy banks of which are not more than 2 yards apart.[25]

In February 1851, José María Martinez, who had moved from Apache-torn Tubac to San Xavier, filed for a small land grant abutting the mission to the east. Martinez began clearing the land in 1849, which was formerly covered with mesquite and sacaton grass and was used by the mission as summer pasture. He also cut a ditch to the spring on the western side of the valley (fig. 4.2). Although the Spring Branch (another name for Agua de la Misión) ran across Martinez's land, the nearby O'odham maintained that he had the right to use "for irrigation [only] the water of the 'rebenton de la Sanja' [Punta de Agua] without using the water of the 'acequia del ojo de agua' [later called Agua de la Misión or the Spring Branch], which flowing from the east irrigated the land of the Mission Indians."[26] The O'odham agreed to the terms of the land grant, in part because they used Martinez's oxen for their plowing.

Under the Treaty of Guadalupe Hidalgo, signed 2 February 1848, the southern boundary of the Mexican Cession was to be surveyed by a joint boundary commission. This southern boundary followed the Gila River and not the present-day location of the US-Mexico border in Arizona. John Russell Bartlett, a prominent

Figure 4.2.
(1912) This upstream view is of the Acequia de Punta de Agua, a streambed spring along the Santa Cruz River south of the San Xavier Mission. The trees are mostly mesquites, although cottonwoods and willows were growing nearby. (Photographer unknown; from Olberg and Schanck 1913.)

bibliophile and amateur ethnologist from Rhode Island, was selected to head the commission. His 1851–1852 survey would be hotly contested in the US Senate, the charge being that perhaps no more than a fifth of the appropriations were actually used for the intended survey. Bartlett used most of the appropriations to fund his research on the Native American cultures of the desert, which he published as a classic tome in American ethnology.[27] In July 1852, Bartlett described his trip along the old trail along the Santa Cruz River south from the Maricopa and Pima villages on the Gila:

> [C]amped eight miles from Tucson [at the Nine Mile Water Hole]. . . . [E]n route to Tucson, wagons mired in crossing arroyos; in Tucson camped on the banks of the Santa Cruz River, where there was an abundance of grass. . . . In addition to the river alluded to, there are some springs near the base of hill [Sentinel Peak] a mile west of the town, which furnish a copious supply of water. . . . [T]he bottomlands are here about a mile in width. Through them run irrigating canals in every direction, the lines of which are marked by rows of cottonwoods and willows, presenting an agreeable landscape. . . . [Left Tucson, heading south and] soon entered a thickly wooded

valley of mesquite. . . . Near [San Xavier] is a fertile valley, a very small portion of which is now tilled, although from appearances, it was all formerly irrigated and under cultivation. . . . Leaving the village, we rode on a mile further and stopped in a fine grove of large mezquit [*sic*] near the river, where there was plenty of grass. . . . [W]e resumed our journey along the valley as before, through a forest of mezquit trees. . . . The rain having continued the whole night, we were much delayed in getting off this morning. The whole country was drenched with water and the road almost impassable for heavily-loaded wagons. After a hard journey of eighteen miles, we stopped at the banks of the river [nine miles north of Tubac] and strange as it may appear, notwithstanding all the rain that had fallen, the river, such is the uncertainty of the streams in this country, was quite dry. Fortunately, in some cavities in the river's bed we found water enough for our present wants.[28]

As it approached Tucson in 1849, the Santa Cruz River "divided to a mere brook, the grassy banks of which are not more than 2 yards apart."[29]

Two serious problems affected Tucson on the eve of its annexation by the United States as part of the

Gadsden Purchase of 1853.[30] The first was the cholera epidemic of 1850–1851 that was spread by itinerant gold seekers. In 1851, one out of ten Tucsonans died from cholera, a devastating waterborne disease. Land-tenure conflicts also plagued the settlement. In 1848, the Mexican government had sought to protect its northern frontier by establishing military colonies. To attract volunteers, a law was passed promising plots of farmland near garrisons to six-year veterans, and several Tucson settlers were dispossessed of bottomland to accommodate them. Traditional patterns of land use, including who controlled and maintained the berms and ditches or what to do when the river flooded, were disrupted. This turnover in land tenure accelerated after the Gadsden Purchase.

Early Observations of the Great Mesquite Forest

Although the significance of the Great Mesquite Forest (fig. 4.1) would not be recognized by scientists for a half century,[31] this bosque, with its northern edge at Mission San Xavier, was noted by Anglo-American explorers by the mid-1800s, particularly by forty-niners on their way west to the California gold fields. In 1849, Judge Benjamin Hayes, with a group of forty-niners from New Mexico, wrote in his journal of the San Xavier bosque that wolves "were howling all around us, and one of very large size, was seen."[32] Robert Eccleston, also in a company of forty-niners on their way west to California, approaching San Xavier from the east, mentioned traversing this large mesquite forest before reaching "an abundant flow of water in the Santa Cruz River."[33]

Early travelers were duly impressed by the extent of this magnificent bosque, with its gigantic mesquite and cottonwood trees. They also wrote in glowing terms of both the quality and the quantity of water in the Santa Cruz River, as well as the beauty and bountiful natural resources of both the river and the Santa Cruz Valley. Another forty-niner, George W. B. Evans from Ohio, entered the upper Santa Cruz Valley in 1849 with a party and "thought they were in the most beautiful valley he had ever seen."[34] In 1852, US-Mexican Boundary Commission surveyor John R. Bartlett wrote in his diary that

> the valley continued about half a mile wide, thickly covered with mesquit [sic] trees of a large size. The bottom-lands resembled meadows, being covered with a luxuriant grass[,] . . . [and the] banks of the

river [Santa Cruz] . . . are lined with cotton-wood [sic] trees of a gigantic size. . . . In some places there are large groves of these trees, rendering this part of the valley the most picturesque and beautiful we had seen.[35]

In 1854, Lieutenant John G. Parke and a military escort of twenty-eight dragoons camped about two and a half miles south of Tucson "on the bank of a clear running brook with an abundance of grass and wood."[36] In July 1855, German adventurer Julius Froebel wrote of the Santa Cruz, "[W]e encamped a few miles above town [Tucson] in a pleasant part of the valley. A rapid brook, clear as crystal, and full of aquatic plants, fish, and tortoises of various kinds flowed through a small meadow."[37] Wild Turkeys were heard calling in the mid-1850s.[38] This species was extirpated from southeastern Arizona by 1930;[39] now, after its reintroduction, it is restricted to montane habitats. Major William Emory, the chief surveyor engaged in determining the location of the new international boundary as prescribed by the Gadsden Purchase, wrote a brief description of the Santa Cruz River south of Tucson in 1857:

> After leaving Tubac . . . the valley expands into a wide open basin, the mountains receding on either hand, and the dry valley[,] now almost exclusively occupied by mesquite, is bordered by a wide stretch of gravelly table land. . . . Approaching the town of San Xavier . . . we again come upon running water, with its constantly associated fertility and verdure. . . . The settlement of Tucson occupies the lowest line of constant running water and consequently, the last fertile basin lying in the course of this valley.[40]

Maps of the vicinity of Mission San Xavier would not be produced until the 1880s and 1890s, but they show conditions similar to those described by visitors after the Mexican War and before the Gadsden Purchase (figs. 4.3 and 4.4). The Santa Cruz River had two branches south of the mission, with maps indicating that the main channel was on the west side in what is now called the West Branch. On the east, a channel with various names, but generally known as the Spring Branch, led to the center of the Great Mesquite Forest, which is roughly denoted with scattered tree symbols. On some maps, the area southeast of Martinez Hill is called a grassland, which is in stark contrast to today's mesquite and creosotebush flats in that area. We infer that this area was a cienega, probably covered with alkali sacaton grasses in a marshy setting, ringed with

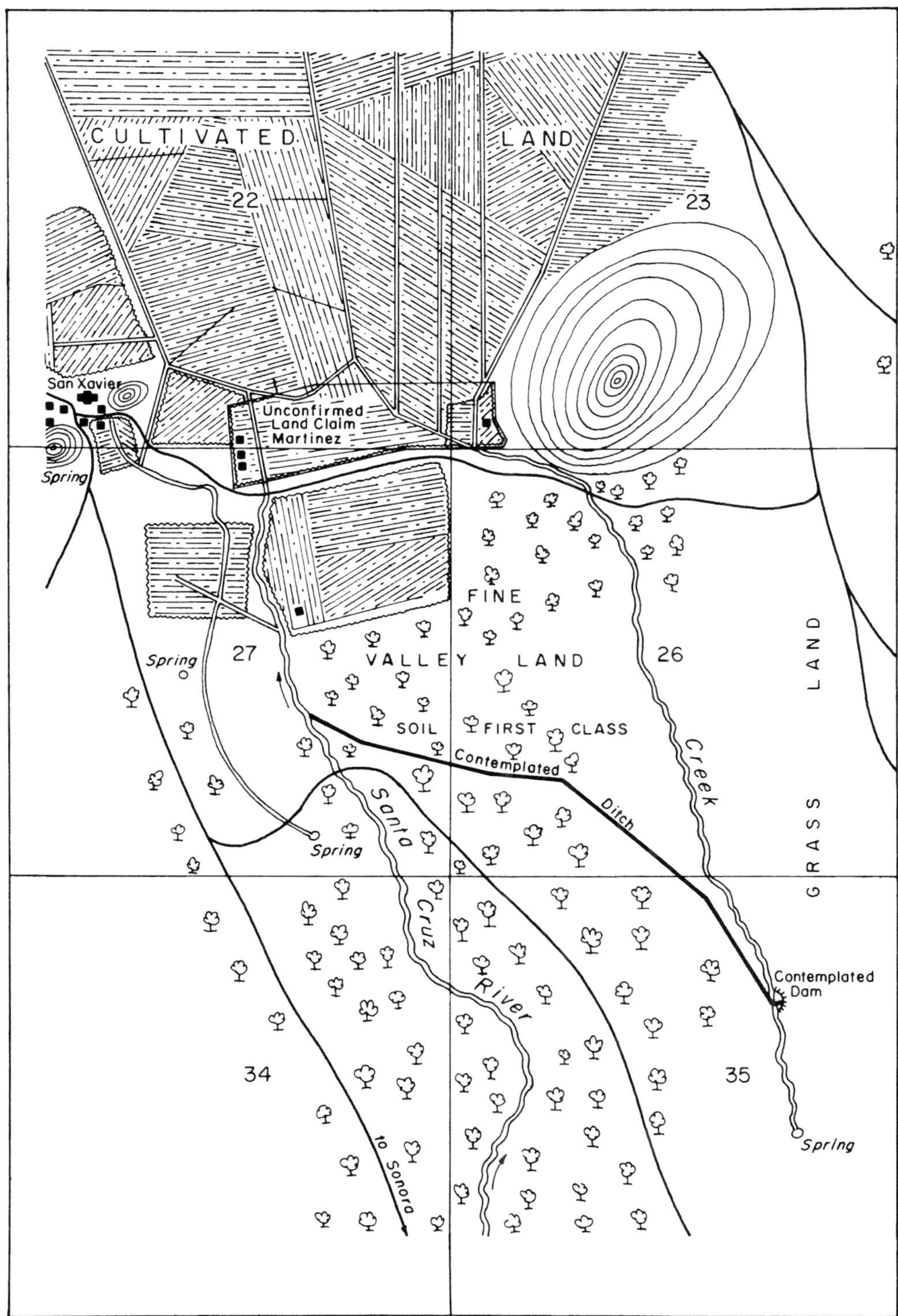

Figure 4.3.
Map of the northeast portion of the San Xavier Indian Reservation in 1882, redrawn by Betancourt (1990: 74) from map prepared in 1882 by George J. Roskruge, while he served as US deputy land and mineral surveyor. The main channel is on the west side of the valley and is now known as the West Branch. The area denoted as "grass land" is probably a cienega with high groundwater levels, and the Great Mesquite Forest is vaguely shown. In this and later maps, the channel of the Santa Cruz River north of Martinez Hill is not indicated, suggesting that the channel had been replaced by ditches in carrying flood flows.

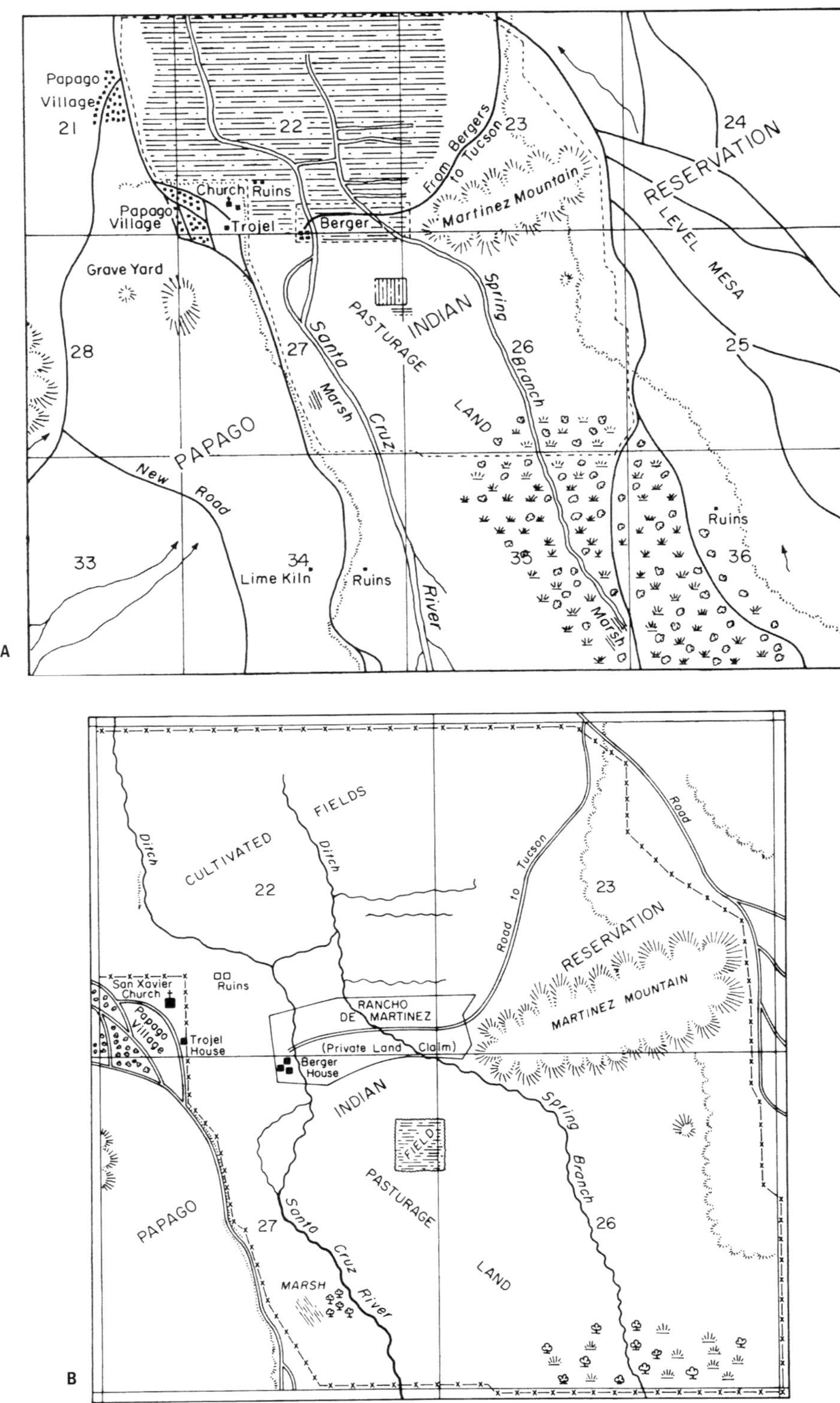

Figure 4.4.

Maps of the San Xavier Indian Reservation, redrawn by Betancourt (1990: 76–77) from originals by **(A)** L. D. Chillson in 1888 and **(B)** the Surveyor General's Office in 1891. In both maps, the Great Mesquite Forest is shown about a mile south of Martinez Hill [Mountain].

dense mesquite. High groundwater in this area would be expected, given the lack of drainage channels immediately west of Martinez Hill. Finally, agricultural clearing had already occurred on the northwest side of the forest as Tohono O'odham utilized the rich bottomlands and abundant water for agriculture.

After the Gadsden Purchase

From the time of initial forging of the Gadsden Purchase, to ratification of the treaty and transfer of funds, to the arrival of official representatives of the United States, Tucson remained under control of the Mexican army but in a local authority answering to no national government. One important change on the landscape was the establishment of the Rancho Punta de Agua south of Mission San Xavier. Fritz Contzen, a German citizen and veteran of Emory's boundary survey, established the rancho with five hundred head of cattle in 1855 in the middle of what would become the San Xavier District of the Tohono O'odham Reservation.[41] His brother, Julius, joined him briefly but died in 1857 of wounds suffered during an Apache raid in Sonora.

Mexican troops remained in Tucson until the US Army took possession in 1856, and the Mexicans were replaced by four companies of the First US Dragoons. Solomon Warner, whose future livelihood would greatly depend on the vagaries of streamflow along the Santa Cruz River, arrived in that year. A native of New York, Warner had worked his way around the continent until 1855, when he helped construct Fort Yuma. In February 1856, he secured a stock of general goods and set out for Tucson, accompanied by William Rowlett of Virginia. Upon his arrival, Warner established Tucson's first mercantile store in partnership with Mark Aldrich. Rowlett and his brother Alfred began construction of a low earthen dam downstream of a spring-fed cienega near Sentinel Peak (now known as A Mountain). The following year, the Rowletts built Tucson's first water-powered flour mill west of the reservoir, in later days known as Silver Lake (figs. 4.1, 4.5).[42]

The millrace probably did not originate at the dam. Juan Romero, a local farmer, gave a complicated, somewhat muddled account of the waterworks at Silver Lake as testimony in an 1885 water rights case.[43] Presumably, three ditches were developed from springs southwest of Sentinel Peak,[44] and all three were gathered up into a single stream that turned the water wheel. The tailrace emptied east into the reservoir, and water would be let out into the main acequias when needed for irrigation

downstream. The purpose of the lake was to avoid having to synchronize operation of the mill with the irrigation schedule for fields immediately west of Tucson.

Continued growth in California increased needs of overland transport from the East, leading to the establishment of a stagecoach service through Tucson in July 1857. Phocion Way, one of the Jackass Mail Route's first passengers, stopped in Tucson, apparently in summer, and wrote this colorful description:

> There is a small creek that runs through the town. The water is alkaline and warm. The hogs wallow in the creek, the Mexicans water their asses and cattle and wash themselves and their clothes and drink out of the same creek. The Americans have dug a well and procure tolerably good water, which they use.[45]

If water from the Santa Cruz River was alkaline in Tucson, it was sweet near Tubac. In 1857, irrigation canals diverted water onto fields that produced vegetables and melons. The produce supplied the town of Tubac and travelers passing through.[46]

Sam Hughes, another of Tucson's early businessmen, arrived on 25 March 1858. A native of Wales, Hughes came via the California gold fields and established a thriving butchering business in Tucson. Here, he met Mose Carson (Kit Carson's brother), who had spent the previous winter trapping beaver on the upper Santa Cruz and Rillito Rivers.[47] According to Hughes, the winter of 1858–1859 was colder than usual and about 1.3 inches of snow fell in Tucson. Recalling that winter, he claimed that "the waters of the Santa Cruz were so deep that a flat boat could be navigated probably clear to the Gila at Maricopa, and that the Rillito was a mile wide."[48] In 1859, a German itinerant described the Santa Cruz just south of Tucson as "a rapid brook, clear as crystal and full of aquatic plants, fish and tortoises of various kinds . . . [flowing] through a small meadow covered with shrubs."[49]

In 1860, the Rowletts sold their mill to William S. Grant and T. W. Taliefero for a considerable profit. By the end of the year, Grant had bought out his partner and the Rowletts were panning for gold in the Cañada del Oro. As a government contractor, Grant supplied the US military at Forts Buchanan, Breckenridge, and Fillmore with flour. He built another mill on the shores of Silver Lake at a cost of $18,000 just before the Civil War started. In 1862, several Mexican farmers enlarged the ditch headings that fed the millrace, hoping to increase flow to beyond what is now St. Mary's Road.[50]

Figure 4.5.
Silver Lake. **A.** (1880s) This view is west across Silver Lake, a small reservoir created by an earthen dam constructed in the 1850s by brothers William and Alfred Rowlett. The structure on the right is a resort hotel. The leafless trees, probably cottonwoods, indicate that the photograph was taken in winter. (Photographer unknown, 18335, courtesy of the Arizona Historical Society.) **B.** (16 December 1981) The dam impounding Silver Lake was damaged by floods in the 1890s, and by 1900, the dam and reservoir were gone. The large trees in this photograph are nonnative Athel tamarisks. (R. M. Turner, Stake 1060, courtesy of the Desert Laboratory Collection.)

46

C

Figure 4.5. (*continued*)

C. (2 October 2001) The ragged natural arroyo banks have been replaced with soil cement (see chapter 9) in this reach. No nonnative species are present in the channel, but Athel tamarisks and palms are cultivated on the distant floodplain. (D. P. Oldershaw, Stake 1060, courtesy of the Desert Laboratory Collection.)

US Military Occupation During and After the Civil War

At the onset of the Civil War in 1861, the Territory of New Mexico, which included both present-day states of Arizona and New Mexico, was sympathetic to the Confederacy.[51] In March 1861, conventions in Mesilla and Tucson voted for secession, although the territorial government in Santa Fe remained loyal to the Union. That year, Union forces arrived in Tucson and set fire to Grant's mill to keep it from falling into Confederate hands. In February 1862, Confederate troops marched into Tucson, meeting no opposition. Union men, such as Sam Hughes and Solomon Warner, were forced to flee. Warner left for Santa Cruz, Sonora, where he erected a flour mill, raised cattle, and operated a freighting business between Guaymas and Tucson. Hughes headed for California, only to return in June with General James Henry Carleton's Union forces. A month earlier, Colonel Joseph R. West had raised the Union flag in Tucson and repaired one of Grant's mills to supply the incoming troops with flour.

To determine which properties should be confiscated, West ordered the town mapped by Major David Fergusson. One of Fergusson's maps shows the ownership of cultivated fields in the bottomlands west of town and shows that river flow vanished underground between present-day Congress Street and St. Mary's Road. Surprisingly, documents from 1862 failed to mention floods along the Santa Cruz River that winter, during which catastrophic floods occurred throughout California, northern and central Arizona, and southwestern Utah.[52]

The presence of Union troops did little to quell hostilities on the landscape. Apaches raided Rancho Punta de Agua in October 1861, taking 376 head of cattle and some horses from the property. After this raid, Fritz Contzen stopped ranching and worked as a livestock broker, buying horses from the Tohono O'odham and selling them to the Overland Stage service.[53] He was briefly caught up in the Civil War, but because he remained neutral and did not pledge loyalty to the Union, he was arrested, taken to Yuma, and forced to sign an oath. Afterward, he served as a military express rider until he and his family abandoned the rancho in 1868.

In October 1862, Fergusson scouted a route from Tucson to the Gulf of California, hoping to find a suitable gateway for moving ship-transported supplies from the Sea of Cortez to mines in southern Arizona. About two miles south of San Xavier, he passed El Rancho Viejo. The road to the ranch was through a meadow with running water 650 feet to the east. For about a half mile south to Struby's Ranch (Punta de Agua), the road ran through dense mesquite, again flanked by running water to the east. Upstream of this point, water could be found only by digging shallow wells in the stream bottom.[54]

Two years later, Joseph Pratt Allyn, a judge appointed by Abraham Lincoln to the first supreme court of Arizona Territory, paid a December visit to Tucson. Traveling south to Tucson, he observed:

> At the point of the mountain range beyond Picacho [Tucson Mountains], you strike the dry bed of the Santa Cruz, a stream that, when it runs, empties into the Gila at Maricopa wells. It is some two hundred and fifty miles in length, and at times huge volumes of water march majestically its whole extent, sweeping and destroying whatever is in its track. . . . The line of the Santa Cruz is marked by its fringe of luxuriant cottonwoods, just now fading into the sere and yellow leaf. . . . After crossing the Santa Cruz the road leaves the river to the left, passing through quite a dense growth of mesquit [sic]; the valley between the two ranges, now on either hand, is of fair breadth and with a good growth of grass.[55]

At Tucson, he found what was left of the Rowletts' mill "where the Santa Cruz runs quite a respectable stream of water." Like Fergusson, Allyn commented on the "thick growth of mesquit" near San Xavier Mission, as well as the beauty of the "broad irrigated fields" adjacent to the church.

Perhaps the most detailed published descriptions of Tucson and the Santa Cruz River and its picturesque valley were those of J. Ross Browne, an Irish immigrant and writer. On his arrival in the Santa Cruz Valley in January 1864, Browne was singularly unimpressed with Tucson, describing it as

> a city of mud boxes, dingy and dilapidated, cracked and baked into a composite of dust and filth; littered about with broken corrals, sheds, bake-ovens[,] carcasses of dead animals, and broken pottery; barren of verdure, parched, naked, and grimly desolate in the glare of a southern sun, . . . [except for] sore-backed burros, coyote dogs, and terra cotta children.[56]

In sharp contrast, Browne presented a glowing description of the river valley, calling it "one of the richest and most beautiful grazing and agricultural regions" he had ever seen: "the grass is abundant and luxuriant[,] . . . mesquit [sic] and cotton-wood [sic] are abundant and there is no lack of water."[57] He added that the watershed outside of the settlements was abandoned in the wake of repeated Apache attacks on isolated ranches.

Traveling with a transcontinental railroad survey party in 1867, William Alexander Bell noted that one hundred acres were cultivated at San Xavier and another two thousand acres were irrigated in Tucson. He noted the river's intermittent flow regime:

> One word about the Rio Santa Cruz, the eccentric course of which can be traced at a glance on the map. For the first 150 miles from its source it is a perennial stream; but four miles south of Roade's Ranch, at a spot called Canoa, it usually sinks below the surface; it then flows underground almost to St. Xavier, and again reappears at a spot called Punta de Agua. . . . Beyond St. Xavier it usually sinks again, rising for a third time as a fine body of water near Tucson, enriching a broad piece of valley for about ten miles around that town, turning the wheel of a fair-sized flour mill, and then sinking forever in the desert to the northwest. During some seasons, it flows further than others, so that the length of the stream above ground is subject to considerable variation; but it never succeeds in reaching the Rio Gila on the surface, although I believe it flows over the bedrock and under the drift which covers it for the remaining one hundred miles from Tucson to Maricopa Wells, where a large spring—the waters of the Rio Santa Cruz, it is believed—comes to the surface and flows into the Gila.[58]

In 1866, the US Army established Camp Lowell on the eastern outskirts of Tucson, overlooking the Santa Cruz River from a distance. The post surgeon also assumed the role of weatherman and maintained daily temperature and precipitation records for the US Signal Corps. John Spring, a member of the Regular Army stationed at Camp Lowell, stated that "the Santa Cruz River . . . had a few places where the water was perhaps a little over four feet deep."[59]

Malaria was common in southeastern Arizona during the nineteenth century, largely because the shallow groundwater fed cienegas that were ideal breeding grounds for mosquitoes. It would be decades before doctors understood that mosquitoes were the vector for malaria; at the time, the fevers were thought to be the result of "bad vapors" and other airborne causes. In 1870, the Surgeon General's Office of the War Department expressed concern over the high incidence of malaria among local Mexican farmers:

> The Santa Cruz . . . runs northward from the Sonora line past the west side of the town and post, and continues its course to a point about four miles below, where its waters cease to run above ground, on account of the porous character of the soil. . . . As cultivation can only be carried on successfully by irrigation, it follows that more or less of the fields are constantly under water, which, combined with the heavy rains in July and August, the tropical vegetation and its rapid decay, favors the development of the malarial poison and accounts for the cases of remittent and intermittent fevers and diseases of the liver which prevail among the Mexican inhabitants during the months of August, September, and October.[60]

Captain John G. Bourke, an aide to General George Crook, passed through Tucson en route to Camp Grant on the lower San Pedro River, where malarial outbreaks occurred in the fall of 1869. His brief stay in Tucson generated considerable commentary about the residents and these brief descriptions of the Rillito and Santa Cruz Rivers:

> Just now I find that all my powers of persuasion must be exerted to convince the readers who are still with me that the sand "wash" in which we are foundering is in truth a river, or rather a little river— the "Riito"—the largest confluent of the Santa Cruz. Could you only arrange to be with me, you doubting Thomases, when the deluging rains of the summer solstice rush madly down the rugged face of the Santa Catalina and swell this dry sand-bed to the dimensions of a young Missouri, all tales would be more easy for you to swallow. . . . [T]hose gently waving cottonwoods outline the shrivelled course of the Santa Cruz. . . . [Tucson] was on the right bank of the pretty little stream called the Santa Cruz, a mile or more above where it ran into the sands.[61]

Bourke's description is the first for the Rillito River, the major tributary of the Santa Cruz River on the north side of the Tucson Basin (see fig. 2.1).

By 1869, there were about five hundred acres cultivated around the Nine Mile Water Hole north of Tucson (see fig. 2.1), with some eighty or ninety pioneers clustering around the stage station.[62] According to Sam Hughes,

the Santa Cruz and other rivers which empty into the Gila were all running high and so great was the snow and rainfall during that season [winter] and the two years following that the Santa Cruz flowed a surface stream from its source to the Gila during 1868, 1869 and 1870, something unheard of since, as the stream is subterranean more than three-fourths of the length of the valley through which it flows.[63]

In the 1885 water rights case involving all of the agricultural lands opposite Tucson, E. N. Fish testified that during the wet years of 1868 and 1869, the river ran clear down to the Nine Mile Water Hole, allowing irrigation of the valley bottom north of town.[64] Along the Rillito River, the stage of a flood in September 1868 was three feet higher than a similar flood in September 1887.[65] According to Hughes, the flood inundated the lower Gila River, destroying a gallery forest of cottonwoods east of Yuma.[66] Possibly in response to these wet years, in 1871 S. W. Foreman recorded an irrigation ditch with its heading at the river about a mile north of the Rillito River confluence.[67] A tentative interpretation is that there was perennial flow at this location, disappearing a short distance to the north.

The Santa Cruz River: The Land of Milk and Honey?

In recent decades, myths about the "good ole days" of desert streams have become public dogma, inspiring talk about restoring the Santa Cruz River to some preconceived past condition, presumably an improvement over its twenty-first-century counterpart. Some people with a skewed view of history say that a perennial stream meandered through grass belly-high to a horse, nothing short of the biblical land of milk and honey.[68] Contrary to this perspective, historical sources indicate that 80 percent of the forty-five-mile stretch between Continental and the Rillito River was mostly dry before 1870 except during floods. Two perennial reaches with a total distance of about nine miles were near Tucson and Mission San Xavier. The characteristics of these short reaches bear closer examination.

At Mission San Xavier, historical perennial flow has been attributed to subsurface basalt that forced groundwater to the surface. Erosion during the Pleistocene cut a wide gap through the Late Tertiary volcanic rocks, leaving Martinez Hill and Black Mountain as remnants on both sides of the valley. The shallow bedrock blocks northward flow of groundwater, a characteristic recognized as early as 1883 by Solomon Warner.[69] At Tucson, Pleistocene terraces from the east nearly converge with the western volcanic outcrops at Sentinel Peak and Tumamoc Hill. An impervious stratum in the narrows of the valley also acts as a barrier, forcing the water upward to the surface.[70] An extremely large cienega with a surrounding halo of mesquite bosque developed here in response to the shallow groundwater.

Other reaches were ephemeral with deep valley fills that quickly absorbed surface water. All accounts agree that the perennial flow of the Santa Cruz first disappeared not far north of Tubac, near the ford at La Canoa. In December 1872, Theodore White noted that "about a mile south of where this line [the southern boundary of T18S, R13E] crosses, the Santa Cruz is a large, ever running stream of water, but [it] sinks into the sand in a short distance."[71] Directly west of this point are the present headquarters of the Canoa Ranch just south of Continental (see fig. 2.1). The flows from the Punta de Agua and Agua de la Misión springs disappeared downstream from Mission San Xavier and the western base of Martinez Hill, respectively. Perennial water reappeared two miles north of Martinez Hill, sinking beneath the surface again in less than a mile. Another brief stretch of perennial flow existed half way to Tucson.[72] The evidence for where the flow disappeared north of Tucson is less clear.

Although cienegas occurred south of San Xavier and at the base of Sentinel Peak, beaver likely were not in the alternating ephemeral-perennial reaches in the Tucson Basin. Likewise, they probably were not present along the Rillito River, despite claims to the contrary.[73] However, they were present in the headwaters of the Santa Cruz River in perennial reaches with suitable habitat, much as they were present where the San Pedro River crossed the US-Mexico border.[74] Carmen Lucero, a Tucson resident in the 1850s and 1860s, reminisced: "I was telling you about the days before the Americans came. At that time the river did not have a deep channel but was just a creek running all over the valley, and there were lots of beaver at Silver Lake."[75] These might have been stocked, or it may be a case of mistaken identification. Remains of beaver have not been found in Hohokam sites along the Santa Cruz River, although evidence of muskrat, an aquatic mammal of marsh habitats, was found; this animal was never mentioned in historical accounts of the Tucson Basin.[76] Muskrats do not build dams and thus would have played a different role than beavers in the maintenance and stabilization of channels and cienegas.

Historical descriptions seldom mention the depth of the channel. In 1849, Clarke describes a small creek with high banks on the western side of the valley south of Mission San Xavier.[77] Some twenty-two years later, Foreman observed that the banks of the Santa Cruz River were 26–32 feet apart and 8–15 feet deep in the same general area.[78] Upstream, about one mile to the south, was the spring at Punta de Agua, which had been developed by Martinez no earlier than 1849. As late as 1915, land surveyor C. F. Leedy noted, "The Santa Cruz River runs through the eastern half of the township mostly on top of the ground with no definite channels before it crosses the Third Standard Parallel South";[79] Punta de Agua is about 650 feet south of the Third Standard Parallel. The springs at Punta de Agua probably issued from a point in the channel where a headcut intercepted the water table.

On the eastern side of the valley, at the present site of the Valencia Road Bridge, Foreman recorded banks 10 feet high and about 65 feet apart.[80] In 1882, the newspapers described a headcut here:

> Some six miles south of the city, the Santa Cruz seems to spring directly from a steep clay bank, rising on the three sides about 20 feet from the bed. Above that point the river is lost under the dry mesa, not reappearing again for miles beyond. Here is found a deep pool of pure cold water, bubbling through the earth from the sides and bottom.[81]

The location of this headcut relative to the mainstem of the Santa Cruz River is uncertain; it was directly in line with the Spring Branch, the stream emanating from the spring at Agua de la Misión, not the mainstem West Branch, which was to the west at that time. The present channel coincides with the former course of the Spring Branch; nineteenth-century maps of this area show that some unknown party diverted the Spring Branch from the base of Martinez Hill west toward the mainstem, probably to supply additional water to the mission and its agricultural lands.[82] North of Martinez Hill, these maps suggest that flood flows followed the irrigation ditches instead of the natural channels. This pattern probably existed back to Kino's time but may have been atypical of conditions about five hundred years ago, when the Santa Cruz probably flowed in a deep channel along the historic course of the Spring Branch.

Alluvial stratigraphic work at the base of Martinez Hill has revealed a large paleoarroyo in the exposed banks of the present Santa Cruz River (see fig. 2.3A).[83] This paleochannel, of comparable dimensions as the modern arroyo, cut and filled between three hundred and five hundred years ago (see fig. 2.3). The Valencia Road headcut developed in a floodplain that was only three hundred years old, at a time when the main flood flows followed the eastern margin of the valley. The age and stability of the Valencia Road headcut is uncertain, but its development may have influenced upstream diversion of floodwaters by the Hohokam or their progenitors. This is suggested by what appears to be the unexpected presence of the main drainage on the western side of the valley and the lack of continuity of the Spring Branch upstream of Ojo de la Misión.

Judging from the accounts before 1870, other headcuts were not present near Tucson. Instead, the river was described as flowing at or near the top of the floodplain or impounded in cienegas. As late as 1907, the riverbanks immediately below the confluence with the Cañada del Oro averaged less than three feet high.[84] In 1871, the channel was about fifteen feet wide at the Nine Mile Water Hole.[85]

Early accounts provide general descriptions of the vegetation in the valley. Dense mesquite occupied the reaches with high groundwater levels upstream and downstream from the perennial reaches. At Tucson, a narrow band of cottonwoods and willows marked the course of the various irrigation ditches. A grassy cienega covering about a square mile spread out on either side of the Spring Branch upstream from Mission San Xavier. An impressive mesquite forest, interspersed with small meadows, may at this time have been restricted to the western side of the valley, between Rancho Punta de Agua and the mission (figs. 4.3 and 4.4). In all likelihood, the extent of this bosque was larger to the east and restricted by the high groundwater levels of the cienegas, which would have limited perennial vegetation to grasses and other emergent aquatic species.

Within the Canoa land grant near present-day Green Valley, surveyors in 1821 referred to "teraques" in listing the local flora.[86] In 1876, the document was translated by the US Surveyor General's Office and *teraque* was taken to mean "tamarisk." Tamarisk, or saltcedar, was introduced to the New World as an ornamental supposedly in the mid-nineteenth century.[87] The non-native saltcedar was first planted at St. George, Utah, in 1896 and along the Salt River in 1901.[88] The large ornamental tree known as Athel tamarisk was introduced to Tucson in the 1920s.[89] Although Athel tamarisk was not known to reproduce or hybridize until recently, saltcedar is an invasive species and a target of eradication efforts. Neither species was present in the United States or the Santa Cruz River watershed in 1821.[90]

5

Land Use, Climate, and Floods, 1873–1888

In 1870, Tucson was a bustling community of 3,224 inhabitants, and the area that would become Pima County accounted for more than half of Arizona's population.[1] Although relations with the Apaches to the east remained hostile, farming and ranching were increasing, and consequently dependence upon surface water in the Santa Cruz River was also increasing. Because channel beds were close to the surrounding land surface, flow was easily diverted onto floodplains for agriculture, and irrigation schemes were devised to convert more and more of the desert landscape into arable lands. At the same time, natural historians began to take an interest in the area's diverse wildlife, particularly its birds. Hydrologic changes that began in the late nineteenth century would continue to affect residents of Pima County into the early twenty-first century.

The 1870s and 1880s would be decades of enormous social and environmental change, when the attention of local residents would shift from defending their families and properties against attacks to extracting resources for a growing population. Climate would both help and hinder, as abundant rain would provide water supplies for agriculture while producing floods that damaged fields and waterworks. A singular earthquake would induce short-term hydrologic changes, while large wildfires swept over the forested highlands. These events were chronicled, not only by newspapers but also by ordinary people who also observed and recorded channel conditions, plants, and animals. During these two decades, the city of Tucson would shift from the Wild West to a semblance of modernity, and the Santa Cruz River would change from having an ill-defined channel to occupying a deep trench. Documentation of these changes began with two disparate groups of observers—land surveyors and naturalists—who had different motives and goals for their work.

The Onset of Modern Land Surveys

Land surveys to establish legal ownership and assess natural resources were extremely important to the development of the western United States, and they were no less important in the Tucson Basin. The lack of up-to-date surveys was causing problems for landowners in the burgeoning community. The earliest known map of Tucson, compiled in 1862 by Major D. Ferguson, served as the only survey of fertile lands in the valley. The lack of adequate surveys and maps posed a significant obstacle to obtaining property titles and precluded establishment of Tucson as a formal town. The first cadastral surveys in the 1870s, commissioned by the General Land Office, provides our first information, not only on the channel geometry of the Santa Cruz River and land uses of its floodplain but also on the status of riparian vegetation along its length through the Tucson Basin. This information helps corroborate the anecdotal accounts of residents, who generally viewed riparian areas as grazing lands, sources of firewood, and impediments to travel.

Levi Bashford was appointed surveyor general of Arizona in 1863 with offices in Tucson, but ongoing skirmishes between settlers and Apaches postponed the sorely needed surveys. In July 1864, Bashford and his records were transferred to Santa Fe, slated as the new headquarters of the General Land Office for Arizona and New Mexico. John Clark, who became surveyor general in 1865, warned that surveying parties would remain inactive until peace could be attained between Native Americans and settlers. Jurisdiction over Arizona shifted to California in 1867, and three years later, Arizona was finally granted its own district with John Wasson as surveyor general.

In January and February 1871, the survey of T14S, R13E, which included the proposed townsite of Tucson, was completed. By 1872, every township in the valley between Rillito and Continental had been surveyed by either S. W. Foreman or Theodore F. White, despite the potential for attack.[2] Indeed, Foreman's 1871 survey of the San Xavier District occurred just after a raid on the Punta de Agua Ranch. Both surveyors recorded site-specific information concerning the course and dimensions of the river channel, vegetation, and anthropogenic improvements in the early 1870s.[3] With few exceptions, the Santa Cruz River was narrow and flowed at or near the land surface through the Tucson Basin.

The First Observations of Natural History

Nearly all of the first ornithologists who visited the Tucson Basin were particularly interested in the "Mexican species" that spilled into southern Arizona across the international boundary from Mexico.[4] Although many of these species occurred in the Madrean woodlands of the "sky islands" in southeastern Arizona, ornithologists quickly learned that birds were particularly dense and interesting in riparian areas along the lowland rivers.

The post of army surgeon attracted men of science in the nineteenth century, and many of these were either trained as or became natural historians. Members of the US-Mexican Boundary Survey and other exploratory expeditions were discovering birds new to the United States and even species new to science during the late 1800s.[5] No fewer than seven subspecies and two species new to science were described from near Tucson.[6] Little early information, however, was collected from the Santa Cruz River, because interest was slanted toward birds in the mountains of southeastern Arizona.[7] Even then, work was transitory, with ornithologists rarely spending more than a few days or weeks in this newly explored region of the American West.

One of the earliest ornithologists to spend a longer period of time in this region was Charles Bendire, a captain in the US Army Medical Corps.[8] Although his main responsibility was the health and welfare of the Fort Lowell garrison, he was a trained naturalist, and birds were his passion. He mentioned the Santa Cruz

River several times in his *Life Histories of North American Birds*,[9] but most of his records were from the Rillito River and undesignated localities near Tucson. Bendire was stationed at Fort Lowell on the east bank of the Santa Cruz near the current location of downtown Tucson for eighteen months from 1871 to 1873. He established a permanent camp on the Rillito River seven miles upstream from the Santa Cruz River near its beginning at the confluence of Agua Caliente, Tanque Verde, and Pantano Washes, where Fort Lowell would eventually be moved. The riparian ecosystem here is only vaguely described but likely was a large mesquite bosque surrounding a core of open water or a cienega.

Bendire found approximately fifty-five bird species nesting along the Rillito River in 1872 in riparian and adjacent ecosystems.[10] He discovered two species new to science from near Tucson: Bendire's Thrasher and the Rufous-winged Sparrow.[11] In addition, he added three species from Mexico to the avifauna of the United States, including the Gray Hawk and Ferruginous Pygmy-Owl, which occurred at or near riparian zones along the Rillito River.[12] The third bird species, the Painted Redstart, is a montane nesting species. Bendire recorded species such as the Crested Caracara,[13] a species not occurring in the Tucson region today, and collected eggs of the Song Sparrow,[14] a riparian and wetland nesting species that was largely extirpated from the Tucson Basin by the early 1900s.[15]

Bendire reported specific birds from the Santa Cruz River in his books,[16] but he never developed a complete avifanua list as he did for Rillito River. For example, Bendire wrote, "I first met with the Gila Woodpecker during the winter of 1871–1872 in the vicinity of my camp on the Santa Cruz River a few miles south of Tucson."[17] Even more specifically, he wrote, "I found it [the Ladder-backed Woodpecker] common in the mesquite groves on the Santa Cruz River, between Tucson and the Papago Mission Church, Arizona [Mission San Xavier], and much less so among the cottonwood and willows on Rillito Creek."[18] The fact that Bendire considered the Ladder-backed Woodpecker to be common at San Xavier is interesting, because some later accounts did not report this species;[19] we cannot determine whether this local population declined after 1872. However, even though Bendire wrote extensively about the importance of the avifauna along the Rillito River, he did not indicate the ornithological importance of the Great Mesquite Forest south of San Xavier, either in his handwritten notes or in his numerous publications.[20]

On 17 April 1872, while returning to Tucson from Picacho Peak, Bendire found a Mexican Spotted Owl nest in a cottonwood tree ten miles northwest of Tucson.[21] This was only the second known record for this species, and today, nesting of a Spotted Owl on the desert floor, even in riparian habitat, is unknown.[22] Bendire climbed a cottonwood tree to collect an egg from a Zone-tailed Hawk nest on the Rillito River on 3 May 1872. While taking the egg from the nest, he saw several Apaches hiding nearby, watching. With egg in mouth, he climbed down to his horse and shotgun and hastily retreated to camp and safety, with the egg still in his mouth. He reported that he suffered from an aching jaw for some time afterward.[23] Zone-tailed Hawks no longer nest on the desert floor.

The next ornithologist to work in the Tucson Basin, Frank Stephens, "found [Gray Hawks] common about Tucson, especially in some of the large mesquite groves on the Santa Cruz River. . . . [N]ests were placed in cottonwoods and large mesquite trees."[24] There is little doubt that Stephens was writing about the Great Mesquite Forest, but like Bendire, he mentioned it only in passing without giving any suggestion of its exceptional value from the standpoint of biodiversity.

Herbert Brown was a Tucson resident who came to Tucson in the 1870s, and he recorded birds along the Santa Cruz River beginning in the 1880s and extending into the early twentieth century.[25] His records show differences between the Santa Cruz and Rillito Rivers and document populations that were vastly different from those of the early twenty-first century.[26] Brown recorded several first records for Arizona from along the Santa Cruz River, including an Anhinga (a cormorant-like bird that he referred to as a "water turkey") on 12 September 1893,[27] a species resident to the southeastern Gulf Coast and accidental in Arizona;[28] a flock of eight Black-bellied Whistling-Ducks on 5 May 1899,[29] a Mexican species that by the mid-1900s was still considered "of sporadic and rare occurrence in summer in southeastern and central Arizona";[30] and a Purple Gallinule on 10 October 1887, another species from Mexico and the Gulf Coast still considered casual in the state.[31] Even more surprising was a flock of seven or eight Scarlet Ibis, a South American species, seen on the Rillito River near Fort Lowell on 17 September 1890.[32] The Anhinga and Scarlet Ibis records have never been duplicated.[33] Brown also collected the first Streak-backed Oriole, a Mexican species, and Scarlet Tanager, a species common in forest and woodlands of the eastern United States.[34] He also conducted some of the early research with the Northern (Masked) Bobwhite,[35] at present a federally listed endangered species.[36]

Perhaps more surprising than the bird observations were records of the California floater, a large edible unionid clam up to three inches across. This species was first described from the Colorado River at the head of its delta in 1852, and it occurred throughout much of the Colorado River basin. A clam specimen from the Santa Cruz River "just outside Tucson" was reportedly taken in June 1880.[37] A May 1968 excavation of a trash dump in downtown Tucson, dated 1880–1885, uncovered numerous valves of this species, and the association of these valves with kitchen utensils suggested that California floaters were used for food.[38] The larvae, or glauchidia, of this clam are free swimming and become parasitic by attaching themselves to the fins of a given host species of fish until they metamorphose into free-living and, finally, stationary adults.[39] The fish serving as the host and dispersing mechanism for this clam in the Santa Cruz River was probably the Gila chub, since the closely related Utah chub serves as a host species for this clam in other locations.[40]

An Unusual Period of Rainfall, Fires, and Floods

The early 1870s were a time of high precipitation and regional flooding in southern Arizona. The year 1873 was eventful for some positive changes in Tucson. In March 1873, Camp Lowell was moved from the military plaza to the confluence of the Tanque Verde and Pantano Washes, which is the start of the Rillito River. The same year, the regional telegraph network was connected to Tucson, bringing a little modernity to this backwater town. After three years of trying to evict Mexican settlers from Tohono O'odham lands in the San Xavier area, Indian Agent R. A. Wilbur and local Tohono O'odham were finally rewarded in 1874 when Congress agreed to set aside a reservation along the Santa Cruz River.

The winter of 1874 was exceedingly wet throughout central Arizona, and flooding occurred throughout the Gila River basin. On the lower Gila River, the floods matched the high water mark reached in the winter of 1868.[41] Storms in July 1874 inflicted damage to crops at San Xavier and Tucson.[42] The unusually high precipitation lured some Tucson residents into a false sense of water security, and plans were developed to convert more desert into agricultural fields.

Solomon Warner returned to Tucson in the early 1870s from his self-imposed exile with plans for a water-powered flour mill at the foot of Sentinel Peak (figs. 5.1

Figure 5.1.
(1880) Pioneer merchant Solomon Warner built his house and flour mill at the foot of Sentinel Peak. The view, from the lower slope of Sentinel Peak, is to the southeast with the Santa Cruz basin in the background. (Carleton Watkins, 14846, courtesy of the Arizona Historical Society.)

and 5.2). In May 1872, Tomás Elías authorized access for an irrigation ditch across his land near the northeast corner of the garden at the San Agustín Mission. Warner faced stiff competition from two other local mills: the Pioneer Mill at Silver Lake and the Eagle Steam Flour Mill at the corner of Main and Broadway. In October 1874, Warner obtained water rights and rights-of-way for his canal system to and from his mill. Bishop J. B. Salpointe, trustee for the Catholic Church, granted Warner the right to build ditches across church lands.

Warner enlisted Alex McKay for construction of the millrace at a cost of over $1,000.[43] Beginning near the Pioneer Mill, the millrace was about a mile long. The tailrace returned the used water to the Acequia Madre directly east of the San Agustín Mission (fig. 5.3). The amount of water claimed by Warner was 0.13 inches (0.01 acre-feet). A portion of the tailrace was diverted to run through Leopoldo Carrillo's cooling house directly across from the mission building. By October 1875, Warner had completed construction of his flour mill at

Figure 5.2.
The Santa Cruz River valley from base of Sentinel Peak. **A.** (ca. 1880) Solomon Warner's mill is visible at the left center of the photograph. The white building at center right is Leopoldo Carrillo's ice house, which was cooled by water discharged from the mill. (Photographer unknown, 6608, courtesy of the Arizona Historical Society.) **B.** (1 December 1981) Both the mill and the ice house are now gone, although remnants of the walls persist. The barely visible road crossing the image horizontally is Grande Avenue, and Mission Lane is at left center. Downtown Tucson appears in the center of the image. (R. M. Turner, Stake 1052, courtesy of the Desert Laboratory Collection.)

Figure 5.3.
The Santa Cruz River valley from Sentinel Peak, looking east. **A.** (1880 or 1882) San Agustín Mission is at the center of this view, with Warner's mill complex at lower left. The Acequia Madre, which was fed by Silver Lake, is marked by scattered trees and runs from right to left across the center of the photograph. The acequia probably followed the mainstem of the Santa Cruz River, which at that time had no discernible channel. (Albert Watkins, 18233, courtesy of the Arizona Historical Society.) **B.** (1 December 1981) The only recognizable feature in both photographs is Warner's house in the lower left corner. Most of the modern floodplain has been elevated a few yards by landfills. The Santa Cruz River channel and Interstate 10 bisect the photograph horizontally, and the Tucson cityscape dominates the midground. (R. M. Turner, Stake 1053, courtesy of the Desert Laboratory Collection.)

an expense of between $15,000 and $16,000, a considerable sum at that time.

The increased rainfall also induced vegetation growth on rangelands in southern Arizona, with both positive and negative consequences. Livestock grazing increased in the watershed, aided by more reliable water and less hostilities with the Apache. Several groups from Tucson began investing heavily on grazing lands in the Santa Cruz Valley. In November, Tomás Ortíz sold his half interest in the Canoa land grant to Fred Maish and Thomas Driscoll for $1,100. Compared with the following decades, stocking levels in the 1870s were relatively low compared with the abundant plant growth, and fires—started by lightning and humans—became important on the landscape. Large fires occurred in the Santa Cruz watershed in 1874, 1877, 1879, 1880, 1882, 1887, and 1889.[44] In the summer of 1877, widespread fires broke out along the eastern margin of the Tucson Basin, consuming valuable forage:

> For the last month the country north, south and east of Tucson has been in a constant blaze. The grasses on the mesas, mountains, and valleys have been eaten up by the flames; during the past two days the fire has traveled over the Santa Catalina Mountains and is burning now miles beyond. It has climbed almost to the summit of the Santa Ritas.[45]

Whether or not fires were unusually frequent and extensive in the 1870s and 1880s, and what impact they might have had on watershed processes, is unknown. The 1877 fires followed a prolonged winter–spring drought, which continued into the late summer, causing much alarm in Tucson. No rain fell in August 1877, an unusual occurrence that greatly concerned Tucson residents, who had grown used to abundant rainfall.[46] Recovery of grasses dependent on summer rainfall would have been delayed until the arrival of late summer rains in 1878. On 11 July 1878, unusual flooding occurred along tributaries that originate in the eastern part of the Tucson Basin and flow through town, fully a month before the burned and drought-stricken grassland could have recovered:

> On Thursday afternoon of last week, the most violent rainstorm that has occurred in Arizona during the recollection of the oldest settlers visited Tucson and vicinity, doing an immense amount of damage to property. The clouds began to gather at about 3:00 in the afternoon and two were plainly visible, coming from an opposite direction, which were heavily laden with water. . . . At about 5:00 the crash came, although rain was falling slightly for a little time before. But darkness came with the two black clouds, and meeting, they burst. Being over an elevated position of the country, the water started within one mile of the city, and soon the streets of Tucson leading to the bottomlands were a roaring sea of water. Buildings were washed out, walls torn down and families were fleeing for safety in great alarm. In the lower end of the city the damage was more severe, as the water came tearing down in great volume among the buildings on the lower side of Main Street. The valley was flooded, the fields being covered to a depth of two and three feet, gardens were destroyed, trees taken out by the roots and swept away with the current, carrying devastation in their course. The vegetable gardens, and they are numerous on the bottom, were destroyed, and the houses flooded with water, so much so that the people had to seek places of safety on more elevated ground. . . . There were forty-two buildings, as far as we have investigated, which have been damaged or washed away, and the scene of wreck and ruin is beyond description. The storm lasted about 3/4 of an hour and rain fell to a depth of six inches on the level.[47]

The 11 July storm contributed to one of the wettest months of July in history,[48] and the interaction between land use, climate, and floods was beginning to be felt downstream in the Tucson Basin. Especially in the coming decade, this interaction would lead to significant changes to the Santa Cruz River, decreasing the water security of the residents of Tucson.

A Dependable Water Supply for a Growing Population

By the end of the 1870s, Tucson was running short of domestic water. Although the fields across from town could be easily irrigated, demand for domestic water had outgrown the days of the presidio. Most of the domestic water then came from El Ojito, an artesian well with groundwater flowing out of the ground,[49] which was located where the floodplain abutted the Pleistocene terrace on the east side of the valley. In the 1870s, Adam Sanders and Joseph Phy owned a hand-dug well on South Main Street, from which they filled a large, square iron tank mounted on a wagon. Each day they would drive through the streets selling water at five cents a bucket.[50] Thomas J. Jeffords, a former Butterfield stage driver and the man who brought the Apache leader Cochise to peace terms, tackled the problem in May 1879. Jeffords signed a contract with the city of

Tucson to develop enough artesian flow to supply local water needs for twenty-five years, but the amount of artesian flow was insufficient.[51]

In the spring of 1879, Leopoldo Carrillo purchased about 12 acres of uncultivated land south of the road to the San Agustín Mission and Warner's Mill near the artesian well at El Ojito on the east side of the floodplain. His objective was to plant an orchard to be irrigated with the groundwater:

> He boldly planted a lot of fruit trees . . . and then set to work to provide them with water. A ditch parallel with and close to the mesa line, with short lateral ditches running from the main ditch into the side of the mesa hill, soon filled with water and showed that the neighborhood there was full of springs. Mr. Carrillo then built a stone tank about 50 feet and added thereto a simple pumping apparatus, and the thing was done. The trees have flourished and there seems to be no reason why the place, which was before a mere alkali patch, may not in the future be a fruitful orchard.[52]

This episode plainly shows that groundwater was very shallow just upstream from Sentinel Peak and west of the river, but it was not dependably artesian.

Early in 1880, Carrillo bought or built a hundred houses to sustain Tucson's growing population, which had doubled during the 1870s. Arrival of the Southern Pacific Railroad in March ensured that Carrillo would turn a large profit and increase his status as Tucson's foremost landlord and wealthiest resident. Carrillo also developed four large springs at his Sabino Canyon Ranch in the foothills of the Santa Catalina Mountains, running the water into a reservoir. Apparently, this was done in preparation for opening a resort the following summer.[53] A similar resort was already available at Silver Lake (see chapter 4), and those facilities were improved in 1881:

> Silver Lake . . . is caused by a dam of masonry in the Santa Cruz River and extends over several acres. Several boats [are available] for sailing and rowing up the river beyond the lake. . . . A row of commodious bath houses for bathers and a stout rope extends across the lake for the convenience of persons learning how to swim. The hotel, bath houses, pavilion, lake and grove occupy a space of 20 acres.[54]

Flooding occurred in July 1881, but its impact on Silver Lake is not clear:

> At Brown's station the crossing constructed by Mr. Brown was carried away and a washout eight feet deep was made. The Santa Cruz below Brown's was a sheet of water half a mile wide and in places 6 feet deep and holes washed out in every direction making the whole country a whole lake.[55]

The organization of a volunteer fire department in 1881 once again highlighted Tucson's pressing water needs. The mayor of Tucson, Robert N. Leatherwood, devised a scheme whereby gravity flow from the river six miles upstream could be piped up on the west terrace to town. In the spring of 1882, he enlisted the aid of Sylvester Watts and J. W. Parker, capitalists from the Midwest who had experience in such affairs, to develop the springs at the Valencia Bridge headcut:

> The first material move toward perfecting the plans adopted by them was the purchase of 1000 acres of land surrounding the source of water supply. By the 20th of May, several carloads of heavy sheet iron were on their way from St. Louis and [in] early June the work of constructing the water mains [was] begun in the old government corral. By August 1, a sufficient number were finished to justify excavation of trenches between this city and the forbey [*sic*; forebay] beyond Gay's ranch and a few days afterward, the first pipe was placed in position. From that time to the 15th of September, the largest available force of laborers were employed in digging tunnels and cementing mains, and on this latter date, as the reader must be aware, the Santa Cruz was forced into Main Street. . . . [From the Valencia Road Bridge headcut] for a half mile northward, a wide ditch has been excavated four feet wide in its bed being clean gravel. Through this, water from the underground springs is constantly rising to the surface. Over the entire length of the ditch a triangular aqueduct of redwood has been constructed, completely watertight, thereby preventing any surface water from entering. The aqueduct gathers all water reaching the surface along the whole distance and pours it into a huge forebay, 20 feet deep. This is a compartment built of strong timbers, which acts as a receiver for the aqueduct water and from which the water is discharged into pipes 20 inches in diameter. These extend probably 100 feet, leading to the corner of Main and McCormick Streets.[56]

By 1883, there had been a noticeable drop in the water level at Silver Lake. According to Solomon Warner, development of groundwater upstream from Silver Lake by Watts and Parker had reduced inflows into the lake. He lamented,

> It is impossible to run a flour mill with the natural flow of the river except two or three months in the

winter. . . . I run my mill from Lee's pond [Silver Lake] and as long as he was running his mill it was very satisfactory. But after he stopped running his mill [ca. 1882], the water came very irregular both in quantity and [in] time and it became necessary for me to get the full benefit of the water.[57]

Early in 1883, Warner began buying up lands along the West Branch of the Santa Cruz River. In summer, he started construction of a large earthen dam that would impound the West Branch cienega at the foot of Sentinel Peak, creating what became known as Warner's Lake (fig. 5.4):

I put the flood gates in my present pond some time in July last year and commenced damming the water that was running from my own ciénega with but little success at first, the water leaking through nearly as fast as it came in. About the first of December, the embankment gave way near the flood gates and let almost all the water out of the pond. In repairing that I succeeded in stopping nearly all the leaks and about the first of January, the water in the pond was on a level with the water in the flume. I then made a connection with the flume and the pond and commenced using the water from them both. The water in the Santa Cruz near Tucson has been much less during the past two years than formerly. Both years being very dry and the Tucson Water Company using a large amount of water in Tucson. They have taken it for about two years. The watershed that supplies the ciénega is quite extensive. It commences on the west side of the Sierritas, 30 or more miles in length and 10–15 miles in width. In some seasons the quantity of water running from the ciénega is equal to one-quarter to one-half enough to run the mill several months a year. The pond is situated at the foot of a mountain [Sentinel Peak] and there are several springs at the base, and for a considerable distance water oozed out so much that the cattle to avoid it made a trail through the mesquite and over the rocks at the base of the mountain. Tullies [or tule, *Typha* sp.] and water grasses grew on all the land the pond covers with the exception of three or four acres on the south and east side. Beside[s] the stream of water that continuously ran in the lower part of the land which the pond covers there were other depressions where the water remained all the time.[58]

The newspaper discussed a different benefit of Warner's Lake:

The result of this big dam has already been wonderful. The waters of the many springs of the different ciénegas on the Warner land have been held back by the dam and have risen till they have covered some 20 acres of land, creating a sheet of water that is beautiful to look upon. Already the wild fawn have made it their resort, and an organization of hunters have obtained the exclusive right to shoot upon its waters. A flat-bottomed boat sails over its surface. The different kinds of ducks killed there are the gray and spoonbill, the green and red winged teal, mallard, canvass back, widgeon, spring tail, the butter and a new kind never seen before called the fish duck. . . . The snipe, curlew and plover appear abundantly. When the dam is completed and the waters have occupied all their space, about 50 acres will be covered.[59]

No sooner had the dam been constructed than Warner began drawing up plans for bathhouses to compete with the resort at Silver Lake. He had completed the foundations for the houses and was about to erect the structures when bad news came. Warner was a creditor in the amount of $2,300 in a company that had just failed, and the loss compelled him to cease all improvements. Foreclosure on his credit marked the beginnings of troubled times. In July 1884, he received legal notice from Hereford Lovell, attorney for the water overseer and landowners immediately downstream of the lake. According to Lovell, Warner was obstructing streamflow into the public acequia without consent of the water overseer. Two weeks after Warner received this legal notice, certain landowners in the valley adopted measures to secure a more efficient distribution of water. This set the stage for a new era of litigation over the diminishing water supply along a river that, even in the 1880s, had a flow that was overallocated.

Too Many Users, Too Little Water

As new lands were cleared for cultivation downstream of Tucson, the irrigation issue escalated into serious litigation between landowners north and south of the hospital road (now St. Mary's Road). Fields north of the road received their water after it had turned Warner's waterwheel. Landowners south of the road thought that the combination of Warner's Lake and the mill deprived the oldest fields of irrigation water. Warner described a power struggle over the water in the lake wherein downstream users opened his hydraulic gates to lower the lake level and provide more irrigation water downstream.[60] The issue at hand was establishment of water rights for the downstream farmers versus Warner's water rights for his lake and mill.[61]

Figure 5.4.
Southeast view of Warner's Lake. **A.** (1883) Solomon Warner constructed this earthen dam to impound Warner's Lake along the West Branch of the Santa Cruz River. The reservoir ensured a continual supply of water for his flour mill and provided a breeding pond for carp. The shallow channel of the Santa Cruz River downstream of the dam is on the extreme left of the photograph. The dam and reservoir lasted less than 10 years. (Photographer unknown, 12565, courtesy of the Arizona Historical Society.) **B.** (31 December 1988) This approximate match, taken 105 years later, shows no sign of the dam or reservoir. The course of the Santa Cruz is obscured by Athel tamarisks at lower left. The roadway in the midground is Starr Pass Road, the westerly extension of 22nd Street. (R. M. Turner, Stake 1055, courtesy of the Desert Laboratory Collection.)

By May 1885, the conflict between farmers north and south of St. Mary's Road was being settled in the county court before Judge Gregg.[62] The court transcript included testimony by W. A. Dalton, one of several landowners north of the road, who testified that about 1,400 acres of irrigated land growing mostly wheat and barley were negatively affected by water diversions. A second crop during the dry summer could not be grown on fields north of the road because of heavy water use upstream and the scant flow of the Santa Cruz River during the hot months. Fergusson's 1862 map of cultivated fields indicated that fields just downstream from Congress Street were worked only when there was an abundance of water.[63]

In the summer of 1884, landowners south of the road provoked the lawsuit by cutting off flow to the north, promising to evenly distribute surplus water in the future. Landowners to the north were skeptical, anticipating heavier use to the south and the unlikely prospect of a surplus in the immediate future. Traditionally, water was diverted to the field that needed it the most, regardless of the owner or his position in the irrigation schedule. This pattern was now being disrupted by unusually heavy water use just downstream of Silver Lake. Since the coming of the railroad, several Chinese families had begun leasing fields from Hughes, Carrillo, and Warner to grow and sell vegetables. By 1885, these gardens covered about 150 acres. Although fields of wheat and barley required irrigation only once a month, the closely planted vegetable gardens needed water every week or daily if water was available. Dalton accused the Chinese of stealing water from the ditches to keep their gardens growing.[64]

In June 1885, Judge Gregg rendered a judgment in favor of Carrillo and the landowners south of the hospital road. Lands north of the road would receive surplus waters only after fields to the south had been fully irrigated, including the Chinese gardens.[65] For the landowners north of the road, the unfavorable judgment was made worse by drought in the summer of 1885.

In February 1886, a newspaper discussed the abundance of groundwater:

> Our ditches—eight streams of water run through the Santa Cruz Valley opposite Tucson. Five of these ditches are 7 feet wide that now contain a foot and a half of running water. The other three are narrower, and contain less. Besides there are many smaller ditches taken from the larger ones and running nearly parallel with them. The source of all this water is Silver Lake and the Santa Cruz, the spring at Carrillo's Garden and the natural springs at War-

ner's Lake. About 6 miles below the city these ditches cease, the water sinking into the plain where it is reached at a distance of from 100 to 300 feet below the surface.[66]

> The water not only shows on the surface for miles in a running stream but also [can be seen] in many bubbling springs at San Xavier, at Carrillo's Garden, and other places, including those ranches where there is no surface water. All along the valley, whether the stream appears or not, are wells, such as those at Verdugo's or Osborn's, giving abundant water at from 4 to 6 feet below the surface.[67]

The owners of the lakes in the desert worked to keep their properties viable or to divest of them. Early in 1886, Warner leased his lake to parties wishing to improve the resort. Maish and Driscoll, who acquired the Silver Lake property in 1884, renovated the hotel that same year. In May 1886, Warner sold all of his land, including the mill and lake property, to Mrs. T. L. Shultz for $6,000. To keep up with the competition, Leopoldo Carrillo improved the gardens on the edge of town and waged an aggressive advertising campaign in the newspapers.[68]

More Floods, an Earthquake, and Sam Hughes's Intercept Ditch

As one newspaper reported in June 1886, "There is a scarcity of water in the Santa Cruz Valley and Silver Lake has spared the settlers all the law allows."[69] Despite this brief drought, long-term residents knew that the summer rains would arrive eventually. The *Arizona Mining Index* warned of the impending danger of building on the floodplain of the Santa Cruz River:

> The bottom of the Santa Cruz is an unsafe place for dwelling houses, and there are a good many such along through it. In 1872, and several times previously, old don Juan Warner of Los Angeles published several articles in the papers cautioning the incoming population about the dangers of building along the river valley and has often advised the city authorities to build certain embankments and take other precautionary measures for protection. To his advice and statements they gave no heed, and within the past three years the floods came again, as of yore, and destroyed many lives, houses and other properties. We state this as a hint to those who build in the Santa Cruz Valley bottom. The upland mesa is the place for homes and hearth.[70]

Even in the 1880s, residents of Tucson were lulled into complacency concerning the relation between their normally benign rivers and washes and the raging torrents these watercourses could become when the monsoon rains were most intense.

Just five weeks later, in August 1886, heavy rain in the upper portion of the Santa Cruz watershed generated floods that wreaked havoc on the floodplain, washing out a bridge and doing considerable damage to the dams at Warner's and Silver Lakes:

> The recent heavy rains in the vast watershed of the Santa Cruz, culminated in a flood near Tucson on Tuesday last [4 August]. The rise in the valley above the city was unexpected and could not be controlled. It appeared with suddenness at Silver Lake and Warner's Lake, two sheets of water above town, and breaking through the dams of each so formed a junction and swept down the cultivated valley for several miles below. When it appeared rolling over the waters of Warner's Lake the parties in charge there attempted to hoist the floodgate at the dam but the accumulated sediment on the upper side prevented them from making use of the safety valve and the rushing flow swept away 30 feet of the dam clear to the bottom level of the lake. At Silver Lake, the flood did not break the stone work, but swept over the bank east of the bath houses and soon cut a wide passage way. Pouring rapidly into the field it swept out a culvert in the graded highway and soon met the flood coming down from Warner's Lake where they united and covered the whole valley, damaging fences and filling the ditches with sand. The flood covered the valley to a point four miles below the town and its mainstream and some of the boards of the fences as far as the Nine Mile Water Hole. . . . The estimates of the total damage to the lakes, the crops and the ditches [are] placed at $4000. Repairs have already been made at Silver and Warner's Lakes as well as on the fences of this valley. It will require some time to restore the ditches to their former capacity.[71]

A week later, another newspaper reported,

> The Santa Cruz is booming again [14 August]. The flow is from the upper Santa Cruz and the Sonoita Valley. The dam at Silver Lake was washed away again yesterday and a great deal of damage has been done to the lake.[72]

The county bridge across the Santa Cruz near Silver Lake was washed away by the roaring flood.[73]

As if floods were not enough of a problem, a powerful earthquake occurred the following spring. On 3 May 1887, a magnitude 7.2 event with an epicenter just north of Bavispe, Sonora, affected an area of nearly one million square miles in the southwestern United States and northern Mexico. This earthquake, one of the largest in Arizona history, produced a thirty-mile fault scarp with three to six feet of vertical offset with its northern terminus five miles south of Douglas, Arizona.[74] The ground's shaking leveled many villages in Sonora and Arizona, including Charleston on the San Pedro River,[75] and several buildings in Tucson were damaged. At Mission San Xavier, the walls of the Spanish cemetery were demolished and the church sustained extensive damage. Viewed from Tucson, the dust raised by rockfall in the Santa Catalina Mountains was mistaken for forest fires.

After the earthquake, significant hydrological changes were reported within a hundred-mile radius of the epicenter,[76] including sudden water level rises or falls in wells,[77] increases in spring discharge, rise of groundwater to the surface because of subsidence or opening up of new fissures, and forcible ejection of water and mud from fissures (fig. 5.5). A fissure twenty miles long was reported along the San Pedro River north of Benson, and a considerable stream of water reportedly flowed from it. Throughout southern Arizona, there were reports of increased stream discharge during the driest month of the year. Nearby, the San Pedro River at Charleston suddenly ceased flowing and for a short time was entirely dry, only to resume flowing with a stage at least two feet higher than before.[78] The Agua de la Misión on the Spring Branch near San Xavier dried up following the earthquake, and water rose to the surface upstream.[79] The O'odham built a dam at the new spring before the onset of the summer monsoon.

Earthquakes are rare in southeastern Arizona, but there is some scant indication of other events that caused hydrologic changes. The Apache chief Esconolea, an old man when interviewed in the 1850s, recalled an earthquake at the beginning of the nineteenth century:

> The whole earth split open from one side of the valley to the other, sending forth a blue smoke [ejection of water?] heavenward for a mile. The same day it began to rain and continued for several days. When the storm ceased all of the earth on this side of the San Pedro River had closed together again, while a crack in the earth about a mile long, five feet wide, and from ten to twenty feet deep remained.[80]

In mid- and late summer of 1887, more flooding occurred, especially along the Rillito River:

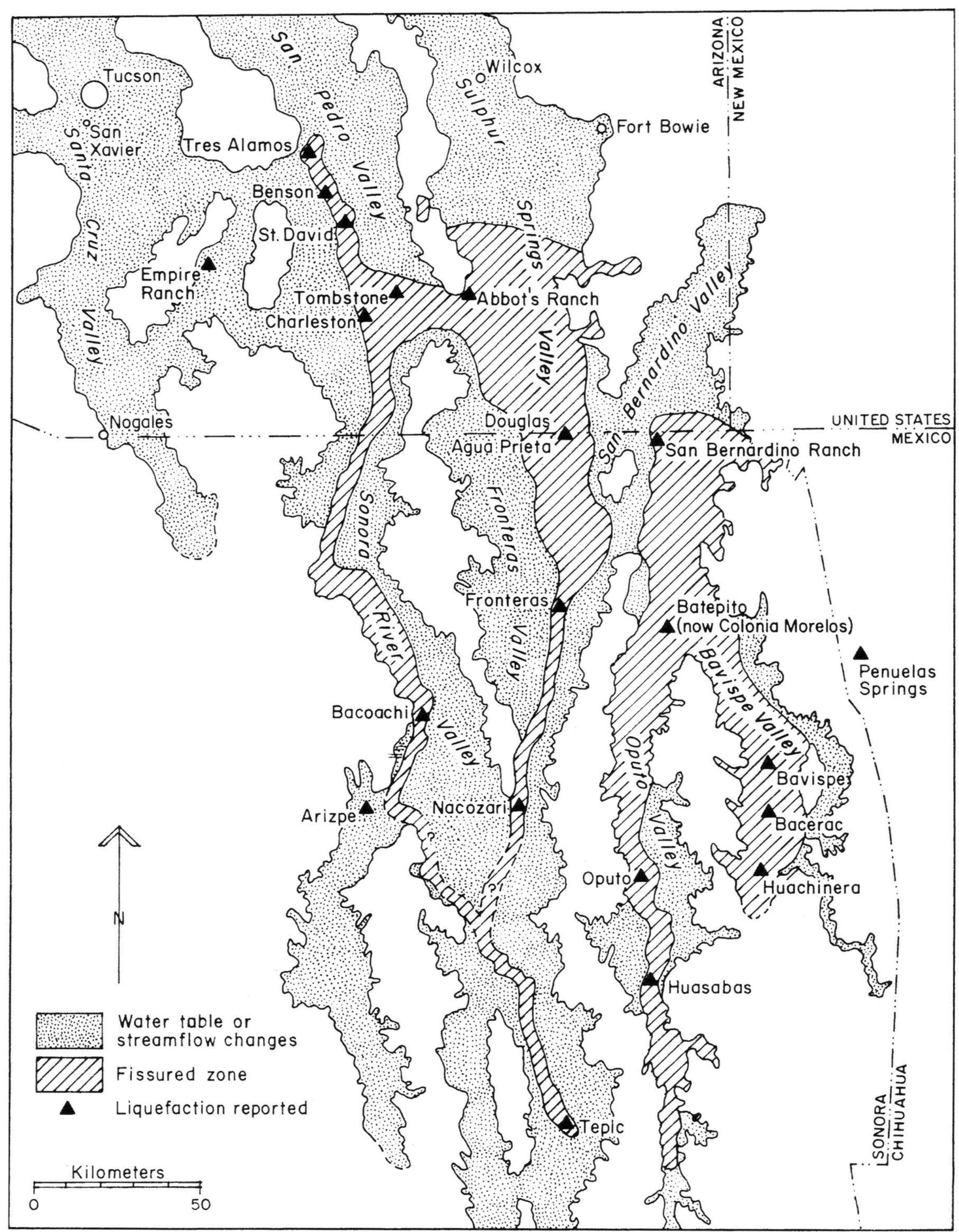

Figure 5.5.

Map of northern Sonora and southern Arizona, showing hydrological effects of the 1887 earthquake (after Dubois and Smith 1980).

[12 July] There is a considerable volume of water coming down the Rillito from the watersheds of the Rincon[s] and Santa Catalinas. It has been reported that the recent flood carried away the bridge between here and Silver Lake across the Santa Cruz River. The late heavy rains to the southward have caused the Santa Cruz to boom. The water is now coming down the valley in three large streams.[81]

[13 July] The Santa Cruz River is more than a mile wide and deep enough to float a mammoth steam boat. . . . The big rain yesterday afternoon filled the arroyo in North Tucson [Tucson Arroyo] full to the brim[;] the angry waters roared like a young ocean turned loose through a new channel. It swept everything before it, flooding the valley below.[82]

[9 September] The flood in the Rillito on Friday last was much worse than was thought. Many cattle being taken unaware by the sudden overflow of the stream were swept away and drowned. Trees of good size, bridge timbers and other debris were found scattered along the banks yesterday. . . . The waters north of town are reported to have stood about 2 miles wide.[83]

[12 September] The Santa Cruz River is running higher now than ever before this season. For the first time in many years, it is navigable from Tubac to the Gulf.[84]

The 1887 floods washed out the new dam on the Spring Branch and scoured a channel immediately downstream. In February 1888, surveyor L. D. Chillson recorded a gulch thirty feet wide and fifteen feet deep near the former Agua de la Misión. The flooding that followed the earthquake and its hydrological effects may have heralded downcutting in some reaches of the San Pedro River. Near Tres Alamos, north of Benson,

the river deepened eleven and a half feet between 1885 and 1889.[85]

The 1887 floods were yet another blow to the irrigation system in the Tucson Basin, and residents sought to improve the reliability of their water systems. In September 1887, Sam Hughes bought a narrow strip of land following the river and centered on what would become St. Mary's Road. North of the road, Hughes planned to excavate a ditch twenty feet wide at the heading and a total of fifteen miles long to intercept and channel groundwater.[86] Apparently, the capital for Hughes's venture was not forthcoming, so he dug a smaller ditch in 1888 that was designed with an artificial headcut that would erode upstream to create the desired ditch. Maish and Driscoll, owners of the Silver Lake resort, had a similar intercept ditch dug at the Canoa Ranch near Continental.[87] This type of ditch, essentially a headcut created for flood extension into a channel, was probably the most popular method to secure near-surface groundwater before the turn of the twentieth century (fig. 5.6). Hughes's intercept ditch (figs. 5.7 and 5.8) and its artificial headcut would later figure prominently in the downcutting of the Santa Cruz River through Tucson.

The Beginning of a Modern Water Distribution System

By the late 1880s, demand for irrigation and domestic water had exceeded the supply afforded by the perennial flow of the Santa Cruz River, in large part because its channel was beginning to change and its flow was less predictable. The dams at Silver and Warner's Lakes

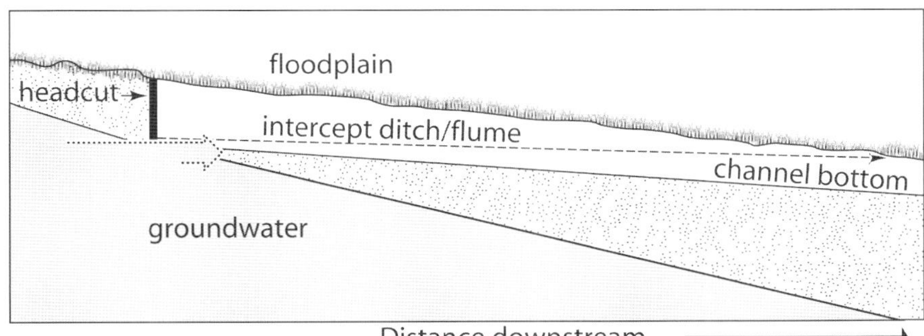

Figure 5.6.
Schematic diagram showing the longitudinal profile of an intercept ditch, which was a nineteenth-century technology for harnessing effluent groundwater for diversion onto farmlands. The *thalweg* is the deepest part of the channel or ditch. This diagram is not to scale; the real slopes would require the horizontal distance to be several miles along the Santa Cruz River.

Figure 5.7.
(October 1889) In 1888, Tucson merchant Sam Hughes excavated an intercept ditch on the Santa Cruz River at the St. Mary's Road crossing. The ditch was designed to actively erode into an irrigation canal, even with minor flooding, as shown in this upstream view. At this time, the river had a shallow channel, and moderate floods would inundate the surrounding broad floodplain. (Henry Buehman, courtesy of the University of Arizona Special Collections Library.)

Figure 5.8.
(October 1889) Taken on the same day as figure 5.7, this image presents a slightly different view of the headcut at the St. Mary's Road crossing, with Sentinel Peak at the upper right. (Henry Buehman, 2922, courtesy of the Arizona Historical Society.)

blocked the normal flow of the river, and water released at their headgates was used to irrigate about 450 acres of land. J. M. Berger, whose wife was heir to the Martinez land grant and farmer-in-charge on the Papago Indian Reservation, estimated that another 790 acres of land were irrigated near San Xavier in 1890.[88] A major conflict over water rights between landowners upstream and downstream of St. Mary's Road eventually was settled in favor of the upstream irrigators. It was time for a systematic solution to the water problem.

In 1884, city engineer J. P. Culver reported discharges at the various springs in the valley, including 17.7 ft[3]/s at Punta de Agua, 4.6 ft[3]/s at the Valencia Road headcut, 0.35 ft[3]/s at a spring in the channel a mile downstream, between 12 and 18 ft[3]/s through the flume at Lee's Mill on Silver Lake, and about the same through Warner's millrace.[89] In 1889, the new Tucson Water Company began to pump groundwater instead of diverting the less reliable surface water into canals; they installed a steam-driven pump on a 40-foot-deep well that produced an astonishing 1,250 gallons per minute.[90] This eased the water shortages temporarily, but the growing city continued to thirst for more water. A gravity-feed system, completed in 1882, moved more surface water into town.[91] Foreshadowing events to come, water-supply engineer J. R. Watts reported a drop in the water table near Mission San Xavier.[92] The next century of Tucson water history would consist of drilling more and more wells in a vain attempt to keep up with the thirsty needs of a growing population.

6

Arroyo Downcutting and Widening, 1889–1915

The late 1880s were a time of numerous hydrologic changes in the Tucson Basin caused by floods, an earthquake, and the quest for reliable water supplies. The stage was set for arroyo downcutting as a result of floods in 1887 and 1889, alterations of the floodplain, the increase in livestock in the watershed, and possibly even the effects of the 1887 earthquake. Events of the following two decades would alter this hydrologic network, changing the drainage pattern in the Tucson Basin into its current form and altering Tucson's water supply from mostly surface water to the more dependable and abundant groundwater.

Tucson was growing more vigorously, placing greater demands on its tenuous water supply. A battle brewed between water users trying to extract surface water from the river and the river itself, which seemed intent on carving itself into its once wide floodplain. The harder farmers and city dwellers tried to secure a water supply, the more the river changed its course and its bed elevation, until groundwater became the only option for a thirsty city and crops. With pacification of the Apaches, ranching expanded, only to be thwarted temporarily by the greatest drought in recorded history. Among the incoming population were more ornithologists, who were attracted to the famous bird habitat along the Santa Cruz and Rillito Rivers, bolstering our knowledge of riverine conditions during these changing times.

Sam Hughes's Ditch and Arroyo Downcutting

In the late 1880s, the channel of the Santa Cruz River of the twenty-first century was transient and poorly defined; instead, the floodplain was characterized by several vaguely defined channels and artificial water bodies (see fig. 4.1). In 1888, Sam Hughes excavated an intercept ditch to capture shallow groundwater at St. Mary's Road and direct it onto the floodplain to irrigate new fields downstream. For all practical purposes, he had built an artificial headcut with an unprotected heading. All that was needed for Hughes's intercept ditch to erode and migrate upstream was a shift in climate and an increase in the magnitude and frequency of large floods.

Floods in the summer of 1890 received a great deal of attention in the newspapers.[1] The season was unusual in the character and amount of precipitation for July and August, anomalously cool summer temperatures, and the repeated flooding. The first flood swept through the valley from 27 July through 1 August, producing overflows more than two thousand feet wide and up to twelve feet deep in the Tucson Basin. On 11 August, the dam at Silver Lake was swept away; on 9 and 15 September, the combined flows of the Rillito and Santa Cruz Rivers inundated the floodplain north of Tucson. On the Rillito, the most recent flood with a higher flood stage had been in September 1868.[2]

Newspapers covered the destruction, which began on 27 July when Silver Lake overflowed and many of its carp were washed downstream. Workers salvaged and marketed some fish in town, and others reportedly were washed downstream as far as Picacho. The rains continued to fall:

The rainfall has been quite general lately in this section. The heaviest rains occurred day before yesterday [July 30]. There seemed to be two storms—one traveling towards the southeast and the other towards the southwest—and [they] seemingly met at the Coyote Mountains, hence the overflow of the Santa Cruz at the present time. The San Pedro, the Gila and the Rillito also [were] reported full. The width of the overflow of the Santa Cruz is about 700 yards and the deepest part is over twelve feet.[3]

The floodwaters damaged gardens, cornfields, fences, irrigation ditches, arable bottomland, and adobe houses. The crossings at the hospital road (St. Mary's Road) and to Warner's Mill became impassable.

In Lochiel, where the Santa Cruz River crosses the border into Mexico, Alice F. Cameron, rainfall observer for the US Signal Service, made the following notes:

> July 31st was very damp and cloudy ranging low all day—very damp and chilly. . . . The rains of August 1, 4, 5, 6, 11, 15, 22, and 24 came from the southwest, as in fact have most of our heaviest rains this year. Heretofore, in other years, since my first coming to this country in 1884, the heaviest rains come from the south and east. . . . Many heavy rains this year have fallen (particularly this month of August— also in July) five miles northeast of here, when not one drop fell here. The rainfall there must have been tremendous for the San Rafael Valley has been flooded very many times this summer, by that fall of water.[4]

The rains were so unusual that E. L. Wetmore, the rainfall observer at Tucson, remarked that July rains resembled winter and spring rains, cooling off the atmosphere. The owner of a popular brewery, Alex "Boss" Levin, was quoted as saying that on 5 August the river was higher "than at any time during the last twenty-five years"; the article continued, "[H]e ought to know as he has lived on its banks during that time."[5] Another newspaper article quipped that steamers would depart daily from "Levin's Landing" bound for Yuma.[6] The *Arizona Daily Star* reported that

> the water in the Rillito was a foot higher last Tuesday night [5 August] than the highest water mark of past years, and . . . there was at least one third more water, as the recent flood had so cleaned out and deepened the channel that a third more water could be carried without it reaching the height of Tuesday night. . . . [T]he sand deposit on the overflowed land is from two to four feet.[7]

On 8 August another observer reported:

> The single channel which was being washed out through the fields of the Santa Cruz by the floods, resulted in considerable damage but this danger has been greatly increased from the fact that the wash or channel has forked at the head, and there are now several channels being cut by the flood, all of which run into the main channel. If the flood keeps up a few days longer there will be hundreds of acres of land lost to agriculture. As these new channels or washes are spreading out over the valley, they will cut through and greatly damage the irrigating canals. Those who have visited the spot say that it is not too late yet to direct the water so as to cause it to cut a single channel and thus confine the flow. In view of the probable great destruction it would be well for some one to lead off in this matter.[8]

Fred Maish tried to lure customers back to his Silver Lake resort:

> [8 August] More than fifty acres of land which has formerly been under cultivation in the Santa Cruz bottom, has been rendered worthless by being washed out so as to form an arroyo. . . . The water has subsided so that travel to and from Silver Lake has been resumed. Mr. Maish will put on a force of men to work this morning repairing that part of the road leading from the county road to the hotel. The bathing pond of the hotel is full of clean, fresh water. This pond is 14 feet deep and a springboard has been erected to accommodate bathers.[9]

Less than two weeks later, some forty feet of earthen dam impounding the lake was washed away.

Perhaps the most important news concerned Sam Hughes's intercept ditch. On 4 August, the floods had widened the heading of Hughes's canal near St. Mary's Road; several days later, it had crossed Stevens Avenue (the future Congress Street) and was still heading upstream.[10]

> A terrible cloud burst is reported southeast of the city, some 20 miles away. Look out for another big flood in the Santa Cruz. Wetmore, the signal service man says that 2 and 1/4 inches have fallen during the last ten days. The floods, however, have come from storms outside of the Tucson Basin. The channel or cut being made by the overflow of the Santa Cruz River, is now one mile and a half long, by from one to two hundred yards wide. . . . Constant and heavy rains reported from Nogales for last ten or twelve days.[11]

By the end of the month, the resulting headcut had eroded two miles upstream to a point near Silver Lake. Tucson surveyor George Roskruge photographed the new channel from the east bank at the hospital crossing

Figure 6.1.

View looking west across the Santa Cruz River at St. Mary's Road. **A.** (August 1890) The Santa Cruz is in flood, and its newly formed arroyo threatens a homestead on the west bank of the river. (George Roskruge, 45854, courtesy of the Arizona Historical Society.) **B.** (4 February 1982) A landfill now occupies the upper three to six feet of the former floodplain, and the arroyo walls support a growth of ruderal plants, probably tumbleweed. The St. Mary's Road Bridge is on the far right. (R. M. Turner, Stake 1065A, courtesy of the Desert Laboratory Collection.)

Figure 6.1. (*continued*)
C. (20 December 2003) The channel of the Santa Cruz is now defined by soil-cemented banks that are lined by the paved trails and cultivated trees of the Santa Cruz River Park. An office complex was built on top of the former landfill. (D. P. Oldershaw, Stake 1065A, courtesy of the Desert Laboratory Collection.)

(figs. 6.1 and 6.2). On 13 August, he described the erosion as "Sam Hughes' ditch taking a walk to Maish's lake [Silver Lake] after water."[12]

> [17 August] Silver Lake dam, gates open, was washed away for about 40 feet in the bank of sand at one side. Many Tucsonans visited scene. Fields flooded, stage passengers detained, fences gone, four acres of fine bottomland in W. C. Davis' field washed down. A man on horseback rode into newly made waterhole on hospital road, going down out of sight for some seconds.[13]

Daily storms kept the river flowing and eroding its bed and banks. On 20 August, the Santa Cruz River at St. Mary's Road was over three hundred feet wide.[14] Toward the end of the month, the river rose even more, washing away houses on the floodplain.[15] By the end of the month, the river had fallen but left behind quicksand in the crossings, and the new head-cut had reached the ruined Silver Lake.[16] One astute observer noted that the deepened channel of the Rillito River could now carry a third more water without reaching the flood stage of a few days before. Ar-royo downcutting, then, was initiated by four major floods spaced three to ten days apart in the summer of 1890.

Several details bear noting here. Groundwater was high in the Tucson Basin, and although it was locally exploited for irrigation, there is no indication that water levels were dropping regionally. Therefore, arroyo downcutting on the Santa Cruz River occurred during a period of high—not depleted—groundwater levels, which is opposite to what some scientists have concluded for other rivers. The erosion occurred during unusual and sustained flooding near the end of a period of notably wet summers in Tucson history. Although we cannot know the magnitude of these floods, the descriptions of width, depth, and damages sustained suggest that they were quite large. There appears to have been some augmentation of discharge owing to the failure of one or both dams upstream from Congress Street, although these lakes did not impound large volumes of water. Finally, there is little question that erosion was focused by human modifications of the floodplain, particularly Sam Hughes's Ditch.

Figure 6.2.
Upstream view of the Santa Cruz River at St. Mary's Road. **A.** (August 1890) This upstream view of the Santa Cruz River was taken on the same day and from the same location as figure 6.1A; the cottonwood tree with the distinctive, asymmetrical crown that appears on the right bank is also visible in figure 6.3. On August 8 or 9, the headcut divided into two channels, and their confluence is shown in this photograph. (George Roskruge, 45852, courtesy of the Arizona Historical Society.) **B.** (20 December 2003) Riparian vegetation on both sides of the channel obscures soil–cemented banks, which were installed in 1983. The bank protection helps to stabilize floodplain infrastructure, including the powerline pylon at left, the bridge, and buildings on the right side. A cable crosses through the center of the view. (D. P. Oldershaw, Stake 1065bx, courtesy of the Desert Laboratory Collection.)

Tucson's Reaction to the Arroyo and Flood Control

Tucsonans viewed the new channel west of town with mixed emotions. Some threatened lawsuits, alleging wrongdoing by those who tapped the water resources for irrigation or domestic supplies. A few residents suggested that there was something to be gained from the erosion: "The big arroyo cut through the Santa Cruz Valley will afford the means of drainage for the city. It is an ill wind that does not bring good to someone."[17] In October, a petition circulated urging the Pima County Board of Supervisors to build a bridge at the hospital road.[18] The petition failed, and the bridge was not built until 1899.[19]

Many Tucsonans came to blame Sam Hughes for the arroyo downcutting in 1890, although he could hardly be blamed for the unusual number of storms that summer. Two decades later, Volney Spalding of the Carnegie Desert Botanical Laboratory on Tumamoc Hill wrote:

> According to statements of residents, this extensive erosion is of recent date. Previous to the advent of cattlemen some 20 years ago, and the destructive effects of over-pasturing, the valley of the Santa Cruz had a luxuriant growth of sacaton and other vegetation, which prevented the cutting of the channels, and the water spread out over the whole valley instead of flowing through the deep cuts it has since made; tules grew thickly in the springy places, and a fine forest of mesquite covered the ground. I have given the commonly received version of the cause of the cutting of the channels of the Santa Cruz, but have since been told by Mr. Herbert Brown, of the Tucson Post, that about 20 years ago, certain old settlers undertook to develop water at a point about two miles down the river, where there were springs, and in order to accomplish this most easily cut a channel for a little distance, expecting the river when it rose to do the rest. Their expectations were fully realized, for the river scoured out the cut and kept on with its work, as already indicated. At present the effects of such erosion are seen most plainly from the point below Tucson already indicated to one about two miles above the city.[20]

In 1934, another pioneer, Rollin C. Brown, recalled:

> Tucson was a great old place when I first came here in 1873. . . . In 1873 Tucson was situated down around Meyer and Main Street extending west as far as the Santa Cruz. And, by the way, there was no big river

Figure 6.3.
(ca. 1900) In this downstream view of the Santa Cruz arroyo between Congress Street and St. Mary's Road, the cottonwood with the distinctive crown, also visible in figure 6.2A, is prominent. (Henry Buehman, 1902, courtesy of the Arizona Historical Society.)

Figure 6.4.
Downstream view of the confluence of the West Branch and the Santa Cruz River from the lower slope of
Sentinel Peak. **A.** (1904) The lower half of the photograph encompasses the former area of Warner's Lake (see fig.
5.4A). A remnant of Warner's Dam is visible at left center, just upstream of the confluence. By 1904, the headcut
from Sam Hughes's Ditch had extended along the Santa Cruz mainstem and the West Branch, as shown by this
photograph. (Walter Hadsell, 24868, courtesy of the Arizona Historical Society.) **B.** (17 December 1981) The
West Branch was filled in artificially in the 1960s and is now marked only by a shallow depression lined with
a few mesquites. The Santa Cruz proper is bordered by taller saltcedars that are visible in the center of the
photograph. The intersection of Mission Road and 22nd Street is visible in the lower right. (R. M. Turner, Stake

Figure 6.4. (*continued*)
1026, courtesy of the Desert Laboratory Collection.) **C.** (17 September 2000) In the nineteen years since the previous image was made, the channel of the Santa Cruz has been spatially anchored with soil cement. Several of the tall saltcedars are now gone, and the Mission Road–22nd Street intersection has been enlarged and a traffic light installed. (D. P. Oldershaw, Stake 1026, courtesy of the Desert Laboratory Collection.)

bed down there then. All that country was a beautiful garden spot, smooth and covered with grass and big cottonwoods. That was where Levin's beer garden used to be [current northwest corner of Congress and Granada Streets]. There was no finer place in the town. . . . Then a man by the name of Hughes had one of these new fangled ideas of digging a ditch in the little brook that was then the Santa Cruz and when the flood season came he could bring the water in by gravity. Well, all in all, his gravity idea did not work so well, and when the rainy season came on that ditch got bigger and bigger until you have that big dry river bed that you see today. I can remember the many times when Main Street was completely covered with water during the flood time.[21]

Cirilio Solano León, who was the son of a lieutenant in the old Mexican garrison and was born around 1850, reminisced:

When I was [a] boy there were no river banks. I remember the time the banks were washed out. It isn't very long that the present channel has been here. Mr. Hughes used to own a small piece of land where the Deaf and Blind School is now [on the west bank, just downstream of Speedway Avenue], and he dug a channel about five or six feet wide, and when the floods came along the waterfall began to cut away the land greatly, clear back to San Xavier. This lowered the water level all over the whole valley. Much of the land that is dry now had water before this.[22]

At the 1924 dedication of Sam Hughes Elementary School, historian Edith Stratton Kitt handed down still another indictment. Speaking of the school's namesake, she stated, "He started one of the early irrigation ditches in this part of the country—the ditch which when the floods came cut back and formed the channel of the Santa Cruz River."[23]

In the summer of 1891, there was more flooding in the Santa Cruz valley. On 2 July, the newspapers reported that the river was full and as high as at any time during the previous summer.[24] In August, Maish repaired the dam at Silver Lake again. Later that month, the river was once again running full and level with its banks, causing great damage to the irrigation network west of town.[25] By 29 August, Silver Lake was "again full of clear and limpid water, and much deeper in places than before."[26]

The 1890–1891 floods extended the headcut from St. Mary's Road upstream and past Congress Street (figs. 6.3 and 6.4), the San Agustín Mission, and along both streams above the confluence of the West Branch and the Santa Cruz River. The damage to agricultural land in the summers of 1890 and 1891 raised special concerns about the pressing need to build flood control structures.[27] The river apparently was not affected during the widespread floods of January 1891, which were devastating in central and northern Arizona,[28] and the following summer was extremely dry, marking the beginning of the early twentieth-century drought. The situation was aggravated by the minimal flow in the fall of 1891. Once again, a conflict developed between landowners in the bottomland. This time a deep channel literally divided the factions on either side of the valley, as those on the west side of the Santa Cruz River claimed that those on the east side received twice as much water as they were entitled to.[29] People feared for the worse as the dry year wore on: "The Santa Cruz River has never been as low since 1872 as it is now. At that time the people had to dig in the bed of the river for water, and barely obtained enough for home consumption. It is feared that the same conditions will come to pass this fall."[30]

The newly eroded channel required stabilization measures as well as a new approach to irrigation.[31] In 1891, W. A. Hartt developed one of the first farms that depended entirely on pump-well technology.[32] S. W. Grossetta followed suit in the same year by installing pump wells on his farm just north of present-day Grant Road.[33] Just south of Hartt's ranch, J. K. Brown constructed a dam thirteen feet high and five hundred feet long that would impound 80 million gallons of water fed into the reservoir by tributaries that drain the northeast slopes of the Santa Rita Mountains.[34] Also underfoot was an impressive plan to dam the Santa Cruz near Nogales, making possible the irrigation of some 300,000 acres of bottomland as far north as the San Xavier Indian Reservation. The plan never came to fruition.

In 1892, Frank and Warren Allison set out to rejuvenate fields west of Tucson that were left dry by entrenchment of the channel and destruction of irrigation ditches. They developed a water source by draining the marsh at the former site of Warner's Lake into a ditch that extended a mile downstream.[35] This project, perhaps the first large-scale extraction of groundwater in the Tucson Basin, was expensive, requiring a large labor force and dynamite to blast out the rocky sections along the base of Sentinel Peak.[36] In January 1893, the ditch extended two miles, terminating in a ten-acre reservoir.[37]

Tucson's industrial technology was powered by steam, and boiling water required an ample and convenient wood supply. Demand for fuelwood impinged upon the vast mesquite growth on the floodplains of the Santa Cruz River. By 1892, the fuel supply on most nearby ranches had been exhausted, leaving the San Xavier Indian Reservation as the only convenient future source of wood in the area:

> Ten years from now [we] will see wood scarce and hard to get. Mr. Shortridge, in the wood business, tells that already for ten miles around . . . the supply is nearly ended. Mexicans are now bringing in roots and stumps, dug up and cut up into stove size. Many Mexicans, he says, go as far out as twenty or thirty miles, taking two or three days for the trip. Others make a precarious livelihood by stealing what they can from ranches or government land. The San Xavier Reservation has a fine wood supply. This the Papagos [Tohono O'odham] are becoming aware of, and are raising the prices accordingly. When the wood from the surrounding ranches is gone, which it will be at the present rate in five years, they will have enough to keep Tucson going for another five years.[38]

The decimation of the ancient mesquite trees of the Great Mesquite Forest then began in earnest, but the woodcutting spurred regrowth of a secondary forest.

The early twentieth-century drought, which persisted until 1904, was far more significant to water supplies than were more-recent events in the middle of the twentieth century and at the start of the twenty-first century, when abundant power and wells offset decreases in surface-water supplies. In 1893, the Tucson Water Company was struggling to supply the town with domestic water, particularly during the dry season. Potable water was sprinkled on Tucson's streets to keep the dust down, consuming 10,000–12,000 gallons of water per day. Drinking water was used to irrigate landscaping installed to beautify the city in the 1890s. Restrictions for the watering of gardens, lawns, and trees only during the evening and nighttime hours were recommended, a tactic still used in the twenty-first century to reduce water demand in the peak of summer.[39]

To counter water deficits, the Tucson Water Company was forced to dig new wells inside the San Xavier Reservation, a short distance upstream of the original gravity-flow system. The new well at San Xavier was only twenty feet deep to bedrock, and the water was pumped into a tank that drained into the water main on demand using a pump capable of supplying one million gallons a day. Tucson residents consumed over 100 gallons per day per person, higher than in other areas because of the large demand for irrigation water, and the company proposed installing flow meters to charge residents for the water they used.[40]

Installation of the new well system began in March 1893, and production began sometime the following month. Nevertheless, water became scarce in June, and it was only intermittently delivered to consumers.[41] Low-flow or no-flow conditions persisted until January 1895, when the Rillito River overflowed its banks:

> The Rillito was the highest yesterday morning that it has been for a long time. It was out of the question for the Mammoth stagecoach to cross it, but the mail was taken on by a Mexican boy who forded the river on his horse. It was said in the afternoon that the water was out of the banks of the stream to the distance of a mile, and in some places more than two miles [wide].[42]

In June 1895, J. R. Watts, superintendent of the Tucson Water Company, discussed the problem of supplying water during a drought:

> This fact is determined by the well from which the city supply of water comes. Originally the well was but 18 feet deep and the process of sinking is still going on. Formerly the city supply came through submerged sluices in the river bed and to some extent these still furnish all that is necessary, but the company has been obliged to run their pump 27 months in the last two and a half years. To do this it required 1,782 cords of wood at an expense of $4500. Tucson uses an average of 13 million gallons of water per month. Recent rains in the mountains will probably help matters with the company.[43]

Later in the fall of that year, wood, mostly transported from the San Xavier Reservation to Tucson on burros, doubled in price as the O'odham turned their attention to planting crops, and the fuel-hungry town turned its attention toward the nearby mountains.[44]

Heavy rainfall on 9 August caused flooding when 2.75 inches of rain occurred over two hours in the middle of the night.[45] This storm brought some relief from the drought, but the resulting flood damaged road and railroad bridges over the Rillito River. Because many houses and walls in Tucson were adobe, roofs and walls in town crumbled as the dusty streets became rivers of mud and debris. Even during the drought years at the end of the nineteenth century, deluges brought significant runoff into the rivers.

In March 1895, the Allisons, who controlled about 1,200 acres of bottomland, expanded their irrigation works:

> The Allison ditch in the valley is full of water which is being used to good advantage for irrigation purposes on the rich land north of Stevens Avenue [Congress Street]. It is said that the Allison Brothers are thinking of taking water out of the river further up than they are now and that they will cut the ditch deeper thus being able to supply farmers in the valley with more water.[46]

By the following year, the Allisons' canal on the west side of the valley had become unusable: "The land down on the west side had too much alkali and was no good, so we finally dug the present canal on the east side of the valley and located the land which is called the Flowing Wells."[47]

At the site of Warner's Dam, the Allisons constructed a flume that carried water across the channel into a newly dug ditch, which they referred to as the East Side Canal. The water from this canal powered the waterwheel at the new flour mill just south of the Tucson smelter.[48] By 1898, they were making considerable money shipping watermelons on the Southern Pacific Railroad. In 1900, the Allisons sold their Tucson property for $60,000 to a group that included Levi H. Manning, the former surveyor general for Arizona Territory and future mayor of Tucson. The Allisons reinvested in land between San Xavier and Tucson, clearing what they called "heavy mesquite" to create agricultural fields irrigated using water developed at Black Mountain southwest of Tucson and transported fourteen miles by ditch.[49]

Others followed the Allisons in developing groundwater for agriculture. For example, in 1902, Manning developed "a large and permanent supply of artesian water" at the base of Sentinel Peak by boring into deeper strata. He hired James C. Fulton, an experienced well driller, to conduct the project.[50] About three-quarters of a mile upstream from Sentinel Peak, he drilled through an impervious layer and got strong artesian flow at a depth of twelve feet. They obtained a flow of 250,000 gallons per day from a four-and-one-half-foot-diameter pipe, and eventually the well field yielded 1.5 million gallons per day. Manning thought that he had struck the proverbial underground river, long sought in the desert, and that he could rival the flow of the entire Salt River in Phoenix.[51] Of course, the underground river did not exist and the artesian flow soon ceased, but optimism for large-scale groundwater development fueled enthusiasm in Tucson in the summer of 1902.[52]

As Tucson grew northward, the Rillito River was looked to as another fertile site for agriculture, and its broad floodplains offered plenty of arable land if only water were available. The present-day arroyo of Rillito Creek began to downcut in 1881, and a particularly severe flood occurred on 11 September 1887 that killed livestock, destroyed property, and inundated the floodplain to a width of two miles in places.[53] By the early 1890s, floods had deepened and widened the channel and the arroyo had become fully entrenched,[54] but groundwater levels remained high. The Rillito River had long served as a source of water for humans and domestic animals, including soldiers at Fort Lowell in the 1870s. By the start of the 1900s, thirty small ditches were used to withdraw water from Rillito Creek during the flood season.[55] Channel widening occurred during the first decades of the twentieth century but ceased temporarily by the early 1940s.[56]

The irrigation ditches were popular with local children for recreation in a time when sufficient water for swimming was uncommon in Tucson. Glenton Sykes, city engineer in the mid-twentieth century, reminisced about his childhood:

> At the present date of 1964 there are reliably reported to be well over 1500 swimming pools in the Tucson environs. In 1907 there were about four, a couple of water holes in the river bottom & the irrigation ditch. The ditch was pretty good, there were several beautiful spots along the ditch from Mission Lane to West Congress St[reet]. Grassy slopes under a canopy of huge Cottonwood & Willow trees. The ditch was about eight feet wide & three feet deep, long trailing moss & slowly moving water. A beautiful place to laze & swim on a hot summer afternoon, there were dozens of kids frolicing [sic] in the water & the older people picnicing [sic] or asleep under the trees during holiday & week ends & no fees or regulations. The ditch system ran North under St Mary's Rd. & on past Speedway. At some places it was slightly above the adjacent property & here the people would install a pipe in the bank & secure enough free water to raise a good vegatable [sic] patch. The small amount of water used was completely negligible compared with the quantity being carried by the ditch & no one said any thing even when they knew it.[57]

Sykes recalled the difficulty of fording the Santa Cruz River because of the changing channel and the dense vegetation that lined it. In the spring of 1907, the ten-year-old Sykes set out in a horse and buggy to ford the river at Alameda Street, downstream from Congress Street:

> There had been a little rain in the evening before & the river had been up enough to impair the crossing, there was quite a cut bank on the West approach & apparently no one had yet attempted to use it. One got to the crossing through a dense growth of arrow weed & willow & before I realized it we were at the cut bank. I tried to back the horse, I tried to turn him round, we all got confused, & finally he jumped, down the bank & into the river. We almost broke the shafts, we almost turned the buggy over but not quite, we got across alright.[58]

Harry S. Swarth, Frank Willard, and the Great Mesquite Forest

Despite a considerable amount of ornithological work conducted in the Tucson Basin during the 1870s and 1880s (see chapter 5),[59] bird populations and the habitats they used along the Santa Cruz River seldom were mentioned. Nonetheless, many visitors to the region chose to single out the Tucson Basin for its ecological resources. For example, from 11 to 13 December 1893, Dr. Edgar A. Mearns, US-Mexican Boundary Survey biologist, visited Tucson. Mearns's description of the rivers of the Tucson Basin and their attendant riparian biota exemplified the findings of early settlers and biologists:

> Tucson, Arizona[:] . . . No [other] place in Arizona has as rich a fauna; and there is considerable variety to the flora. The Santa Cruz Valley is well wooded with cottonwood, willow, mesquite, and cultivated fruit trees. . . . The vegetation in the region about Tucson presents an extremely picturesque appearance. The streams—Rillito Creek and the Santa Cruz River—are well wooded with screw bean, mesquite, cottonwood, willow, boxelder, and ash, groups of which are often converted into fragrant bowers by climbing grape.[60]

By 1900, the focus of ornithologists had shifted largely from the Rillito to the Santa Cruz River, and especially to the Great Mesquite Forest.[61] Herbert Brown collected the state's first Pacific Loon in 1900 at San

Xavier,[62] and he published the state's first records for an Anhinga and Purple Gallinule,[63] both from the Santa Cruz River. Occasional sightings aside, Harry S. Swarth and Francis (Frank) Willard were the first ornithologists to publish their findings from the Great Mesquite Forest in scientific journals. Swarth was a professional ornithologist with institutional backing, mostly from California institutions. Willard was a resident of Tombstone and an amateur. Both left important records of the Great Mesquite Forest and its birdlife at the start of the twentieth century.[64] Willard may have been an amateur, but he was highly respected in the ornithological community, and his assistance during later fieldwork inspired A. C. Bent to write a glowing account of Willard's ornithological abilities after his death.[65]

With approximately fifty-six nesting species recorded at that time (more were recorded later) and nearly a dozen that foraged in or over the forest during the breeding season, it was unequaled in the region for species numbers and nesting densities. During the early 1900s, as many as sixty-five species were known to nest in the Great Mesquite Forest and adjacent ecosystems,[66] and several additional species nested opportunistically. For example, on 5 May 1899, six of a flock of eight Black-bellied Whistling-Ducks were shot on the Santa Cruz River south of Tucson.[67] This species now nests regularly in nest boxes in southern Arizona,[68] and it has even been known to nest in agricultural areas.[69]

By the 1890s, the cienega upstream from Mission San Xavier, which was dominated by perennial grasses such as alkali sacaton, was affected by arroyo downcutting through this reach, which was not as significant as it was near downtown Tucson. Groundwater levels dropped in response to the downcutting to a new base level, which drained much of the marshlands. Mesquite, which ringed this cienega with enormous trees, then became established in the unsaturated rich soils of the former marsh.[70] Instead of destroying the bosque, as some ecologists have intimated, arroyo downcutting enhanced it by creating a large area of aerated soil over a shallow groundwater table.

Swarth was the first to attempt to document the bird populations in this bosque. In 1896, he began his birding career in the Huachuca Mountains,[71] southeast of Tucson, during the first of six collecting trips he made into southeastern Arizona.[72] In 1902, he started working in the Great Mesquite Forest, and his research was the first of its kind in the Southwest. Interestingly, Swarth worked during the same decade that long-term

botanical research projects began at the Desert Laboratory on Tumamoc Hill, west of downtown Tucson (see fig. 4.1).[73] One of the botanical publications from the Desert Laboratory during this time discussed unusually large graythorn, a plant normally occupying the margins of dry washes, in the Great Mesquite Forest.[74]

In 1902 and 1903, Swarth described the forest:

> South of Tucson, Arizona, along the banks of the Santa Cruz River, lies a region offering the greatest inducements to the ornithologist. The river running underground for most of its course, rises to the surface at this point, and the bottomlands on either side are covered, miles in extent, with a thick growth of giant mesquite trees, literally giants, for a person accustomed to the scrubby bush that grows everywhere in the desert regions of the southwest, can hardly believe that these fine trees, many of them sixty feet and over, really belong to the same species. This magnificent grove is included in the Papago Indian [Tohono O'odham] Reservation, which is the only reason for the trees surviving as long as they have, since elsewhere every mesquite large enough to be used as firewood has been ruthlessly cut down, to grow again as a straggly bush.[75]

Swarth's first foray was with an experienced field assistant named O. W. Howard from 17 to 23 May 1902. Howard had accompanied Swarth on his first Huachuca trip and later published several papers on the birds and nests of Madrean birds.[76] The following year, from 3 to 13 June 1903, he finished his work in the Great Mesquite Forest with field assistant Frank Stephens, who had collected birds in Arizona beginning in the 1880s.[77] Swarth published his observations based on thirty-six person-days of fieldwork in 1905,[78] which was the first in a series of avian research publications on the bosque that would become the most thoroughly documented riparian ecosystem in the American Southwest. In particular, he noted that the White-winged Dove was "by far the most abundant bird in the mesquite forest" and observed that "the noise they made morning and evening was such as to almost entirely drown the notes of the other birds."[79]

In 1911, Willard spent two days in and around the Great Mesquite Forest. He traveled 375 miles during a week-long egg-collecting expedition, during which he traveled from the Huachuca Mountains, where he lived, to the Tucson area. This was a considerable distance, considering the conditions of automobiles and roads in those days. Willard reported:

The mesquite trees are wonders of their kind. . . . There were some whose trunks at the base scaled over four feet in diameter. The large bases branched a few feet from [the] ground into several limbs fifteen or eighteen inches in diameter. The tallest reaches a height of over sixty feet. The undergrowth is a thick mass of hackberry, etc. with various thorny bushes growing close to the ground. Meandering wood roads lead in every direction and one can never be quite sure that he is on the right one.[80]

He also observed that "White-winged Doves outnumbered any two of those [other fifteen species] mentioned."[81]

Swarth's and Willard's observations indicate changes in bird populations that occurred even while the bosque was expanding. Their observations suggest that several breeding species may have been extirpated from the Santa Cruz during the thirty years between 1872, when Bendire worked along the Rillito, and 1902, when Swarth first visited the Great Mesquite Forest. Some of these species include the Common Black-Hawk, Ferruginous Pygmy-Owl, and Song Sparrow.[82] This trend apparently continued into the mid-twentieth century and may have been related to the twin effects of woodcutting and aquifer dewatering, which was favoring woody vegetation over that of the cienegas.

Declining Groundwater

Optimism over the apparent abundance of groundwater diminished as the early twentieth-century drought wore on. In 1903, the city, which had bought the Tucson Water Company in 1900, urged citizens to schedule irrigation of lawns and gardens to save water.[83] The drought abruptly ended in 1905, when Tucson received 24.16 inches of rainfall, the wettest year in its 140 years of recorded rainfall at the University of Arizona (see fig. 2.4). More than half of this rainfall fell from January through April. In early March, about fifty feet of grade at the west end of the Jaynes bridge over the Santa Cruz River was swept away by floods.[84] A couple of weeks later, a half acre of land washed away near the bridge.[85] This flood, and the ongoing shortage of irrigation water, may have spurred the Irrigation Department at the University of Arizona to install a streamflow gaging station at the Congress Street Bridge in November 1905.[86]

In June 1905, an anonymous observer perceived the connection between the recent groundwater shortages and the newly developed arroyo, noting that the wash "was drawing off the underflow which previously formed a vast underground reservoir and from which the city water wells drew their supplies."[87] Similar claims were made in the 1910s by hydrologists and engineers working in the Tucson area, who predicted that groundwater effluent to the Santa Cruz River eventually would lower the regional aquifer.[88] Again, these observations reinforce the idea that declines in groundwater levels followed arroyo downcutting, instead of preceding it. Groundwater lowering was neither the cause for channel erosion nor even a contributing factor in downcutting. Floods, however, were the problem, and in a 1913 report, hydrologists Olberg and Schanck succinctly observed, "The two subjects of water development and flood action are so intimately associated that it is necessary to consider them in conjunction."[89]

The 1905 floods continued the downcutting and widening initiated in the 1880s and 1890s. Diversion of surface water was made more difficult, increasing the frustration of local residents who depended on irrigation for their livelihoods. By the early 1910s, there were four main ditches that irrigated lands west of Tucson and along the Santa Cruz River floodplain. The oldest of these, called the Farmer's Canal, had headgates on the east side of the river upstream from Tucson and tapped effluent groundwater at the site of Silver Lake, but by the early 1910s, the headgate was ten or twelve feet above the channel. Water was fed into the ditch by diversion along the riverbed, which washed out whenever the Santa Cruz River flooded. Because each flood lowered the channel bed and the diversion point moved farther and farther upstream, the owners of the canal built a flume across the river to transport water from what was called the Manning Ditch (fig. 6.5) from the west bank to the eastern Farmer's Canal. The channel deepened with successive floods, and groundwater effluent rose farther and farther upstream. Ditches in the channel, made temporary because they were destroyed during each flood, were the solution to keeping irrigation water flowing toward Tucson.[90]

The struggle to sustain a reliable water supply made small-scale agriculture difficult, and consolidation of farmlands was one solution to this problem. In 1910, a group of Chicago and British businessmen pooled their capital and formed the Tucson Farms Company. In 1911, the company purchased some 1,200 acres from Manning and the Allison brothers, who no doubt needed funds to maintain their water supply. By 1913, the company's holdings included about 6,000 acres of valley land. Manning reinvested in the Santa Cruz Reservoir Project, a $10-million boondoggle to dam the confluence of the Santa Cruz River at the Aguirre Valley in Pinal

A **B**

Figure 6.5.

The Manning Ditch near Sentinel Peak. **A.** (1907) In 1900, Levi H. Manning acquired land and waterworks from brothers Frank and Warren Allison, including this ditch built in the late 1890s near the former site of Warner's Lake. The men in the photograph are dumping copper sulfate into the ditch, presumably to retard accumulation of moss. The Santa Cruz River and Sentinel Peak are visible in the background. Even though the stream downcut through this reach in the 1890s, it remained perennial as the alluvial aquifer drained; flow may actually have increased with deeper intercept of the water table. (Photographer unknown, 2709, courtesy of the University of Arizona Special Collections Library.) **B.** (4 February 1982) Seventy-five years later, all traces of the ditch are gone. Now, this view shows the river from the east bank. (R. M. Turner, Stake 1073, courtesy of the Desert Laboratory Collection.)

County (see fig. 1.1). Both the Tucson Farms Company and Santa Cruz Reservoir Project were investment schemes designed to attract farmers from the Midwest to the Tucson Basin.

The Tucson Farms Company project called for additional groundwater development at the base of Sentinel Peak, near San Xavier, and upstream on the Santa Cruz River at Sahuarita, increasing the total area subject to irrigation. Developed parcels of land with partial water rights were sold to farmers for $1,200–$1,900 per acre.[91] At San Xavier, the Tucson Farms Company installed a number of pumping plants powered by electricity, a new development. From south of Sahuarita to the north end of the Canoa land grant, they drilled twenty-four wells, most of which were 200–500 feet deep, but one

was drilled to a depth of 900 feet.[92] In 1912, Manning bought the Canoa land grant and diverted flow from just upstream of the point where the perennial flow of the Santa Cruz River disappeared into the deep sands of the Tucson Basin. He dug an "artificial ravine" starting with a deep ditch into the higher ground near the Canoa Ranch headquarters at a point deep enough to intercept the shallow alluvial aquifer; the depth of the ditch decreased downstream to where it was used for irrigation.[93]

The Tucson Farms Company implemented its most extensive water-development system beginning about 300 feet downstream of the former dam at Silver Lake. Nineteen wells were drilled to depths ranging from 45 to 150 feet, forming a line across the valley as part of a "crosscut system" (figs. 6.6–6.9). The wells were connected by a reinforced-concrete canal, 4,740 feet long and 5 to 12 feet below land surface, that fed into a 4-foot-diameter 1,500-foot-long pipe. This pipe fed into a siphon under the Santa Cruz River and into about seven miles of main ditch, of which 12,650 feet were lined with concrete. This in turn fed twenty-one miles of lateral ditches, of which about 3,000 feet were concrete lined. Numerous concrete flumes and weirs regulated the flow, and smaller siphons conducted water under the Santa Cruz and Rillito Rivers farther downstream from Tucson. The project included one short-lived timber dam across the Santa Cruz River.[94] The main concrete-lined canal was about nine miles long, the first three miles of which followed the old Manning Ditch (fig. 6.7).

Percy Jones, who was the concrete inspector and instrument man during construction of the crosscut system, related that the "trouble with building the damn thing was too much water, and then it was never a success because they never had enough."[95] The project cost almost a million dollars, in part to pay more than five hundred workers. One effect of the Tucson Farms Company project was the displacement of the confluence of the West Branch and the Santa Cruz River, which was moved upstream to protect the West Branch outlet against floods on the mainstem. By 1915, the system delivered almost 30 million gallons of water a day.

The downstream counterpart to the Tucson Farms Company was the Santa Cruz Reservoir Project, an elaborate scheme started by Colonel William C. Greene near present-day Redrock in Pinal County (fig. 6.10; see fig. 1.1). This project diverted streamflow from the Santa Cruz River into an artificial ditch known as Greene's Canal and then into Aguirre Wash, terminating in a reservoir with a capacity of 300,000 acre-feet. The reservoir was impounded in a natural depression upstream from the confluence of the Santa Cruz River and Aguirre Wash. Colonel Greene died in 1911 before the project was completed. In 1913, P. E. Fuller, the project engineer, reported that an earthen-fill diversion dam 2,000 feet long and 10 feet high and equipped with a spillway was built across the Santa Cruz River. Flow was diverted into a 20-foot-wide, 5-foot-deep ditch, which conducted water to the reservoir. A levee across the Aguirre Valley fed runoff from this ephemeral channel into a 6-mile-long ditch of about the same width but only 1.0 to 1.5 feet deep. The reservoir drained into an earthen canal initially capable of delivering $600\,\mathrm{ft^3/s}$ that flowed nine miles to the north and supplied numerous lateral ditches delivering water to farmlands.[96]

Like Sam Hughes's Ditch, Greene's Canal was designed to be eroded by floods, as was the ditch from Aguirre Valley. Furthermore, the ditch was expected to change its course to match that of the natural channel it replaced, especially since it had a higher slope. The designers knew that these artificial channels could not convey the largest flood expected for this reach of the Santa Cruz River. They assumed the risk that high flows would erode enough to further their design but not so much that the floods would destroy their waterworks. .

Channel Manipulation in the Great Mesquite Forest

By 1912, the headcut that started at Tucson had joined the one at the Valencia Road Bridge, and the continuously entrenched reach extended three miles south into the San Xavier Reservation and the Great Mesquite Forest to the base of Martinez Hill (see figs. 2.1 and 4.1). The downcut channel, about 30 feet deep, was within a half mile of capturing the entrenched segment of the Spring Branch (see fig. 4.4B). At that time, the Spring Branch

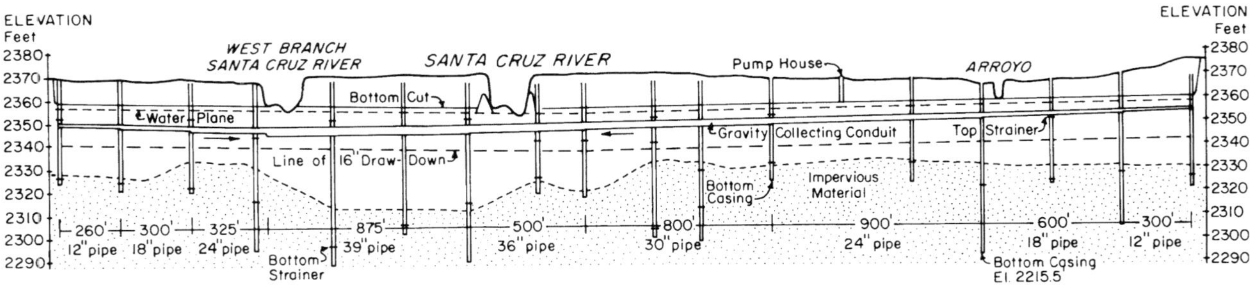

Figure 6.6.
Plan of the Tucson Farms Company Crosscut and distribution system (redrawn by Betancourt [1990: 136] from Hinderlider 1913).

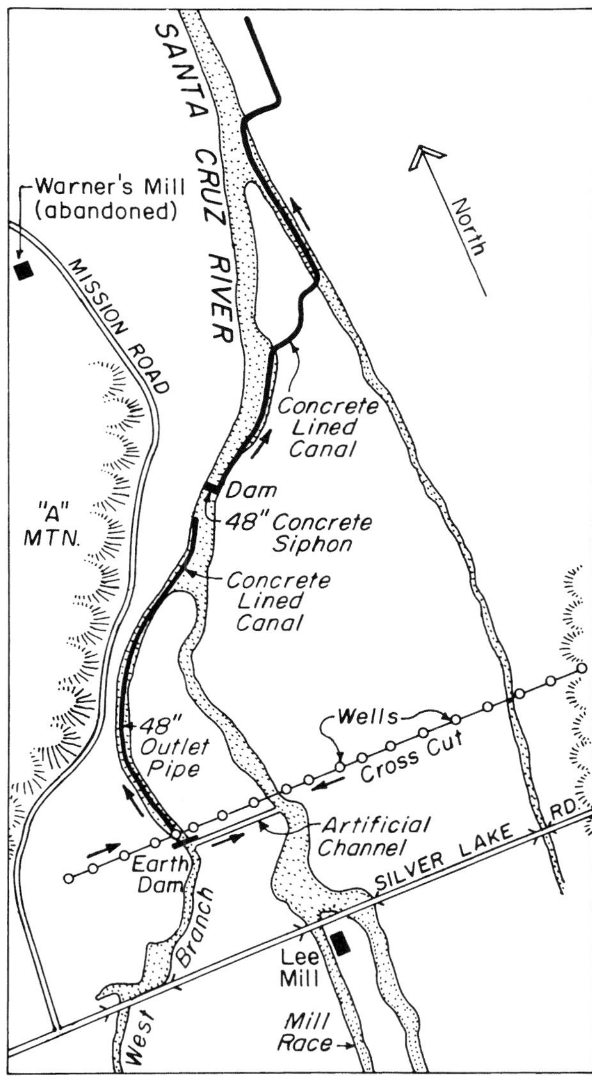

Figure 6.7.
Vertical profile of the line of wells, pump houses, and gravity collecting conduit associated with the Tucson Farms Company Crosscut. Note the intercept of the water level of the aquifer (the "water plane") by both the West Branch and the Santa Cruz River (from Hinderlider 1913).

had a channel 100 to 200 feet wide and 16 to 20 feet deep at a point two miles downstream of the spring. To collect water and to try to prevent further erosion, an earthen dam was built across the Spring Branch to divert water for five hundred acres of agricultural lands. This dam washed out with every flood. The slope from the base of this dam to the Santa Cruz River was large, although outcropping beds of clay minimized the potential for downcutting over this distance. Concern that

downcutting was inevitable prompted plans for erosion control.[97]

To the west, the channel downstream of Punta de Agua (see fig. 4.1) had downcut as early as 1849 as part of the discontinuous arroyo system in this part of the Tucson Basin. By 1912, ongoing erosion along this reach had forced farmers to construct a ditch 60 to 100 feet wide and 6 feet deep at the mouth of the channel, which increased up-gradient to about 20 feet deep. Outcropping calcrete in the bed halted further downcutting but not widening, and a broad sandy channel continued upstream. Water flowed over the impervious calcrete, but floods deposited 6 to 8 feet of sand over the hardpan, requiring considerable Tohono O'odham labor to reinitiate surface flow.[98]

The most serious erosion on the reservation resulted when overbank flow in the Spring Branch crossed from the west to the east side of the valley, cascading into the downcut channel of the Santa Cruz River at the base of Martinez Hill. In July 1908, a flood with a peak discharge estimated at 6,700 ft^3/s at Congress Street had significant overland flow near Martinez Hill, resulting in gullies with headcuts propagating to the west through farmlands, rendering them useless for agriculture. The headcut propagated a short distance upstream as well, into the mesquite bosque.[99] These hydrologic changes ultimately contributed to the destruction of the Great Mesquite Forest (see chapters 7 and 8).

In 1909, J. H. Quinton, a consulting engineer, submitted a plan designed to control the erosion to the chief engineer of the US Indian Service. Quinton proposed that a dike and canal be constructed across the upper end of the valley near present-day Pima Mine Road, joining the west and east barranca above the reservation's best agricultural lands. This dike was at what likely was the upstream end of the Great Mesquite Forest and would divert surface water from the headwaters into the east side channel and through the bosque south of Martinez Hill. A major obstacle to this strategy involved litigation between the federal government and the Tucson Farms Company, which had bought the Martinez land grant and exploited groundwater from several wells on their property that could be affected if the canal was constructed. Quinton's proposal threatened the company's investment, and they responded with the threat of a lawsuit.

An alternative erosion-control plan was proposed in 1913 and enacted in 1915. Quinton's group built a dike across the Spring Branch channel, diverting flow eastward to a point a short distance north of present-day Pima Mine Road.[100] In addition, the dike across the

Figure 6.8.
(1912) This west-facing view shows the crosscut under construction, just downstream of the former dam at Silver Lake. (Percy Jones, 2758, courtesy of the University of Arizona Special Collections Library.)

Figure 6.9.
(1913) A sector of the finished concrete-lined canal inside the east bank of the Santa Cruz River, which formed a portion of the crosscut water distribution system. (Percy Jones, 2713, courtesy of the University of Arizona Special Collections Library.)

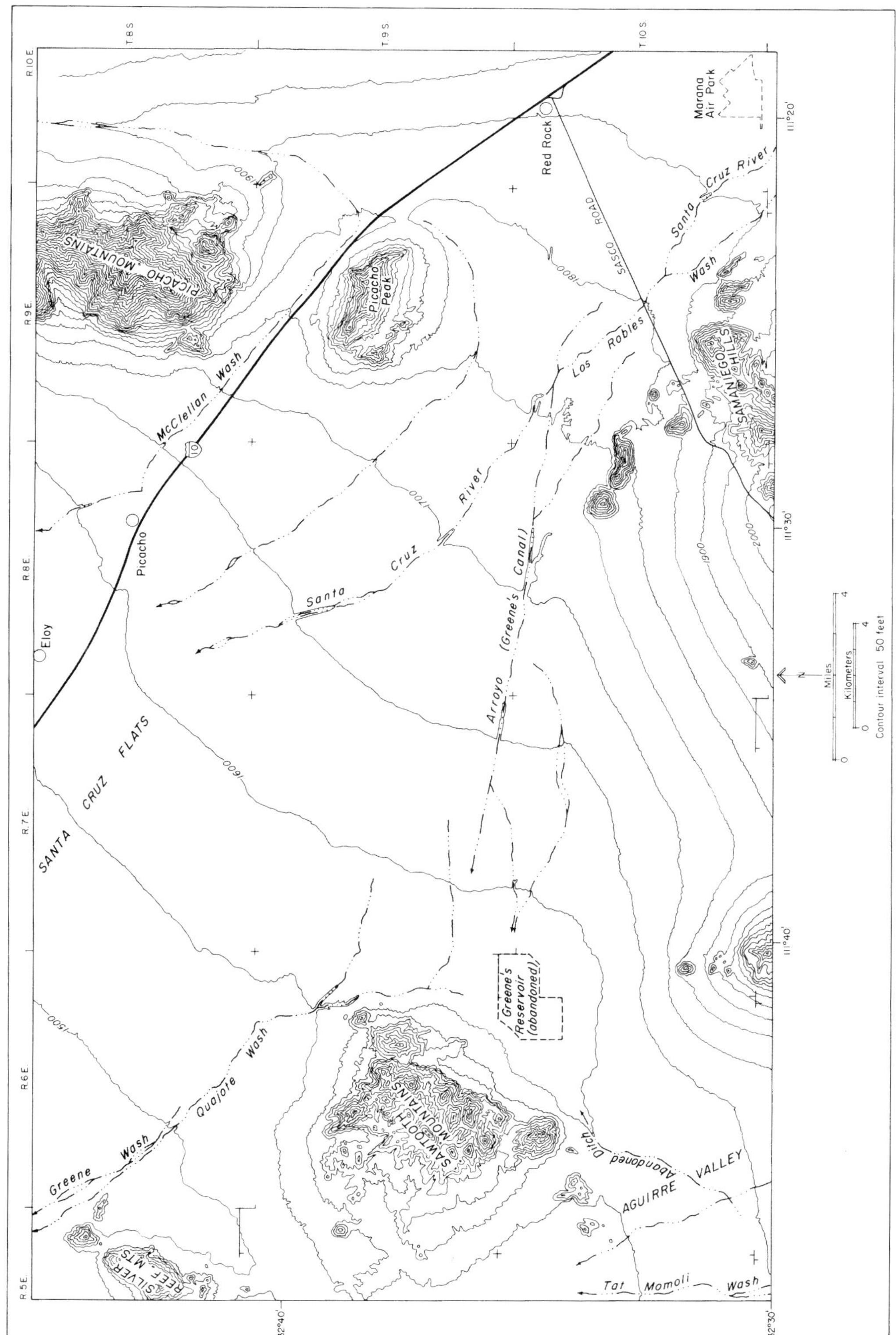

Figure 6.10.

Map of Greene's Canal and the lower Santa Cruz River, based on USGS 15′ quadrangles (Betancourt 1990: 142).

Spring Branch channel, about 7 feet high, would impound a temporary reservoir with a capacity of 350 acre-feet. An artificial channel directing flow eastward would be confined between two dikes, 4 feet high and 300 feet apart. One problem was that this artificial diversion channel had a total drop of 50 feet over 1 1/3 miles,[101] a steepened gradient that inevitably would lead to accelerated downcutting. Today, the channel of the Santa Cruz River, 20–25 feet deep and more than 300 feet across, follows the route of the 1915 dike into the former course of the Spring Branch (see fig. 4.1).

The Floods of 1914–1915

The winter rains of 1914–1915 generated the largest flood to date on the Santa Cruz River, where 55,150 acre-feet of flow passed the Congress Street gaging station in December alone. A peak discharge of 15,000 ft^3/s oc-

curred on 23 December (fig. 6.11), and a University of Arizona group measured a flow velocity of twelve miles per hour at Sahuarita, where houses were flooded, people were stranded on rooftops, and two people drowned. The flood damaged or destroyed $10,000 worth of wells and pumps operated by the city, and the Tucson Farms Company sustained another $10,000 in damages to ditches and wells. Twenty-five miles of railroad track between Marana and Cortaro were destroyed, and part of the Twin Buttes railroad bridge washed out and struck the bridge at Congress Street. Numerous bridges and reportedly all of the dams were washed away in the flood. Twenty acres of prime farmland planted in alfalfa on the west side of the river were eroded, with losses estimated in the many thousands of dollars.[102]

The bridge and abutments at Congress Street were damaged during the 1914 flood but remained intact. As early as 1902, a meander had developed on the east

Figure 6.11.
(1915) In this northwest (downstream) view of the flood, taken from just downstream of the Congress Street Bridge, onlookers stand perilously close to the eroding east bank of the Santa Cruz River. The east bank (at right center of the photograph) is undercutting. (Henry Buehman, B38373, courtesy of the Arizona Historical Society.)

Figure 6.12.
The Santa Cruz River at the Congress Street Bridge. **A.** (January 1915) This northwest view shows the bridge after erosion of the east bank during the 1915 flood. The cottonwood stand present in 1902 (see fig. 6.3) was destroyed in the 1915 flood. (Photographer unknown, courtesy of the University of Arizona Special Collections Library.) **B.** (July 1915) This image, taken six months later, shows the berm that was built to join the east approach of Congress Street to the remaining part of the bridge. (Forrest Shreve album, courtesy of the University of Arizona Special Collections Library.)

A

B

Figure 6.13.

Downstream view of the Santa Cruz River from Rillito Peak. **A.** (12 March 1910) Ellsworth Huntington, the noted geographer, took this photograph and described it as follows in his unpublished journal: "[L]ooking northwest from end of Tucson Mountains [Rillito Peak] at Santa Cruz Valley, now dry, near where this river finally merges into a large playa. . . . The dry channel of the river . . . here possibly 5 feet below the terrace." (Ellsworth Huntington, courtesy of the Beinecke Library, Yale University.) **B.** (30 November 1983) The Santa Cruz River has a narrow channel, visible at the extreme right of the photograph, and a wide floodplain that was eroded during the flood of October 1983. Most of the left bank of the river has not been developed. (R. M. Turner, Stake 1104,

c

Figure 6.13. (*continued*)
courtesy of the Desert Laboratory Collection.) **C.** (20 December 2010) The left side of the view is blocked by a young paloverde tree, but the Santa Cruz River and its floodplain remain visible. The low water channel, filled with wastewater effluent discharged into the channel several miles upstream, is readily apparent. The widened floodplain has become revegetated and is not well defined. Agricultural clearing has proceeded on the left side, altering the desert to create farmland that is fallow at this time. (H. A. Raichle, Stake 1104, courtesy of the Desert Laboratory Collection.)

bank of the river upstream from the bridge. This meander did not move significantly during the floods of 1904–1905 or on 23 December 1914. Things changed on 31 January 1915, when the meander migrated during renewed flooding and eroded more than five acres of land, breached the east approach to the bridge (fig. 6.12), and created a two-hundred-foot gap in the eastern approach to the bridge. The Menlo Park neighborhood on the west side of the river was cut off from the rest of town, and continuing meander migration threatened houses and eroded agricultural lands.

Thousands of people, kept more or less in line by police cordons, gathered to watch the channel erosion and bridge collapse, while workmen dumped dirt into the raging current in an attempt to stop the bank erosion. Those who happened to be on the west side of the river suddenly found themselves isolated from the main part of town proper when the embankment gave way. A railroad employee insistent on catching a train jumped over the rapidly widening gap and was the last person over before the crossing became impassable. Homes near the east bank were evacuated, and Tucson Farms Company

management feared for the safety of the crosscut outlet ditch, just thirty-five feet away from the flood.[103]

The floodwaters had receded sufficiently by 2 February that the river could be forded, at least by those who needed to cross to get to their property. The threat temporarily was over for owners of homes near the east bank and the Tucson Farms Company. A newly formed association of property owners planned to raise enough money "to install the necessary works to defend their lands against any further ravages of the treacherous Santa Cruz." Newspapers further reported damages amounting to at least $50,000 and stated that the Congress Street Bridge had sunk about two and a half feet.[104]

Downstream of Tucson, in the alluvial plains of Santa Cruz Flats (see fig. 1.1), the floodwaters of the Santa Cruz River spread out to a width of more than five miles. Near Casa Grande, the overflow is said to have extended from the outskirts of town west to the alluvial fan of the Table Top Mountains, implying an inundated area of up to ten miles wide. As predicted by the engineer for the Santa Cruz Reservoir Company,[105]

Greene's Canal widened to two hundred feet in places, but much to the dismay of company officials, it deepened twenty feet or more to form a discontinuous arroyo between the Santa Cruz River and the reservoir. The flood transformed a sinuous, narrow channel (fig. 6.13) into a broad system of braided channels near present-day Marana, in some places widening the channel by as much as six times its former width.[106]

Upstream of the city near San Xavier, the diversion dam along the old Spring Branch sustained considerable damage.[107] The effects of the flood were recalled in 1937 by the supervising engineer of the Office of Indian Affairs Irrigation Service:

> It should also be kept in mind in connection with the San Xavier irrigation situation that the so-called diversion dam [the Indian Dam] is not really a diversion dam but is a soil conservation dam built with funds allocated for the purpose of preventing further erosion and deepening of the San Xavier River channel in the immediate locality upstream from the dam. We were hopeful, of course, that it incidentally would act also as a diversion dam but in this particular there has been disappointment due to the fact that the largest flood in the history of the Santa Cruz River had to come within a week or ten days after completion of the dam and before the river channel had been adjusted to the change of the grade by normal flood flows. The erosion immediately below the dam as a result of this excessive flood, cut through the underlying clay stratum with a result that a considerable part of the water that heretofore would have been diverted into the canal by this dam flows through the gravel stratum which is below the clay and comes into the river channel below the dam.[108]

In other words, the dam failed and the arroyo through the Great Mesquite Forest deepened as a result of the 1915 floods.

Observations of Desert Laboratory Staff

Since its inauguration in 1903, the Carnegie Desert Botanical Laboratory on Tumamoc Hill (see fig. 4.1) was frequently cut off from most of Tucson when the Santa Cruz River was in flood. Volney Spalding, one of the principal researchers at the Desert Botanical Laboratory, wrote about the Santa Cruz River and its riparian vegetation in his monumental work on desert plants in 1909:

> The banks of the river are lined with willows . . . and cottonwoods . . . which constitute their conspicuous vegetation, while the arrow-weed . . . is of frequent occurrence. . . . It is the cottonwoods and willows that constitute the typical vegetation of the river-banks.[109]

He noted that the riparian areas of Arizona were what he referred to as "mesophytic forests," unlike the widespread forests of the eastern United States. He stressed the locally high densities of mesquite, annotating a photograph of large mesquites with a location of "Edge of mesquite forest, on [flood]plain of Santa Cruz River south of Tucson."[110] In this reach, he writes,

> the mesquite ranges in size from a mere shrub a few feet high to a tree 2 feet or more in diameter and upwards of 40 feet in height. Such trees grow thickly on the bottom-land near the old mission of San Xavier, forming the fine forest that stretches for miles up the river.[111]

The St. Mary's Road Bridge had washed out in 1905 and was temporarily replaced with a ford. After the Congress Street Bridge was made impassable in 1915, the laboratory staff had to ford the river to get to and from town. The correspondence of D. T. MacDougal, director of the laboratory, reveals some interesting information, including rapid recharge of the water table during the winter of 1914–1915:

> [T]he Santa Cruz has been in flood for a week and taking out bridges, and our only communication with town has been by rider and by wagon fording the river. The thing that concerns you, however, is the rise of the water table, which in this part of the valley amounts to from four to six feet.[112]

Several weeks later he wrote:

> The Santa Cruz River in all of this stretch between the two tips of the mountains has broadened itself this winter and filled its channels with the floods. Do you remember the little old narrow bridge that we crossed in going in and out of town [Congress Street], and the comparatively narrow channel here? This channel [near Congress Street] was widened this winter by a cut which gives it a width of about three times the old one, and the banks are now not as high. This is true of pretty nearly all of the stretch between here and the tip of the mountain [near Rillito]. We were down to the extreme tip of the mountain two weeks ago and at that time a stream in two channels was going on out into the desert with a comparatively large volume of water, and the cutbank

which you photographed and published,[113] giving a record of its height is now no more than four or five, or possibly six feet at any place in that region.[114]

In the summer of 1916, MacDougal was stationed at the Carnegie installation in Carmel, California, but Desert Laboratory geographer Godfrey Sykes kept MacDougal informed on events about Tumamoc Hill. Sykes was the father of Glenton Sykes, a well-known figure in Tucson history who served a long term as city engineer. The floods of that summer were frequently the topic of their correspondence.[115] In mid-August, floods generated upstream in the watershed were not particularly large, but the river crossing became treacherous for horse-and-buggy teams as laboratory staff attempted to ford the river to get supplies.[116]

Wet years continued into the twentieth century, but large floods like those from 1904 through 1915 were singular events heralding the change from downcutting to widening. The artificial diversion channel upstream from San Xavier led to creation of a continuous arroyo that extended from north of Marana at least to the small town of Sahuarita upstream (see fig. 2.1). The Santa Cruz River through the Great Mesquite Forest was now fully channelized but not deeply incised, and groundwater levels declined to the new channel level, no doubt stressing many of the numerous trees in the bosque. The 1914–1915 floods had spurred interest and action toward flood control and floodplain modification, and the ongoing quest for water for the thirsty town and fields eventually would doom the bosque upstream from Mission San Xavier.

7

Water Development and the Great Mesquite Forest, 1916–1942

The Santa Cruz River adapted to its new channel, passing relatively small floods in the late 1910s and early 1920s. The artificial diversion channel upstream from Mission San Xavier led to creation of a continuous arroyo that extended from north of Marana at least to the small town of Sahuarita (see fig. 1.1). The channel through the Great Mesquite Forest had become fully entrenched, and groundwater levels declined to the new channel level, no doubt stressing many of the numerous trees in the bosque while enhancing habitat for others. At the same time that Tucsonans were desperately trying to create a sustainable water supply for their town and fields, the Great Mesquite Forest was surviving despite the channelization and related reduction in groundwater levels. Increasingly, the bosque was under stress from the combination of woodcutters, agricultural clearing, and water development. With the implementation of newer pump technology, however, the life expectancy of the Great Mesquite Forest as a functioning ecosystem would soon diminish.

Although the flood peaks in the river were smaller, they continued to widen the raw and barren channel through Tucson, prompting the initial stages of channelization in the urban area. Sediment eroded in the upstream reaches would be deposited in the Tucson reaches and farther downstream, and this sediment would cause temporary increases in the bed elevation at crossings in Tucson, such as at Congress Street and St. Mary's Road, as the sediment wave moved through Tucson and downstream to the Santa Cruz Flats. Although channelization was complete, further adjustments would occur that required Tucson floodplain managers to control the river for good.

Increasing Agricultural Water Demands

In the 1920s and 1930s, the Santa Cruz River changed to a regime of more frequent, smaller, mostly summer floods that continued to widen and deepen the arroyo through the Tucson Basin. The relatively wet summers of 1919, 1921, and 1923 were followed by drought years through 1930, although occasional floods occurred, such as in November 1926 and late September 1929 (see fig. 2.7). On the San Pedro River, what may have been a dissipating tropical cyclone in September 1926 produced the largest peak on record in Arizona south of the Gila River, estimated at 98,000 ft^3/s near Charleston.[1] Except in the headwaters, this storm had minimal effect on the Santa Cruz River. Floods in September 1929 destroyed the bridges at St. Mary's Road and at Continental despite a relatively low peak discharge of 10,400 ft^3/s; at Congress Street, the crest of the flood rose within three feet of the top of the bridge.[2] The same tropical storm produced the second-highest peak of record on the Rillito River (24,000 ft^3/s).

Residents continued to pursue their livelihoods while further developing water resources, and the focus shifted from surface-water diversion to groundwater extraction. New ideas for crops, in addition to ongoing expansion of land under cultivation for more traditional ones, kept the pressure up for finding new water supplies. Southern Arizona residents continued to experiment with crops in the desert, particularly given the national demands of World War I. Two resources that southern Arizona residents thought they

Figure 7.1.

The Santa Cruz River at St. Mary's Road. **A.** (1 August 1927) This upstream view of the Santa Cruz River appears to be from the left (west) abutment of the St. Mary's Road Bridge, although the lack of background mountains or other features makes identification of the camera station uncertain. A small summer flood is passing down the river through an open gallery forest of cottonwood trees. (J. K. Doutt, 238, copyright Carnegie Museum of Natural History.)
B. (1 August 1927) This downstream view of the Santa Cruz River shows a small summer flood flowing in a channel lined with dense cottonwoods. A bridge appears in the distance, partially hidden by the trees. The appearance of the Santa Catalina Mountains, above the trees on the skyline, and the bridge and the right-turning meander suggest that this photograph was taken from the left (west) abutment of the St. Mary's Road Bridge looking downstream to the Speedway Boulevard Bridge. Arrowweed appears in the foreground; this species is now rare in the Tucson Basin and unknown along the Santa Cruz River. A group of children appears under the cottonwood trees on the far (right) bank. (J. K. Doutt, 239, copyright Carnegie Museum of Natural History.)

could contribute to the war effort were rubber and cotton.

In July 1916, L. H. Manning had sold the north half of the Canoa land grant to the Continental Rubber Company, which planted guayule for the manufacture of synthetic rubber. By 1920, approximately 1,100 acres of guayule were under irrigation. Later that year, the drop in the price of rubber after the end of World War I cut into the guayule profits and the farm was abandoned.[3] Manning also encouraged the McGee colony of Mormons to settle on tillable lands in the south half of the land grant. The remaining parts of the Canoa property became a renowned breeding ranch for thoroughbred beef stock and Arabian horses in the 1920s.

In the 1910s, Edwin R. Post purchased a large amount of acreage in the floodplain between present-day Marana and the Rillito–Santa Cruz confluence. The Post Project, with headquarters at Cortaro, emulated the Tucson Farms Company in an attempt to lure immigrant farmers by drilling ten new wells and emphasizing the lucrative market in cotton. At the end of the decade, cotton prices plummeted, and several Post Project farmers went bankrupt. The project was eventually transferred to the Pima Farms Company and later to Cortaro Farms. By 1930, the cotton market had improved, and land was leased to farmers on a sharecrop basis. Pumping of groundwater in the northern part of the Tucson Basin lowered water levels dramatically in the 1920s, but this decline was slowed by reduced farming in the early years of the Great Depression:

> On the Cortaro Farms Project around Cortaro, Rillito and Marana, the water supply has apparently been overdrawn, as the water level has lowered since irrigation was started. The acreage under cultivation on this project has been decreased recently, and if it is held at or below the present acreage, the water supply will be sufficient.[4]

In 1920, domestic demand for water was growing, and the City of Tucson was forced to place restrictions on irrigation, again asking its citizens to water only early in the morning and late in the afternoon.[5] To meet the increasing demand, in 1921 the Tucson Water Company expanded its system by drilling three wells east of the University of Arizona, which at that time was east of Tucson. In June 1922, the newly formed Flowing Wells Irrigation District assumed control of the Tucson Farms Company Crosscut Canal. The company's land north of San Xavier was sold to Midvale

Farms, and the stage was set for an expanded water distribution network from the abundant water beneath the Great Mesquite Forest to the multitude of users in and downstream of Tucson. Even so, in the 1920s, riparian gallery forests still thrived along the Santa Cruz River in Tucson despite more than thirty years of water development (fig. 7.1).

Water Development on the San Xavier Reservation

Anticipating increasing demand from the City of Tucson and agricultural users downstream, the Tohono O'odham attempted to secure their water supply on the San Xavier Reservation in 1925:

> Completion of [a] new gravity irrigation system will enable Papagos [the Tohono O'odham] to abandon [the] temporary pumping system installed by the government after 1914 when the erosion of the Santa Cruz River lowered the water level of the stream until the water would no longer flow over the Mission fields by gravity. Installation of a 30 in. infiltration gallery . . . will deliver 4500 gpm [gallons per minute]. Prior to the erosion of the river bed in 1914 the Santa Cruz was a very small and narrow channel, in many places not even well defined, and Indians secured gravity water for the irrigation of their lands by diverting the steady flow of the small stream above the Mission into irrigating ditches.[6]

On the San Xavier District reservation, groundwater discharge into the channel still produced enough water to make surface-water diversion for agriculture.

> Above Tucson the Papagos [Tohono O'odham] annually constructed an earthen dam [visible in fig. 7.2A] with which to irrigate their fields near San Xavier Mission. During the 1920s, this was replaced by a supposedly superior concrete dam, "Indian Dam," which promptly silted full. The Santa Cruz, however, continued to flow below the dam and was diverted for irrigation. This flow finally ceased about 1945.[7]

Indian Dam continued to be referenced by those who studied the Great Mesquite Forest into the 1960s, largely because it was central to prime riparian habitat. One of the last descriptions we have for the Great Mesquite Forest states:

> Approximately 12 road miles south of the University of Arizona Campus[,] . . . Indian Dam is on the Santa

Cruz River which runs through extensive mesquite bosque, with several scattered cottonwood groves located along the sides of the river. These groves also contain willow, mesquite, catclaw and other shrubs, usually with grass completing the vegetation.[8]

In 1931, the US Senate conducted extensive hearings concerning Tucson's plans to draw from the water supply on the San Xavier Indian Reservation. Development of water by other than Tohono O'odham would require an act of Congress, but the idea was rejected on the strong objections of the chief irrigation engineer at San Xavier. C. K. Smith, then mayor of Tucson, testified:

> The city of Tucson has been scouting for some time to get a larger and more available water supply for the city. Our engineer employed for the purpose of finding what the available sources of water were came to the conclusion that the Santa Cruz Valley carries from its watershed the largest and most valuable source of water for Tucson. We are a growing community. We have an adequate supply for the present but we must look forward to the future. Eight or nine miles up the river is the Indian Reservation. . . . It is the most available place for water in the entire river course. Now, I want to offer a tentative plan that might be of benefit to the Indian Service and also to Tucson. Our engineers have investigated the claim that there is more water than the Indians can ever use and more than Tucson can use for 50 years to come.[9]

That same water sustained the Great Mesquite Forest; fifty years later nothing would remain but dead stumps. Environmental issues would be of little concern in the face of water-development plans in the United States for another half century.

Channelization Begins

Even during the 1920s and 1930s drought, and perhaps because of frequent summer floods, the Santa Cruz River continued to change its channel. As a result, in the early 1930s, the Works Progress Administration (WPA), a Great Depression–era jobs program, focused its workforce on flood control along the river, mostly following the recommendations of H. C. Schwalen, a hydrologist at the University of Arizona. Schwalen wanted to smooth the sharp meanders in the channel, which were subject to erosion, by excavating new channels across the floodplain and piling the overburden on the lower side of the bars. Projecting points along the channel

were smoothed, trees and other vegetation growing in the channel were removed, and trees were planted behind metal bank protection to stabilize banks.[10] This activity, which preceded by a half century the extensive bank protection and channelization of the 1980s and 1990s, resulted in a straighter, steeper channel that, ironically, would lead to more erosive flood flows by increasing flow velocity.

The WPA implemented these recommendations for the Santa Cruz between San Xavier and Congress Street:

> Six long pilot channels were constructed across some of the more severe bends, the current being deflected into these channels by means of deflectors or revetments. These revetments were constructed of automobile frames in instances where the pressure was excessive and of double and triple lines of hog wire fence with posts of boiler tubes filled with cement. Current is deflected away from the big bends by means of boiler tube and hog wire fences placed in such a manner that the river current would do the greater portion of its own cutting, thus eliminating a great deal of the unnecessary labor. Projecting points were shaken up with dynamite so that they will be carried away by the first heavy flood. All heavy bends have been protected with tree planting behind jetties and revetments. Trees have also been planted along the entire river at points where it was desirable to maintain and hold existing banks. A recent survey indicates that about 95 percent of the trees planted have taken root and are growing.[11]

The revetments built by the WPA were successful, and in several cases, they are now buried by point bars as originally intended (fig. 7.3).[12] The WPA channel and bank protection, combined with low-flow conditions during the Dust Bowl years, probably explain why the stage-discharge ratings at the Congress Street gaging station were nearly constant from 1929 to 1946, indicating that channel downcutting did not happen during this period.[13]

In 1936–1937, the Soil Conservation Service conducted a survey of the entire Santa Cruz Valley, acquiring aerial photography to create a primary database of land use and condition in the region. The preliminary results launched yet another investigation of water resources within the San Xavier Reservation. Because of a lack of water and accelerated erosion in agricultural fields, many Tohono O'odham had turned from crops to fuelwood as a source of income. That fuelwood was mesquite harvested from the Great Mesquite Forest. A concerted attempt to develop new water and arrest erosion was aimed at encouraging a return to farming on the reservation.

Figure 7.2.

Upstream view of the Santa Cruz River from Martinez Hill. **A.** (1913) This upstream view of the Santa Cruz River, from near the summit of Martinez Hill, shows the river flowing in a narrow, cottonwood-lined channel through the Great Mesquite Forest. Open water is visible in the center of the view, showing that the alluvial aquifer here was high. At this time, a channel thirty feet deep marked the course of the Spring Branch, with a steep headcut terminating just below the dam in the center of the photograph. This was a temporary dam that was replaced later, in the 1930s. by the concrete "Indian Dam." Punta de Agua, the water source for the Mission San Xavier del Bac, is out of the view to the right, and the mission would be visible if the camera were turned 90° to the right. The bosque persisted into the 1940s but declined steadily thereafter. (Unknown photographer, Senate Document 973, 62–3 Plate I, courtesy of the National Archives.) **B.** (15 December 1981) Groundwater overdraft, which became a significant problem just downstream from here in the 1920s, combined with floods, such as the 1977 event, to completely change riparian vegetation in this reach. The remains of a failed agricultural project appear in the center. The corridor for Interstate 19 crosses the upper

Figure 7.2. (*continued*)
right of the view, and a thin haze partially obscures the Santa Rita Mountains in the background. (R. M. Turner, Stake 1057, courtesy of the Desert Laboratory Collection.) **C.** (5 April 1989) The 1983 flood caused substantial widening in this reach, causing failure of one of the Interstate 19 bridges (out of the view to the right). More agricultural clearing is apparent on the extreme right to the left of the freeway corridor. (R. M. Turner, Stake 1057, courtesy of the Desert Laboratory Collection.) **D.** (25 November 2002) Other than the establishment of some desert shrubs and the shifting of the channel thalweg to the left (right in this upstream view), little has changed in this reach. Decline in groundwater levels, coupled with ephemeral flow in the river, precludes establishment of riparian vegetation, even mesquite. (R. M. Turner, Stake 1057, courtesy of the Desert Laboratory Collection; this photographic match was previously published in Webb et al. 2007b.)

Figure 7.3.

The Santa Cruz River near 22nd Street. **A.** (1935) In 1935, the Works Progress Administration (WPA) constructed several flood-control features along the Santa Cruz River. In the reach just south of Sentinel Peak (at left), the river's flow was deflected into pilot channels by means of revetments, in this case fashioned from salvaged automobile frames (right center). By the following year, summer flows had filled the area behind the revetment with about three feet of sediment. The intent was to eliminate sharp meanders and to reclaim the areas they incorporated for cultivation. (R. C. Baker, 211, courtesy of the Arizona State Archives.) **B.** (11 May 1982) The WPA measures were largely effective in eliminating the sharp meanders. The 22nd Street (now Starr Pass Road) bridge is visible in the center of the photograph. Saltcedars are growing both in the channel and on the right bank. (R. M. Turner,

Figure 7.3. (*continued*)
Stake 1074, courtesy of the Desert Laboratory Collection.) **C.** (2 October 2001) In the nineteen years since the previous image, the banks of the river have been stabilized with soil cement. Desert broom and other shrubs are growing in the channel, while the trees along the bank are mostly cultivated. (D. P. Oldershaw, Stake 1074, courtesy of the Desert Laboratory Collection.)

The Soil Conservation Service was encouraging the growth of agriculture in the lower Santa Cruz valley while minimizing soil erosion. In 1937, irrigated land along the Santa Cruz River from the international border to the confluence with the Rillito River was only 22,000 acres, compared to 100,000 acres from the Rillito River to the Gila River.[14] This was a complete reversal from the agricultural dominance of the upper Santa Cruz River before the turn of the twentieth century and underscored the growing influence of wells powered by jet-driven pumps.[15] At about this time, monitoring of wells began on the San Xavier District reservation north of Valencia Road, and water levels in the late 1930s were relatively high (fig. 7.4). In the following decades, the monitoring wells would document an enormous fall in water levels that would cause a precipitous decline in riparian vegetation.[16]

By 1936, the old wagon road between Tucson and Nogales had eroded into a deep channel on the west side of the valley south of San Xavier (fig. 7.5).[17] The headcut of this channel eventually eroded upstream and intercepted the channel of the Santa Cruz River flowing north from Tubac, completing the rerouting and entrenchment of the river channel to its current position entering the Tucson Basin. Although some channel geometry information remains hazy, the Santa Cruz River appears to have had a continuous arroyo from the San Rafael Valley to downstream of Marana by the mid-1930s.

Development of a well-defined channel in and upstream from the Tucson Basin shifted the potential for inundation during floods downstream toward the farmlands around Cortaro, Rillito, and Marana. In the late 1930s, F. H. Knapp, a local engineer, reported,

> The Santa Cruz above its junction with the Rillito has a deep, well-defined channel still in process of some bottom cutting in the immediate vicinity of Tucson. The Rillito, on the other hand, and the Santa Cruz below its junction are undergoing deposition of sand in the stream bed and, consequently, widening of the channel and changing of course—This is aggravated by poorly located bank defense works built for local protection without regard to a comprehensive plan, and a crooked and meandering alignment. . . . As to deposition of the silt burden, most of the fine silts of the main stream find their way to the [Santa Cruz Plains], a part going with diverted water to

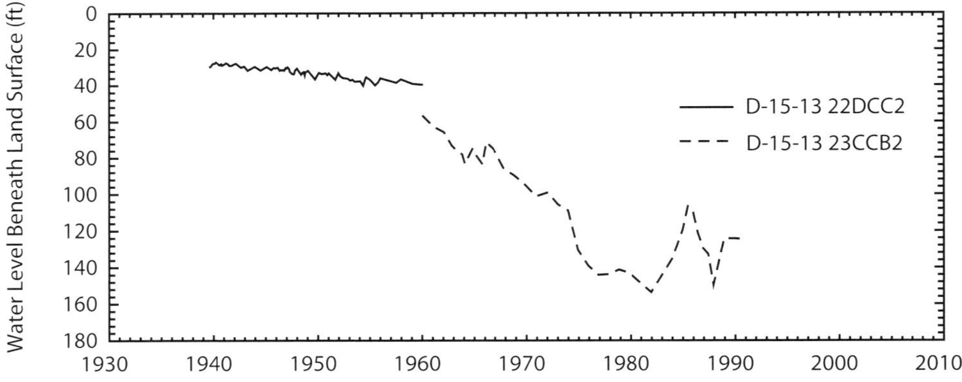

Figure 7.4.
Groundwater levels for two wells—D-15-13 22DCC2 (old well) and D-15-13 23CCB2 (new well)—along the
Santa Cruz River north of Martinez Hill near Tucson (from Webb and Leake 2006 and Webb et al. 2007b).
Excessive withdrawals are responsible for water-level declines after about 1950, and rises in the early 1980s and
early 1990s reflect recharge during large floods.

the irrigated lands, the rest spread over the uncultivated plains. The larger part of the coarser sands are
being deposited at a point about 2 miles upstream
from the road bed of the old Silver Bell Railroad near
Sasco. General leveling of the plains is taking place
here, resulting in the termination of the well-defined
channel [the arroyo at Greene's Canal,] which moves
upstream (report from 1/4 mile to 1 mile in past 10
years). Additional heavy silting is taking place upstream from this channel end [headcut] all the way
to the mouth of the Rillito.[18]

In other words, most of the channel change in Tucson either naturally ceased or was halted by the WPA
revetment work, while other reaches, particularly on
the Rillito River and downstream from its confluence
with the Santa Cruz River, were continuing to widen or
aggrade from all the sediment eroded from upstream.

On 21–22 September 1937, the War Department held
flood-control hearings at Casa Grande, Tucson, and Nogales related to nationwide efforts to curb flood losses
and soil erosion during the Depression years. The principal issues were the increase in flood hazards, particularly related to agricultural infrastructure in the lower
Santa Cruz Valley, and the recognition that future
floods like those in 1914–1915 would cause millions of
dollars in damages. Knapp and other local engineers
recommended construction of a large dike across the
valley that would intercept flood flows with sufficient
spillway capacity to divert excess water down the
arroyo at Greene's Canal to what used to be the Santa
Cruz Water Company reservoir. One feature of the
project was to prevent further headcut migration on
Greene's Canal.

Storms in the basin seemed to want to emphasize the
potential flood hazard. On 13 August 1940, a storm of
wide areal extent, uniformly heavy rainfall, and high
intensity affected the entire Santa Cruz watershed.
Considerable damage was reported along Tucson
Arroyo and in downtown Tucson. The Crosscut Canal,
which ran parallel to the Santa Cruz River west of
downtown Tucson, was destroyed by the flood and sub-
sequently was abandoned. This flood was an aberration
in a time when the arroyo mostly seemed to freeze its
cross section, still passing relatively small floods but
neither widening nor deepening while the region
passed into the midcentury drought.

An Ecological Wonderland

The Great Mesquite Forest Becomes Famous

Following the work of Charles E. Bendire and Harry S.
Swarth, professional ornithologists flocked to Tucson to
conduct studies,[19] and by the early 1900s the Great Mesquite Forest had become the Southwest's most famous
bosque and one of the best studied bird localities in the
region. The unparalleled avian habitat of the forest was
so impressive that descriptions by ornithologists waxed
superlative: the exceptional number of species, avian
densities, number of nests, and size of trees inspired
descriptors using even magical or spiritual terms. Im-
portantly, ornithologists recognized that the Santa
Cruz River, flowing northward from Mexico, was an
important migratory corridor for birds, particularly

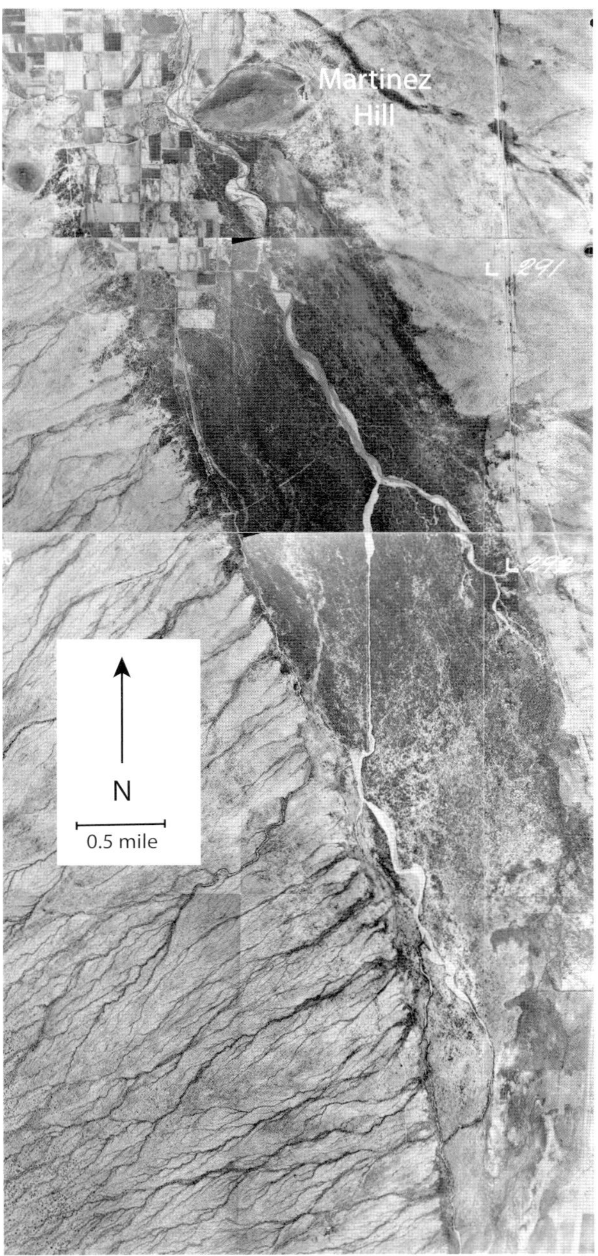

Figure 7.5.
Composite aerial photographs from 1936 showing the extent of the Great Mesquite Forest from Martinez Hill to Pima Mine Road. Agricultural clearing south of Mission San Xavier (upper left) removed what reportedly was the densest part of the bosque by this date. Numerous logging roads are present, affirming reports of those who left written accounts concerning the ongoing wood harvesting.

neotropical migrants and waterbirds. The Great Mesquite Forest served as ideal stopover habitat for resting and feeding by these migrants.

Publications on the bosque increased, which attracted still more ornithologists, but no one group of ornithologists created a comprehensive bird list. For example, in their time in the Great Mesquite Forest in the 1910s and 1920s, neither Harry S. Swarth nor William L. Dawson found nests of the Green Heron or Pied-billed Grebe. In the 1930s, C. T. Vorhies, an ornithology professor at the University of Arizona, found nests of the heron and heard Pied-billed Grebes calling during the breeding season.[20] Compilations of the disparate lists help us understand what birds used this magnificent resource (see appendixes A, B, C, and H).

Although the Great Mesquite Forest was the best-known birding locality in the Southwest for the first half of the twentieth century, estimates of its precise location and size vary. The bosque was large, observers agree, but its dimensions varied widely, ranging from a starting position between nine and twelve miles south of Tucson.[21] The northern boundary varied because woodcutting and agricultural clearing increased the distance from town. Its exact area and extent along the Santa Cruz River were only vaguely described. Volney Spalding, a researcher at the Desert Laboratory, wrote that it "stretched for miles up the Santa Cruz River near the old mission of San Xavier."[22] F. C. Willard spent "the whole day in this fashion, skirting along the edge of the mesquite forest a distance of some six miles."[23] Herbert Brandt reported that it "bordered both banks of the Santa Cruz for a number of miles and appeared to be four or five miles in width at its broadest part, tapering back to the river on either side."[24]

The Soil Conservation Service obtained aerial photographs of the Tucson Basin on 22 February 1936 as part of their overall documentation of arable lands in the West (fig. 7.5). Using the photographs that show the Great Mesquite Forest, we estimate that its area was 7 square miles at that time and that, after allowing for agricultural intrusions at its north end, its maximum area likely was 7.8 square miles. It extended about 6.25 to 6.75 miles along the Santa Cruz River. The dense core of the bosque appears to occupy about 6 square miles, with the remainder of the area either agricultural lands or an open, savanna-like forest. The size of the Great Mesquite Forest likely fluctuated in response to groundwater conditions, floods, woodcutting, and agricultural clearing, and we have no way of knowing whether 1936 represented its maximum extent.

Observers of the Great Mesquite Forest: 1916–1940s

Because of the Great Mesquite Forest and the bird populations along the Santa Cruz and Rillito Rivers, Tucson became the "ornithological capital" of Arizona, if not the entire Southwest,[25] even though Phoenix had natural and cultural resources that equaled or surpassed those of Tucson. Although the Santa Cruz and San Pedro Rivers are similar hydrologically,[26] and both provide ideal pathways for migrating birds, no large cities comparable to Tucson developed along the San Pedro River, so no university or citizens conducted studies on the local bird populations. Elevation differences also are important: the San Pedro River is approximately 1,000 feet higher in elevation, which apparently is too high for some lowland species such as the Ferruginous Pygmy-Owl.[27] Brandt found approximately sixty species of breeding birds along the San Pedro in the mid-1940s,[28] and his work in the 1930s and 1940s along the Rillito River compared the avifauna between the San Pedro and Rillito Rivers, the latter of which also had sixty species, including six species that did not occur along the other watercourse.[29]

Ornithologists trekked to Arizona in the late 1910s, lured by what was reported from the bosque south of Tucson. For example, A. Brazier Howell spent time from 7 December 1915 to 27 March 1916 camping along the Rillito River near Fort Lowell.[30] Howell's published work only mentions the Great Mesquite Forest in the context of a visit with F. C. Willard on 20 March 1916, when he saw a "forest of giant mesquites twelve miles southwest of Tucson" where they observed "two or three" Gray Hawks.[31] In his field notes on another trip to the Tucson region in 1918, he wrote, "[S]topped to hunt for rarities only for an hour and a half at the giant mesquite forest . . . some [trees] of which must have been 50' tall."[32]

William L. Dawson led a collecting party of five scientists who spent most of their time from 7 May to 14 June 1917 in the "great Mesquite Forest" during a regional tour of avian sites.[33] The expedition created a list of birds they saw in the bosque (see appendix A) and described the habitat they were exploring:

> The real objective of our expedition was the famous Mesquite Forest. This lies, or did lie, some twelve miles south of Tucson. Fed by secret springs of the Santa Cruz River, which has been alternately flowing and trickling underground ever since it left the Mexican border, some sixty miles to southward, the humble mesquite . . . rises here to the dignity of a

real tree, and this forest, once matchless for size and extent, contained trees sixty and seventy feet in height and up to three feet in diameter.[34]

During much of the spring and early summer of 1922, A. C. Bent took time from his writing of *Life Histories of North American Birds* to tour southern Arizona, accompanied by Willard.[35] They spent several days in the Great Mesquite Forest, and Bent's accounts for several species contain information from his field notes.[36] Like others before and after, Bent found the avifauna to be inspirational. Descriptions abound with phrases such as "in the early morning the medley of bird songs was absolutely confusing, and the number of individuals of the many species found in this region was far beyond what is usually the case in the lowlands of Arizona" and "the forest was rich in bird life and the air was filled with their music."[37] Other descriptions included "birds fairly swarmed,"[38] "the forest fairly teemed with bird life,"[39] and "no book on Arizona birds would be complete without the mention of the Tucson mesquite forest . . . of great mesquite trees, all of them forty feet or more high."[40]

In 1935, Herbert Brandt began a study that would last through eight field seasons until 1948, accompanied in the field by other noted ornithologists.[41] Brandt wrote eloquently of the place he called the "Grand Mesquite Forest,"[42] which occupied two chapters of his book, *Arizona and Its Bird Life.* L. W. Arnold, who conducted his master's degree work at the University of Arizona from 1938 to 1940, compiled a year-round bird list for the bosque. Arnold's work is the only study that examined all vertebrates in the forest at that time: amphibians, reptiles, birds, and mammals.[43]

In 1931, Allan R. Phillips began extensive ornithological studies over the next three decades in the Santa Cruz River drainage.[44] He lived in Tucson until the 1950s, becoming Arizona's premier professional ornithologist,[45] and pursued degrees in ornithology, attaining a master's degree at the University of Arizona and a doctoral degree at Cornell University, and he produced a complete bird list for the state and the authoritative *The Birds of Arizona* in the 1960s.[46] The Great Mesquite Forest was one of Phillips's favorite places in the Tucson Basin to observe birds.[47] Although several of his publications contained observations and data from the Great Mesquite Forest, he never published a paper solely on its avifauna or the regional importance of the bosque.

Biologists were not the only professionals who wrote of the Great Mesquite Forest and the biological richness of its flora and fauna. Early anthropologists wrote of its value to humans, along with the deep, rich soils of the Santa Cruz Basin.[48]

We created lists of plants and animals known from the Great Mesquite Forest and the Santa Cruz–Rillito River ecosystems for the late 1800s and early 1900s using many sources (see appendixes A–M). The primary references are publications by biologists, especially ornithologists, who conducted specific surveys of the avifauna and other vertebrates and published those findings.[49] Finally, we use unpublished, archived information on specific species or with lists for specific localities such as the Great Mesquite Forest and Fort Lowell.[50] A full list of publications appears in the footnotes of the appendixes and in the reference list.

Plants

As noted in chapter 5, the mesquites were the main botanical feature of the Great Mesquite Forest. The terms that have been used to describe it include a "magnificent grove,"[51] a "magnificent forest,"[52] and "the Grand Mesquite Forest" filled with "mighty mesquite trees"[53] that were "wonders of their kind"[54] and "matchless for size and extent."[55] Henry Dobyns, a social anthropologist, described an older Tucson home with a door made from a single board cut from one of the forest's gigantic mesquites.[56] In addition to huge mesquites, the forest contained a variety of trees and shrubs (fig. 7.6). Large netleaf hackberry trees were present, and the understory had extremely large graythorn and many other species. Near the channel of the Santa Cruz River, netleaf hackberry trees grew along with Mexican elders, Frémont cottonwoods, and black willows. The dry surroundings were largely Sonoran desertscrub on the surrounding alluvial fans and mesquite-paloverde savannas on the adjacent fine-grained river alluvium.[57]

Brandt wrote eloquently of the bosque:

> This virgin forest bordered both banks of the Santa Cruz for a number of miles and appeared to be four or five miles in width at its broadest part, tapering back to the river on either side. Here we enjoyed the only important trace of semitropical forest cover that we encountered in southeastern Arizona. It reminded me very much of a semiarid, hotland Sinaloa jungle, with its lively community of strange animals and plants. This woodland of giant mesquite trees developed so many distinctive natural features differing from the usual cottonwood streamside association, and drew to itself such a fine list of unusual birds, that I feel it merits designation as a separate type of desert area. . . . Of all their widespread kind, these mesquites are the noblest. In 1935 many a grand old patriarch still ruled here that had evidently already looked down on several centuries of desert droughts and savage storms. From my estimates, there should be trees in this wonderful grove that were more than saplings when Francisco de Coronado pioneered in this vicinity about 1640. . . . Here there are, indeed, trees of heroic dimensions; the bole of one stately specimen that we measured reached a girth of 13 feet, 6 inches, and a diameter of more than 4 feet, 3 inches; while the height of another capitol-domed giant was calculated to be 72 feet. . . . Nowhere were the big mesquites closely assembled but grew in spacious reserve. As there was but little underbrush, the embowered aisles were parklike and lured one on and on, in exciting anticipation. Nor was your glorious woodland ramble blocked by fallen or uprooted monarchs, emphasizing the depth and spread of their ample root structure.[58]

Species Richness of Bird Populations

Approximately seventy-three species of birds nested in the Great Mesquite Forest, in addition to numerous species that nested in the surrounding desert and foraged over it, including several raptors and other large birds (see appendix A). Many of these birds typically occurred in Mexico with ranges that lapped over the international boundary into southern Arizona and extended no farther north. Six, approximately 18 percent, of the thirty-three "Mexican" species were recorded here (see appendix D). Some of the rarest birds in the United States nested in the Great Mesquite Forest, including the Gray Hawk, the Common Black-Hawk,[59] the Northern Beardless-Tyrannulet, and the Rose-throated Becard,[60] the latter two species related to the Tyrant Flycatchers. Birds that have been recorded as migrants along the Santa Cruz River near Tucson include such unusual species as Flammulated Owl, Whiskered Screech-Owl, and Spotted Owl.[61]

The Rose-throated Becard provides an example of these losses. This species was found in 1958 and 1959 in the Great Mesquite Forest. An old nest was found in a cottonwood tree on the bank of the Santa Cruz River in the fall of 1958.[62] On 1 July 1959, a male was heard calling, and this bird was again observed on 18 July 1959.[63] Most of the riparian trees along the Santa Cruz River were dead by 1959, and the few remaining cottonwoods were senescent.[64] Only two decades earlier, in July 1938, a small colony of the Tropical Kingbird had been discovered as the first breeding record for the United

Figure 7.6.

View looking south across the Santa Cruz River from Martinez Hill. **A.** (June 1942) A gallery of cottonwoods flanks the river channel, and dense mesquite occupies the bottomlands, which were then a haven for nesting and roosting White-winged Doves. As late as 1942, one could dig by hand and find water in the streambed. (Photographer unknown, courtesy of the Arizona Game and Fish Commission.) **B.** (29 May 1981) The river channel has widened, and the bosque has been eliminated, leaving the dead trunks of large mesquite trees that are not clearly visible from this distance. The tree mortality resulted from a considerable drop in the water table beginning in the 1920s (see fig. 7.4). The 1977 flood was largely responsible for the channel scouring apparent beyond the foreground rocks and vegetation. (R. M. Turner,

Figure 7.6. (*continued*)
Stake 937, courtesy of the Desert Laboratory Collection.) **C.** (5 April 1989) The 1983 flood widened this reach considerably (also see fig. 8.8E), scouring away the floodplain and widening the arroyo. Woodcutting continues to reduce the small trees and dead trunks in the background. (R. M. Turner, Stake 937, courtesy of the Desert Laboratory Collection.) **D.** (25 November 2002) The 1977, 1983, and 1993 floods widened the channel significantly, removing floodplains. By 2002, the active channel was reduced in width and floodplains were deposited, particularly in the area behind the paloverde tree at left. (R. M. Turner, Stake 937, courtesy of the Desert Laboratory Collection. This photographic match was previously published in Glennon 2002: 48, National Research Council 2002, and Webb and Leake 2006.)

States along this stretch of the Santa Cruz River.[65] However, "flow [of the Santa Cruz] finally ceased about 1945,"[66] and the resulting degradation and loss of riparian habitat seem to have prevented the possibility of a similar colonization by the Rose-throated Becard.

The Great Mesquite Forest apparently had one of the highest densities of birds in the region, especially for certain species. The richness of the nesting avifauna and the dense population of nesting birds can be judged, in part, from numerous sightings. Although densities will never be known with any precision, inferences can be drawn from the bird lists that were published. The Dawson expedition of 1917 recorded ninety-five species of birds in May and seventy species in June, including five owls, nine hawks, and several species of migrating birds. Many of these, such as the Gray Hawk and the Crested Caracara, no longer occur along the Santa Cruz River. They recorded 572 nests and collected 165 sets of eggs from forty-two species. They also found fifty nests of Lucy's Warbler and sixty nests of Vermilion Flycatcher and collected a number of each.[67] Since Lucy's Warbler is known as "the mesquite warbler,"[68] its large numbers in the Great Mesquite Forest were to be expected. In addition, they found that "the third commonest species of these woods was, perhaps, the Bell's [Arizona] Vireo,"[69] a species whose numbers have been decreasing and is now listed as endangered under the California Endangered Species Act (CESA).[70] To find such large numbers of Bell's Vireo in a mature mesquite forest seems highly unusual since it is commonly a bird of "early successional stages"[71] that is often in scrub vegetation (such as riparian scrub) or shrubby grasslands.

A similar group of species nested in the Great Mesquite Forest and along the Rillito River, although there is no historic suggestion that the population densities were similar. During a trip to the Great Mesquite Forest in 1918, A. Brazier Howell wrote, "[B]irds, although quite common, were in only about a quarter the numbers as in March 1915, when they fairly swarmed."[72] In the spring and early summer of 1922, Arthur C. Bent wrote, "The forest fairly teemed with bird life, from the graceful Mexican goshawks [Gray Hawk] soaring overhead to the Gambel's quails whistling on the ground."[73] In another publication, he recounted:

> White-winged doves fairly swarmed through the thickets, and their tiresome notes were the dominant sounds, mixed with the softer notes of mourning and ground doves. The forest was rich in bird life and the air was filled with their music, rich-voiced cardinals and hooded orioles, mockingbirds,

desert [Cactus] wrens, Arizona [Bell's] vireos, Lucy's warblers, phainopeplas, and noisy Gila woodpeckers. Overhead, turkey vultures soared lazily."[74]

Successive ornithological work showed that some species observed or collected in the nineteenth century apparently were no longer present in the Great Mesquite Forest. For example, Dawson failed to collect eggs of species that Harry S. Swarth considered breeding birds, including the Willow Flycatcher and Blue Grosbeak. These species are all obligate or preferential riparian nesting birds.[75] Although extirpations in plant and animal populations occurred during the decline and demise of the Great Mesquite Forest, no changes are better known than with the bird populations (see appendix H).

White-winged dove. The Great Mesquite Forest was probably the best-known locality in Arizona for nesting White-winged Doves in the early 1900s.[76] This bosque and Komatke Thicket, both on the Santa Cruz River (see fig. 1.1), were also the best-known Arizona localities for hunting this species.[77] Both bosques were on Indian reservations, complicating access, although White-winged Dove flights to and from the Great Mesquite Forest allowed hunters to hunt on lands adjacent to the reservation. Arnold reported that this bosque had one of the largest known nesting colonies of White-winged Doves in Arizona and that "perhaps one of the best known white-winged dove flights from the standpoint of Arizona hunters is that which parallels the Santa Cruz River about eight miles south-southwest of Tucson and just east of the San Xavier Mission."[78]

In 1922, Bent noted that "White-winged doves fairly swarmed through the thickets, and their tiresome notes were the dominant sounds,"[79] and his comment is particularly telling. As a game species, the White-winged Dove is sometimes accentuated in avifaunal studies. Because Bent was no more interested in doves than in other species, his comment suggests that an unusually dense population of the White-winged Dove lived in the bosque. Other studies suggest that dense mature mesquite bosques were favorite White-winged Dove breeding habitat. Arnold, for example, emphasized the importance of this type of habitat to White-winged Dove after his study in the Great Mesquite Forest.[80] Similarly, the Komatke Thicket had huge numbers of these birds.[81] A definitive book on the species mentions both bosques and states:

> As recently as twenty-five to thirty years ago, bottomland mesquite jungles provided the best and most concentrated nesting habitat for Arizona

whitewings [White-winged Doves]. The simple fact is that the mature mesquite thickets that once supported huge nesting colonies along the Gila River and tributaries are gone.[82]

Cavity nesters on the margins. The abundant resources of the Great Mesquite Forest spilled out into the surrounding desert, enhancing bird populations in more xeric habitat. Several species of birds nested in saguaros, chollas, and other desertscrub species on the margins of the bosque. Most species that nested in the desertscrub also nested in cottonwood-willow and (or) mesquite, or at least at the interface (ecotone) where the Great Mesquite Forest abutted the surrounding desert (fig. 7.5). Cavity nesters that generally nested in either cottonwood-willow stands, saguaros, or mesquites included the American Kestrel, Western Screech-Owl, Elf Owl, Gila Woodpecker, Ladder-backed Woodpecker, Gilded Flicker, Ash-throated Flycatcher, and Brown-crested Flycatcher (table 7.1). Three exceptions to this are

Table 7.1. **Cavity-nesting birds of the Santa Cruz–Rillito River system**

Common Name	Scientific Name	Saguaros Only	Saguaros and Trees	Trees Only
***Primary Cavity Nesters*[1]**				
Gila Woodpecker	*Melanerpes uropygialis*		X	
Ladder-backed Woodpecker	*Picoides scalaris*			X
Gilded Flicker	*Colaptes chrysoides*		X	
***Secondary Cavity Nesters*[2]**				
American Kestrel	*Falco sparverius*	X	X	
Barn Owl	*Tyto alba*			X[3]
Western Screech-Owl	*Otus kennicottii*		X	
Ferruginous Pygmy-Owl[4]	*Glaucidium brasilianum*		X	
Elf Owl	*Micrathene whitneyi*		X	
Ash-throated Flycatcher	*Myiarchus cinerascens*		X	
Brown-crested Flycatcher	*M. tyrannulus*		X	
Purple Martin	*Progne subis*	X		
Cactus Wren[5]	*Campylorhynchus brunneicapillus*	X		
Bewick's Wren	*Thryomanes bewickii*			X
European Starling	*Sturnus vulgaris*		X	
Lucy's Warbler[6]	*Oreothlypis luciae*		X	
House Sparrow[7]	*Passer domesticus*		X	

Sources: Scott and Patton (1975), Scott et al. (1977).

Notes

All species were recorded in the Great Mesquite Forest and (or) in adjacent ecosystems except Ferruginous Pygmy-Owl, which was common along the Rillito River.

1. Species that excavate nesting cavities; all are woodpeckers in the desert lowlands of the US Southwest.

2. Species that nest in cavities excavated by other species, particularly woodpeckers.

3. Because of its large size, the Barn Owl nests in cavities larger than those excavated by woodpeckers, usually cavities caused by falling limbs and (or) rotting wood.

4. Herbert Brown took a Ferruginous Pygmy-Owl south of Tucson in 1884, the only Santa Cruz record (Johnson et al. 2003). The Arizona population was listed as federally endangered in 1997 but delisted in 2007 and extirpated in the Tucson Basin by 2009 (R. R. Johnson, pers. comm., 2012).

5. Cactus Wren usually nests in a large, global nest with a side entrance but occasionally uses saguaro cavities (Proudfoot et al. 2000).

6. Lucy's Warbler usually nests in a small, inconspicuous nest under loose bark but occasionally uses a cavity in a saguaro or tree (Johnson et al. 1997b); Bendire was the first to report the nest and eggs of this species (Coues 1872c).

7. House Sparrow usually nests in a freestanding nest or in a nest built in an opening near or in a building but occasionally builds a nest in a cavity in a saguaro or tree.

the "saguaro" Purple Martin, which nests only in saguaros; Bewick's Wren, which at lower elevations nests
only in riparian trees; and, occasionally, Lucy's Warbler,
which usually nests under loose bark but occasionally
nests in a cavity in desert or lowland riparian environments.[83] At least one new subspecies of Albert's Towhee
was described from type specimens taken in the Great
Mesquite Forest.[84]

Other Vertebrates

Approximately eighty-nine species of vertebrates, other
than birds, and one clam have been recorded in the
Great Mesquite Forest and bordering ecosystems along
the Santa Cruz River. This list includes five species
of native fishes and one introduced fish, approximately forty-three native species of amphibians and
reptiles, one introduced frog, and thirty-nine species
of mammals. Several other species that have been recorded in the Santa Cruz–Rillito River drainage since
the arrival of Euro-Americans in the Tucson Basin
may also have formerly occurred in the bosque (see
appendix H).

The Santa Cruz River had sections of perennial flow
separated by long reaches of dry sand, so it is somewhat
surprising that fish lived in the Tucson Basin. Presumably, fishes arrived in the Tucson Basin from the Gila
River during Pleistocene climates, when flow likely was
perennial from headwaters to the confluence. The five
native fishes that once inhabited the middle Santa Cruz
River included the Gila chub and Sonora sucker, both
medium-sized fish of more than one foot in length, and
the longfin dace, the Santa Cruz (Monkey Spring) pupfish, and the Gila topminnow, three small fish measur-
ing four inches or less.[85]

This bosque was home to a large number of rare and
unusual animals in addition to species that continue to
be common in the Sonoran Desert. White-tailed deer,
javelina, coyote, bobcat, and other mammals were in
the bosque and remain in the area, but other large predators, such as jaguar, may well have used the Great
Mesquite Forest. One of the unusual animals was the
mesquite mouse, which occurs only in the states of
Arizona and Sonora, Mexico, and is one of the rarest of
native mice in the United States.[86] The largest known
population of this mouse lived in the Great Mesquite
Forest before it was destroyed.[87] We will never know
what other animals may have used this unique ecosystem, but regional species lists provide gist for some
speculation.[88]

Comparison with Riparian Resources of the Rillito River

Less is known about the riparian ecosystems along the
Rillito River. By the 1930s, riparian vegetation in the vicinity of Fort Lowell consisted of seepwillow near the
active channel and Frémont cottonwood, Arizona ash,
Arizona sycamore, Arizona walnut, and "willow" on
the adjacent floodplains.[89] A grove of cottonwoods was
present at the confluence of the Rillito and Santa Cruz
Rivers, and groundwater levels were high along the
length of both. Some obligate riparian species have been
extirpated, notably the Mexican garter snake, which
once lived on the floodplains of the Rillito River,[90] and
may have occurred in the Great Mesquite Forest as well.

The Beginning of the End: The Long Decline of the Great Mesquite Forest

We cannot determine from the historical accounts when
the Great Mesquite Forest reached its zenith, in terms of
spatial extent, size or number of riparian trees, or especially in its animal biodiversity. We know that numerous
accounts warned of the forces threatening to destroy it,
and we can assume that its long decline began in the
1930s or 1940s. Woodcutting in the early 1900s resulted
in the loss of the bosque's towering old-growth trees,
some of them more than four feet in diameter and
more than sixty-five feet high.[91] By 1917, 2,500 cords of
mesquite were being cut and removed annually,[92] and a
few years later it was noted that "woodchoppers had
been cutting down the larger trees all over it and making a network of cart roads all through it."[93] Most of
the mesquite had been cut to sustain Tucson's power
needs.[94] The forest resprouted from the stumps, creating a shorter-stature woodland still dependent on high
groundwater levels.

In 1936, woodcutting roads were present in the
bosque, but the Great Mesquite Forest was still dense
with mesquite trees although the arroyo had downcut
and widened by this time (fig. 7.7). When studied in
1940, the second-growth trees were said to be twenty to
twenty-five feet tall,[95] and, by the 1950s, all the older
mesquite trees had been cut, leaving nothing but smaller
second-growth trees that generally were fifteen to
twenty feet tall.[96] At least one remaining mesquite measured more than four feet in diameter.[97] This situation
prevailed into the early 1960s,[98] and by that time the extent of the forest had also been reduced, largely by agricultural fields and roads.[99] Woodcutters kept cutting,

Figure 7.7.
(1940) Inside the Great Mesquite Forest. This photograph shows what probably is a woodcutter's road in the Great Mesquite Forest. Most of the plants appear to be secondary-growth mesquite (from Arnold 1940).

but the trees regrew; woodcutting did not kill the trees but altered the structure and function of the bosque.

Some ornithologists, who considered the bosque a national treasure, were apoplectic about the changes they saw and wrote scathing accounts of the land uses they witnessed. In 1917, Dawson pointed out what he considered a grave situation with this magnificent forest:

> A ruthless policy of deforestation, which was culminating at the time of our visit, had reduced the heavier timber to about four-fifths of its former abundance and the destruction was going on . . . at the rate of 2500 cords per annum. At that rate the forest could not have held out above two years longer. Without doubt the intensity of the nesting operations which we found in progress during our stay was due, chiefly, to the congestion caused by the rapid decrease of available nesting area. As an oological opportunity, the occasion was unique; but as an example of governmental waste, the situation was appalling.[100]

Five years later, during his visit to southeastern Arizona, Bent expressed his disappointment about the on-going destruction to the Great Mesquite Forest. He had obviously been told earlier of this outstanding avian paradise by Willard, his guide, who published one of the earliest accounts of the forest.[101] On seeing the forest, Bent wrote:

> But this magnificent forest did not long remain in its pristine glory. When I was in Arizona with Frank Willard in 1922 . . . [we] were disappointed in the forest, for . . . [they] had been cutting down the larger trees unmercifully and had made a network of cart roads all through it for hauling out the firewood. There were only a few large trees left, more or less scattered, and between them many open spaces in which were thickets of small mesquites and thorn or patches of medium-sized mesquites and hackberries.[102]

In 1935, amateur Ohio ornithologist Herbert Brandt first visited the famous forest. Calling it the Grand Mesquite Forest, he wrote:

> The wood itself is heavy, of magnificent knotty strength . . . It is the best of all the desert firewoods, and herein lies a fatal scourge. In Tucson for years

there has been a profitable demand for small, round mesquite firewood . . . In order to cart out the wood, wagon roads serpentined almost everywhere through the forest, so it was an easy matter for a traveler to go far astray.[103]

In 1940, Arnold recorded 111 species of birds in and adjacent to the Great Mesquite Forest, including 74 summer species, the remainder consisting of migrants and wintering species. Arnold, the only person to conduct a survey of all the terrestrial vertebrates of the forest, wrote:

> Fortunately part of the area which we selected for the more detailed work and a rather extensive section immediately adjoining it more closely resemble the original condition as described by former writers. Here the mesquite attain a height of some twenty or twenty-five feet. They are of sufficient density to form a dense canopy of branches overhead during the summer season and the ground is well covered with litter formed by the falling mesquite leaves.[104]

Arnold also pointed out that although there was little loss in numbers of species of birds, with the loss of larger trees there had been a shift in population densities to fewer individuals of forest species and larger numbers of brushland species such as Vermilion Flycatcher, Yellow-breasted Chat, and Northern Cardinal.

Ten years after he first visited the Great Mesquite Forest, Brandt wrote:

> In 1945, with Doctor Oberholser, I returned to what was left of the grand mesquite forest. With sad heart, I found that every noble tree had been hacked away for firewood, leaving only the tall naked stumps as sad monuments to greed and ignorance. Only brush heaps remained where once grew the mighty mesquites. . . . Again the heedless hand of man destroys an immeasurably old climax growth, ruining forever an exceptional monument of nature.[105]

Some of the intentional destruction was to discourage recreationalists, which is ironic considering Pima County's current emphasis on ecotourism and green space, especially with shade during the hot summer months. Carl Lumholtz, a Norwegian explorer and ethnographer, wrote about the San Xavier Reservation:

> Part of the reservation consists of a large and very fine forest of mesquite [*sic*] trees, some of which have grown to considerable size. The Indians help to supply Tucson with wood from these trees. . . . The

mesquites were at their best in their light green, fresh-looking foliage, and on Sundays these woods, which in less arid regions would not be valued very highly, are used as picnic grounds by the hard-working people of Tucson. Lovers of nature find comfort among the trees although they give little shade, but they are the only woods within a reasonable distance of town. Formerly some magnificent cotton[wood] trees, near the Indian agency of the reservation, used to be the objective point of picnic excursions. There were no less than twenty of these old trees that gave splendid shade during the fierce heat of the summer. The crowds as is their careless habit in America, used to leave newspapers, baskets, and peanuts strewn about the ground, and they would throw empty cans into the alfalfa fields beyond the fence. The owner, much annoyed, posted warnings and prohibitions which were apparently taken lightly; for, finally, to prevent this nuisance, he resorted to the incredibly drastic measure of cutting down these superb trees and making firewood of them."[106]

An excellent summation of conditions before 1940 was published in *The Birds of Arizona*. All three authors of this book had visited the bosque on numerous occasions after 1931, when Allan R. Phillips made his first visit. Gale Monson had also visited the bosque during this same period,[107] and Joe T. Marshall, an ornithology professor at the University of Arizona, published scientific papers based on his work in the bosque during the 1950s.[108] In *Birds of Arizona*, these ornithologists summed up their impressions of the Great Mesquite Bosque:

> Prior to World War II, the river at Sahuarita Butte [Martinez Hill] (between Indian Dam and San Xavier Mission) was a paradise for birds. There were fine groves of cottonwoods, and . . . above these bottoms was a dense but not tall mesquite thicket, of a general level of fifteen feet or more, with much understory of vines and flowers in the summer rainy season. . . . No book on Arizona birds would be complete without a mention of the Tucson mesquite forest [Great Mesquite Forest]. This famous bosque of great mesquite trees, all of them forty feet or more high, occupied the river valley on the San Xavier Indian Reservation before it became channeled. The trees were cut down for firewood in the early decades of the Twentieth Century, and would probably have died in any case with the drastic lowering of the water table. . . . Today in this area it is difficult to find even a single healthy cottonwood along the Santa Cruz.[109]

As a result of flood-control dikes, the entire flow of the Santa Cruz River was diverted east into the former

Spring Branch, creating an abnormally straight channel for about 1.3 miles and then, just upstream from Martinez Hill, a wide and meandering reach (figs. 7.2, 7.6). In the last third of the twentieth century, that meandering and widening would increase, leading to the greatest channel erosion along the length of the river as well as the loss of bridge spans. Channel downcutting drained the alluvial aquifer to the level of the channel, which helped the bosque for decades. Mining of groundwater accelerated the fall of the water levels, and the shallow aquifer sustaining the bosque drained well beneath the level of the arroyo channel through the Great Mesquite Forest. Between 1930 and 1940, the water table dropped nearly 30 feet in this reach (fig. 7.4).[110] Woodcutting, agricultural encroachment, and roads decreased the acreage and overall extent of the bosque, but it was groundwater withdrawal that finally killed the once-magnificent trees.[111]

8

The City and the Arroyo, 1943–1975

Drought is pretty easy to recognize, particularly when it persists for many years, but the drought that impacted the southwestern United States in the middle of the twentieth century had a fuzzy beginning and an even more ambiguous end. Some have referred to this event as the 1950s drought because the drought was most severe early in that decade, with drastic reductions in winter rainfall from 1952 to 1956. In southern Arizona, however, this drought was more persistent, extending from the late 1940s until the early 1960s and, depending on definitions, to about 1975. Whereas the winters were generally dry during this period, the period was punctuated by wet El Niño winters, such as 1957–1958. Summer rainfall was another matter: small floods frequently passed through Tucson in this period, particularly during the mid-1950s. The problem river became a quiescent channel, dry most of the time, and residents of the rapidly growing city began seeing the once unstable and vulnerable floodplain as an opportunity for development, converting the former farmlands to neighborhoods.

The channel and its floodplain continued to serve as a municipal resource, providing residents with an array of products ranging from firewood from the Great Mesquite Forest to water for their gardens. When the channel narrowed in the absence of large floods, the citizens of Tucson began to use the floodplain to dispose of garbage, narrowing the channel further. When increasing construction required sand and gravel for concrete, Tucsonans directly mined it from the channel. When they needed more water, what better place to drill wells than the floodplain aquifer, presumably replenished with those now-smaller floods? Many decisions regarding Tucson's municipal functions in relation to its view of the Santa Cruz River had a lasting impact evident today. To understand these decisions, we need to look back at water development, beginning around the turn of the twentieth century, and how it blossomed into the complex system of pipes and canals that affected the arroyo and its restoration potential.

The Growing City

In the 1900 census, Tucson was a small town of 7,531 residents (fig. 8.1A).[1] A transcontinental railroad had passed through town for twenty years, although the roads coming and going to town were still dirt and the automobile remained a novelty. Then there was a problem river, which seemed to guarantee water while occasionally cutting the town off from its western expanses. Tucson residents wanted to live close enough to the Santa Cruz River to get water for farms and houses while staying far enough away to avoid floods and malaria. At that time, malaria was thought to result from "bad air" (the Italian meaning of the word); its association with mosquitoes would not be determined until 1898, and its eradication from the United States would not occur until 1951.[2]

After the floods and channel widening that marked the first third of the twentieth century, the river seemed less of a hazard, and the floodplain represented an opportunity for growth. The marshes were drained, reducing the malaria problem. The land was no longer suitable for agriculture irrigated from surface-water canals, but its proximity to the inner city made it valuable real estate for both housing and industrial development. From 1940 to 1977, Tucson grew from a sleepy city of 35,752 residents using about 7,000 acre-feet of water annually to a thriving metropolis of 304,600 people (fig. 8.1A) using 100,000 acre-feet.[3] In the same period, the

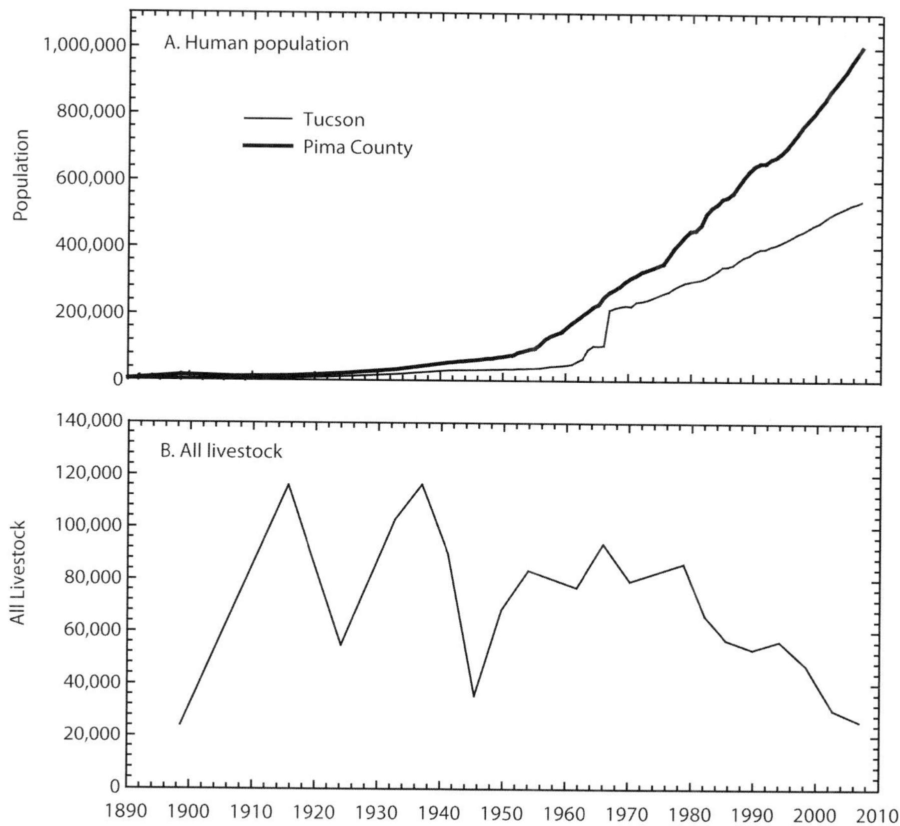

Figure 8.1.

Census and livestock data for Tucson and Pima County, 1880–2007 (see chapter 8, note 1 for sources of data). **A.** Human population of Tucson and Pima County. The unusual jump in Tucson's population in 1960 was the result of annexation of adjacent county lands. **B.** Total numbers of all livestock in Pima and Santa Cruz Counties, including cattle, sheep, horses, and hogs. The drops between 1910 and 1920 and 1938 and 1945 are in response to the late nineteenth-century drought and the Great Depression, respectively.

number of people in Pima County—mostly the Tucson metropolitan area—increased from 72,838 to 468,500 residents (figs. 8.1A and 8.2).[4] This growing population needed property for houses, and the best place to build these houses was on the flat and well-drained bottom-lands next to the entrenched channel. Moreover, most of the wells at that time tapped the shallow aquifer near the Santa Cruz River, making water distribution relatively easy.

Development along the Santa Cruz River proceeded without significant planning for flood hazard during a period of low-flow conditions. Federal legislation concerning floodplain hazards and insurance would not be enacted until 1968.[5] Collective memory of bridge failures and erosion of agricultural fields faded during the midcentury drought, which would foreshadow what was to come later in the century. The floodplain of the Santa Cruz River accommodated the growing popula-

tion of the rapidly expanding city (figs. 8.3 to 8.5). No consideration was made for riparian vegetation, even if the association between high groundwater tables and riparian vegetation was generally known at the time. Native riparian vegetation lined the channel and tena-ciously regrew after repeated decimation by floods, woodcutting, and agricultural clearing, but it finally expired, in part to be replaced by nonnative Athel tam-arisk. Many of the trees planted in this period along the banks of the Santa Cruz River persist into the twenty-first century.

The booming town was not the only change that would affect the Santa Cruz River. During the mid-century drought, livestock numbers remained rela-tively constant following a rebound after the Great Depression (fig. 8.1B). Floodplain agriculture in the vicinity of Tucson, estimated to cover about 4,000 acres and use 47,000 to 48,000 acre-feet of groundwater

Figure 8.2.

Congress Street from Powderhouse Hill. **A.** (1890s) When George Roskruge took this photograph, about 5,000–6,000 people lived in Tucson, the Santa Cruz River floodplain was still active, and Congress Street was a narrow dirt lane. The trees along the West Branch are probably cottonwoods. The Rincon Mountains appear faintly on the horizon. (Roskruge, 46397, courtesy of the Arizona Historical Society.) **B.** (1930s) This approximate match shows that in the forty years between views, entrenchment of the Santa Cruz arroyo turned the floodplain into a terrace, which enhanced drainage and encouraged urbanization. About 32,000–35,000 people lived in the town. (E. Ronstadt, courtesy of the University of Arizona Special Collections Library.) **C.** (17 December 1981) Approximately 343,500 people lived in the city of Tucson at

Figure 8.2. (*continued*)
the time of this view. The former fields are completely urbanized along the reach through downtown Tucson. The river, too far away to see in this view, would be channelized a year later. (R. M. Turner, Stake 1061, courtesy of the Desert Laboratory Collection.) **D.** (20 June 2007) Tucson's population at this time was more than 800,000, and maturation of the foreground neighborhood is emphasized by the increase in the size of trees growing along streets and on private property. Rio Nuevo, a redevelopment project for downtown Tucson, will eventually occupy much of the bare area in the right midground. (R. M. Turner, Stake 1061, courtesy of the Desert Laboratory Collection.)

Figure 8.3.
View looking east across the Santa Cruz River from Sentinel Peak. **A.** (30 May 1927) The east bank of the Santa Cruz arroyo is visible across the bottom of the photograph. Mesquite has reclaimed the formerly cultivated fields in an area that was a bosque in the late nineteenth century. (Norman Wallace, 518, courtesy of the Arizona Historical Society.) **B.** (6 October 1987) The banks of the Santa Cruz River, reinforced with soil cement five years before this photograph was taken, are visible across the bottom of the photograph, and 22nd Street (now Starr Pass Road) is in the center. The mesquite thickets have been replaced by residences and businesses; many of the foreground trees are nonnative Athel tamarisks. (R. M. Turner, Stake 1307d, courtesy of the Desert Laboratory Collection.)

Figure 8.4.
View looking east-northeast across the Santa Cruz River from Sentinel Peak. **A.** (30 May 1927) This image shows the Santa Cruz River in the foreground, flanked by large trees that probably are cotton-woods. Agricultural fields are at the center and on the left. The city of Tucson is in the midground, and the Rincon Mountains are on the horizon. (Norman Wallace, 522, courtesy of the Arizona Historical Society.) **B.** (6 October 1987) The channel of the Santa Cruz is now deeper and lined with soil cement. The city now occupies the former agricultural fields. (R. M. Turner, Stake 1307c, courtesy of the Desert Laboratory Collection.)

Figure 8.5.
View looking northeast across the Santa Cruz River from Sentinel Peak. **A.** (30 May 1927) The Santa Cruz River runs from right to left through several cultivated fields. Large trees, probably cottonwoods and willows, line its banks. The Santa Catalina Mountains are on the horizon, with the city of Tucson in the midground. (Norman Wallace, 502, courtesy of the Arizona Historical Society.) **B.** (6 October 1987) The course of the Santa Cruz has been straightened and lined with soil cement. A landfill, dating to 1950, has replaced the fields on river left, while city development dominates on river right. The large buildings at left center are part of downtown Tucson. (R. M. Turner, Stake 1307b, courtesy of the Desert Laboratory Collection.)

in 1937,[6] may have been at its maximum at the start of the drought (fig. 8.3). More and more wells were installed in the Tucson Basin, and groundwater was mined as if it were an infinite resource. Groundwater levels fell several hundred feet until water managers realized that growth would stop if they did not secure more water.

Water-Level Lowering and Aquifer Compaction

Falling water levels and the increased cost of pumps and electricity were not the only problems faced by water managers. Tucson had drilled many wells into the aquifer within the city limits, and each year wells were pumping more and more water out of the ground. Unfortunately, as the water levels fell, the land surface began to subside in response. Land subsidence caused by aquifer compaction commonly occurs in areas subjected to groundwater overdraft or hydrocarbon extraction.[7] After California and Texas, Arizona has the largest problem with land subsidence, and all of Arizona's subsidence problems are associated with groundwater overdraft in either agricultural or urban areas.[8]

As groundwater pumping increases, land subsidence increases, given certain alluvial basin characteristics. In alluvial aquifers, groundwater is pumped from the pore space between grains of sand and gravel, and if an aquifer has interbedded clay or silt, the lowered water pressure in the sand and gravel causes slow drainage from the clay and silt beds. The reduced water pressure is a loss of support for the fine-grained strata, which compact by decreasing the pore space and packing the fine particles closer together. This compaction results in a land surface permanently lowered in most cases without the possibility of reversal by recharging the aquifer.

In the Tucson Basin, groundwater withdrawals increased from 35,000 acre-feet in 1920 to 240,000 acre-feet in 1970, remaining approximately constant through 1996. This rate of extraction exceeded recharge rates, and the water level declined beneath the city. From 1950 to 1980, the maximum measured land subsidence was only half a foot,[9] with an additional 0.15 to 0.20 feet of subsidence through 1996.[10] The resulting changes in land surface affect flow in storm drains, sanitary sewers, canals, and levees, damaging urban infrastructure.[11] Ironically, subsidence damages the wells and casings that are helping to create the problem. Land subsidence has had direct, indirect, and as yet unknown effects on Tucson. Cracks in

walls and foundations in central Tucson have been attributed to subsidence. Sinkholes, which may be precursors to subsidence fissures, developed on the Tohono O'odham Reservation southwest of Martinez Hill, affecting the ability of farmers to use arable land.[12]

Land subsidence caused by groundwater withdrawals tends to be greatest where pumping and water-level declines are highest. This occurs in the middle of the Tucson Basin, where numerous small streams drain the urbanized area; the Santa Cruz and Rillito Rivers both mostly flow around the center of the Tucson Basin. There is a possibility that gradients for these small streams could be reduced or reversed, with concomitant changes to flood hazard, erosion, and deposition. This effect also was thought to have been one response to the 1887 earthquake (see chapter 6). On the margins of subsiding aquifers, earth fissures open up to accommodate the overall strain on the land surface. Fissures can be more than one hundred feet deep and several thousand feet in length; one extraordinary fissure in central Arizona is ten miles long.[13] These fissures can alter the course of small drainages, some of which may discharge into the fissure. Although subsidence is greater in the Eloy Basin and the Phoenix metropolitan area north of Tucson, it also is causing changes in the Tucson Basin.

Landfills and Sand-and-Gravel Mining

A growing city produces ever-accumulating trash. Tucson burned its garbage during the 1940s at an incinerator on the east bank of the Santa Cruz River at St. Mary's Road. After the incinerator closed in 1950, several million tons of garbage were dumped either in the channel or on the adjacent floodplain. Between 1953 and 1962, the main landfill was on the west side of the Santa Cruz River at the base of Sentinel Peak, extending from the former site of the old San Agustín Mission upstream to the former site of Warner's Lake (fig. 8.6). This landfill covered about twenty-two acres to an average depth of forty-five feet, in one case completely filling in the channel of the West Branch near its confluence with the main channel. Another landfill was created on the west bank upstream from Congress Street, using a section of channel widened by the 1915 flood (fig. 8.7). Landfills have also been operated near the former site of Silver Lake, near the confluences of the Rillito River and Cañada del Oro with the Santa Cruz River, and in the vicinity of Marana.

Figure 8.6.

The Santa Cruz River just south of the Congress Street Bridge. **A.** (31 January 1915) This upstream view, looking southeast, shows the sweeping meander along the east bank as it eroded during the 1915 flood. The trees in the background are cottonwoods. (Photographer unknown, 6518, courtesy of the University of Arizona Special Collections Library.) **B.** (26 February 1982) Landfill operations, which began in 1950, narrowed the channel and caused additional downcutting, especially during the 1977 flood. The trees are mostly Athel tamarisks with a few mesquites. (R. M. Turner, Stake 1067, courtesy of the

Figure 8.6. (*continued*)
Desert Laboratory Collection.) **C.** (19 November 2008) The camera station is located just north of a small arroyo entering the
Santa Cruz on river right. The match is not exact because of changes in the riverbank during channelization, as well as the
dense midground vegetation that blocks key features. The Santa Cruz channel is now defined by soil cement and contains a
great deal more vegetation than it did twenty-six years earlier. Much of the foreground vegetation consists of landscape plants,
such as Mexican paloverde and nonnative mesquites. (R. M. Turner, Stake 1067, courtesy of the Desert Laboratory
Collection.)

Landfills, bridge construction, urbanization of the floodplain, and sand-and-gravel mining of the riverbed constricted the channel and promoted bed degradation.[14] Sand and gravel, needed for construction activities, were listed as a significant mineral resource of the Santa Cruz River in 1917.[15] Sand-and-gravel mines were excavated along the river channel—some wildcat, and some with municipal links—and these caused significant problems. Essentially, a sand-and-gravel mine in the center of a channel creates an artificial headcut that can propagate both upstream and downstream during floods. Although better known along the Rillito-Pantano River system,[16] sand-and-gravel mines also caused problems along the Santa Cruz River. As the value of floodplain land increased and as periodic flooding caused erosion of these mines, they were moved out of the channel onto the adjacent former floodplain, separated from the channel by only a thin wall of unstabilized alluvium. Many of these mines ultimately greatly exceeded the depth of the adjacent channel, creating an erosion hazard if the bank were to be breached during a flood. They also intercepted the groundwater level, creating new lakes in the desert.

Water Use: Surface Water to Groundwater

In response to arroyo downcutting and widening, which destroyed every attempt to harvest a seemingly abundant water supply, Tucson switched from being a small town with extensive agriculture dependent primarily on surface water to being the largest city in the world, at least for a time, wholly dependent on groundwater for domestic consumption. As has been discussed, a few wells were dug in the nineteenth century, but this water had limited usage; wagons delivered river water, some impounded in ill-advised lakes, to individual houses. The switch from isolated domestic well use to groundwater supplying homes, subdivisions, golf courses, and businesses required a concerted effort from public and private institutions to construct

Figure 8.7.

East view of the Santa Cruz River valley and Tucson from Sentinel Peak. **A.** (1932) The river runs from right to left across the center of the photograph, with the Congress Street Bridge at far left. The broad, entrenched channel is lined with cottonwoods. Solomon Warner's house and the ruins of his mill are in the lower left corner. (Photographer unknown, 26758, courtesy of the Arizona Historical Society.) **B.** (8 July 1981) Since 1950, landfill operations and construction of Interstate 10 have constricted the channel. Much of the floodplain surface has been elevated by landfill, in some places by six to nine feet. The only nonelevated part of the floodplain is the former mission garden in the lower center of both photographs. Warner's house is still visible. (R. M. Turner, Stake 1044, courtesy of the Desert Laboratory Collection.)

the necessary infrastructure to distribute water. It also required a new kind of pump, developed elsewhere, to lift water from increasing depths beneath the city.

Lifting water out of the shallow alluvial aquifers that were present at the start of the twentieth century was relatively easy. The centrifugal pump typically was used in shallow wells, but it had limits on the height to which it could efficiently lift water from its position on the land surface. That height generally is twenty feet above water level,[17] with an absolute limit of thirty-two feet before cavitation interrupts flow; these limits were severely tested as the water level in the alluvial aquifer fell.[18] Beginning around 1920, centrifugal pumps were replaced with deep turbine pumps that could be lowered into wells below water level. Instead of sucking water up through a well stem, the turbine pumps pushed water up from below. The limitation on well depth was removed; well pumping was dependent only on the depth to which one could drill a well and the power required to run the turbine pump.

The mayor and his engineers were optimistic about the long-term availability of groundwater extracted using this new technology. Pumping of the alluvial aquifer accelerated, particularly in the vicinity of Martinez Hill and the Tohono O'odham Reservation. Groundwater levels plummeted (see fig. 7.2), and perennial flow in the Santa Cruz River ceased about 1940 in this area.[19] Most wells were only about fifty feet deep, but as pumping accelerated, more wells drilled to deeper depths were required. By the middle of the twentieth century, an array of wells punctured the alluvial aquifer of the Tucson Basin, and a basin-wide decline in groundwater levels was under way. For the entire state of Arizona, water use in 1950 was 5.37 million acre-feet, two-thirds of which came from groundwater, and only 100,000 acre-feet was used for agriculture.[20]

Depletion of the alluvial aquifer had another effect well beyond the economic cost of the installation of deeper wells. The twin springs of Punta de Agua and Agua de la Misión, so important to the early history of the Tucson Basin, were the result of the generally high groundwater levels that sustained the Great Mesquite Forest (see chapter 7). Arroyo downcutting drained the aquifer down to the channel bed, but that only encouraged the growth of woody riparian vegetation in the newly aerated sediments still within reach of permanent water. Pumping continued to decrease groundwater levels under the Great Mesquite Forest until water levels bottomed out in the 1970s (see fig. 7.2). The once-thriving bosque slowly died (fig. 8.8; see also figs. 7.3 and 7.5), leaving a forest of stumps. In 1961, riparian

vegetation was estimated to consume a paltry 1,800 acre-feet along the Santa Cruz River, a water-use level that was expected to go to zero as water levels continued their precipitous decline.[21]

As Tucson continued inexorably in its transformation from frontier town to modern city, the water-distribution system became increasingly complex. Between 1945 and 1950, water connections in Tucson increased by nearly 50 percent.[22] In the boom years after World War II, all water-intensive land uses—mining, agriculture, and municipal uses—expanded their needs. Surface-water diversions on the Tohono O'odham Reservation within the dying bosque ended in 1947,[23] making Tucson wholly dependent on groundwater for domestic supplies.

Systematic study of water resources in southern Arizona began with a series of US Geological Survey reports on the locations of water holes and tinajas in the early 1920s.[24] In the Tucson Basin, it gained some sophistication from the work of hydrologist George E. P. Smith, who studied water supplies—particularly groundwater supplies—during the first third of the twentieth century.[25] Systematic study of groundwater systems really started when hydrologists Harold C. Schwalen and Richard J. Shaw, who were employed by the University of Arizona's Agricultural Experiment Station, collated a considerable amount of data gleaned from the extensive well fields in the Tucson Basin.[26]

Schwalen and Shaw's work was not good news for water users in southern Arizona. Groundwater levels were dropping precipitously in the 1950s, a phenomenon that in no small part could be attributed to the midcentury drought and the increase in agriculture, especially cotton farming in the Marana area. They reported that groundwater had dropped as much as 70 feet along the Santa Cruz River—50 feet in the vicinity of A Mountain—and noted the extensive agricultural water use. Despite the overall bad news, they noted something else: the large number of floods in 1954 and 1955 (fig. 8.9) had induced short-lived rises in groundwater levels along the river.[27] This flicker of hope, that the alluvial aquifer potentially could recharge to near historic levels, was extinguished as the regional aquifer system was mined, reducing water levels another 100 to 150 feet.

River Flow in a Period of Drought

Not all droughts are created equal. When it comes to ecological and hydrological droughts, such as the

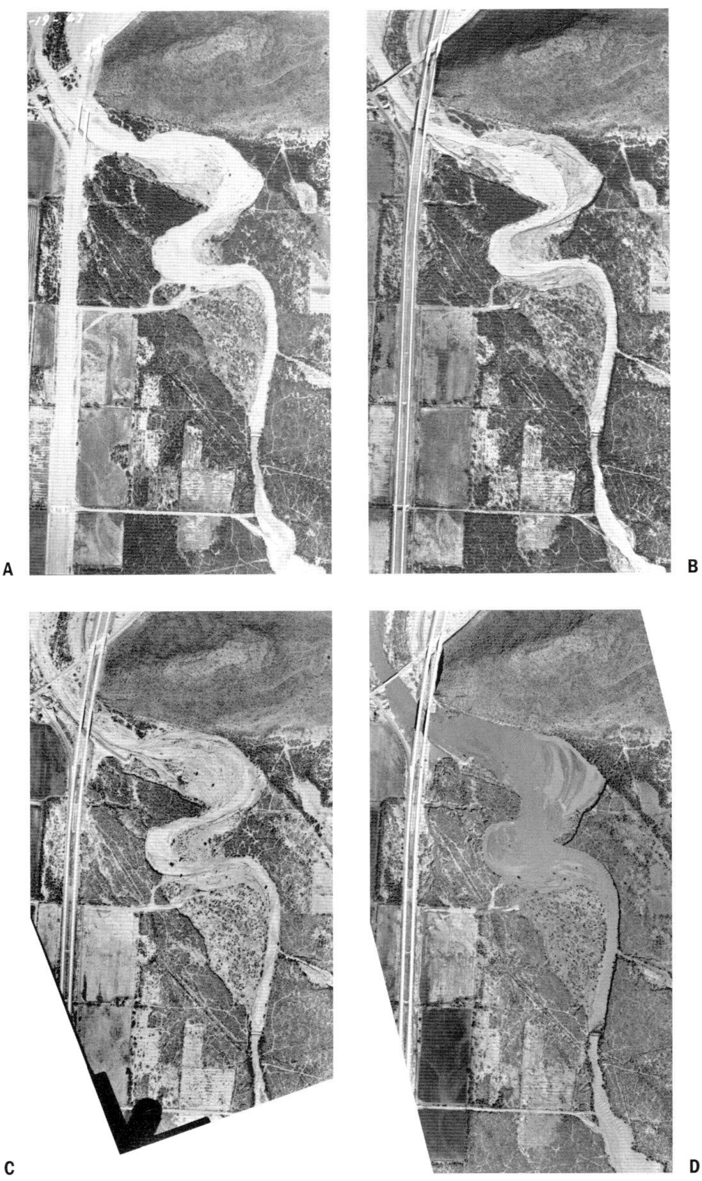

Figure 8.8.

Aerial photographs of the Santa Cruz River south of Martinez Hill. **A.** (19 August 1967) This series of composite aerial photographs shows the Santa Cruz River upstream from Martinez Hill (top). The river flows from bottom to top (south to north), and the Interstate 19 bridges (under construction) appear at the upper left. A distinctive S-loop in the river, which represents an abnormally wide section of channel apparently created by the river's interaction with Martinez Hill, appears in the upper center. Widening of the channel in this loop occurred steadily from 1936 to 1967 (according to Parker 1995b: 29). The dam associated with the surface-water diversions to the west (left) and onto Tohono O'odham Reservation fields appears at the lower right. East-west roads lead to sand-and-gravel mines in the channel that supplied fill material for the freeway construction. **B.** (7 January 1971) In only four years, the channel has narrowed significantly and straightened in the upper part of the S-loop, and riparian vegetation is becoming established on the relatively wide, inset floodplain. **C.** (3 October 1976) At this time, the channel is relatively narrow, and vegetation is well established on the floodplain between the arroyo walls. A small grove of cottonwood trees appears upstream from the Interstate 19 bridges. **D.** (10 October 1977)

Figure 8.8. (*continued*)
The 1977 flood swept through the S-loop, increasing its width to pre-1967 conditions and scouring most of the riparian vegetation that became established on the floodplain. **E.** (4 October 1983) This view, taken during the recession of the 1983 flood, shows a close-up of the Interstate 19 and San Xavier Road bridges, both of which failed during the flood. The dam was also destroyed. **F.** (7 June 1985) The Interstate 19 bridges were quickly repaired, restoring rapid travel between Tucson and Nogales, but the San Xavier Road Bridge remained closed in 1985. The channel, dry at the time of this view, is much wider and devoid of vegetation. **G.** (2005) By the early twenty-first century and its severe drought conditions, the low-flow channel of the Santa Cruz River had narrowed within its overwidened banks. Xerophytic riparian vegetation colonized the mostly dry floodplain, stabilizing it against shallow flows that occasionally overtopped its surface. (Imagery courtesy of Google Earth, accessed 15 January 2009.)

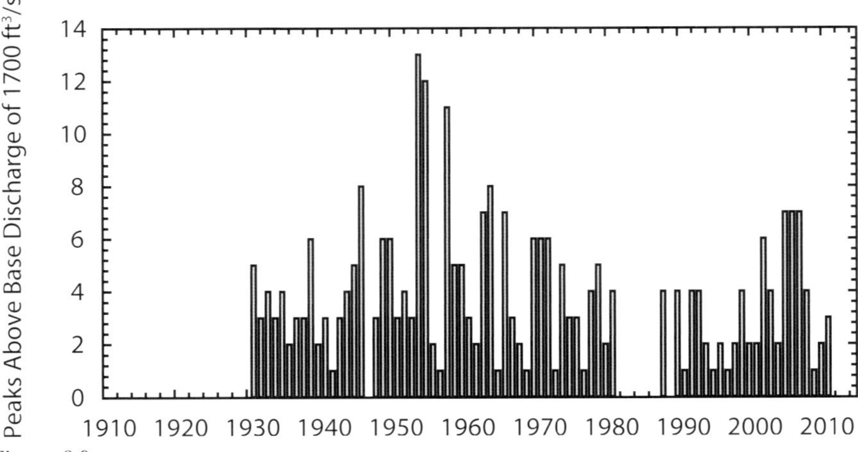

Figure 8.9.

Number of floods that exceeded a base discharge of 1,700 ft³/s for the US Geological Survey gaging station Santa Cruz River at Tucson. No peaks above base discharge occurred in 1948 or 1989; peaks above base discharge were not measured between 1982 and 1987.

midcentury drought, the reduction in precipitation primarily was during the winter months (see fig. 2.4C). Summer rainfall was another matter, and during the midcentury drought, all of the annual flood peaks resulted from summer thunderstorms (see fig. 2.7). Between 1940 and 1977, the largest flood was 16,600 ft³/s on 23 August 1961. This flood resulted from a thunderstorm over Tucson that produced over two inches of rainfall in one hour. The heavy runoff produced by this storm caused approximately $100,000 in damage to city streets, $200,000 in damage to county roads, and $25,000 in damage to automobiles caught in flash floods.[28] From the perspective of the twenty-first century, this was not a very significant flood for the Santa Cruz River, but at that time it was a large outlier in what had been a quiescent flood regime.

Although floods were not extraordinarily large during the midcentury drought, they were numerous. In terms of flood peaks above base discharge, also known as the *partial duration series*, the mid-1950s were far and away the years with the most floods (fig. 8.9). A total of thirteen and twelve floods above the base discharge of 1,700 ft³/s that occurred in the Santa Cruz River in 1954 and 1955, respectively, were the highest totals in the period of record. The third highest total—eleven—occurred in 1958, making the 1950s the decade of frequent small flood in the Santa Cruz River. The numbers of floods in 1954 and 1955 appear to have at least some precedent in the summer flooding of 1890 (see chapter 6).

While Tucson and the Santa Cruz River did not experience extremely large floods during the midcentury drought, several near misses occurred, all related to tropical cyclones (see fig. 2.6A). In August 1951, an unnamed hurricane made landfall in Baja California, Mexico, and sent considerable moisture into western and central Arizona, causing significant flooding in the middle Gila and Verde Rivers.[29] This event was the only significant flood to occur in the state of Arizona at the height of the midcentury drought.

Another storm on 26–28 September 1962, related to Tropical Storm Claudia, brushed by Tucson but caused extensive flooding on the lower Santa Cruz River, particularly downstream of Santa Rosa Wash in Green's Arroyo (see fig. 1.1).[30] Peak discharge at the Congress Street Bridge was a mere 4,980 ft³/s, but the peak discharge in Brawley Wash west of Tucson was 38,800 ft³/s, which was higher than the 1993 flood in Tucson (see chapter 9). For the same storm, the peak discharge in Santa Rosa Wash was 53,100 ft³/s, which was higher than the 1983 flood in Tucson (see chapter 9). These impressive floods did not have a high flow volume, because flow in the Santa Cruz River at Laveen peaked at only 9,200 ft³/s, suggesting that most of the 1962 flood attenuated on the Santa Cruz Flats and dispersed in the broad floodplains south of the Gila River. Finally, Tropical Storm Norma, which set records in September 1970 for rainfall and flooding in central and northern Arizona as well as New Mexico,[31] missed Tucson but generated a flood that peaked at 38,000 ft³/s in Brawley Wash. In the years that followed, the near misses would become direct hits, but despite the climatic complacency,

Tucson managers were readying the city for the floods to come.

Lateral channel change and downcutting were by no means solved by the installation of crude revetment in the 1930s. The elevation of zero flow at Congress Street dropped ten to fifteen feet between 1946 and 1980,[32] in large part because of the 1977 flood (see chapter 9) but also because of the channel constrictions imposed by bridges and their abutments. Similar downcutting is evident in comparing 1940s and 1980s photographs from the Rillito confluence to Continental.[33] Besides the Works Progress Administration (WPA) work and the ad hoc channelization created by landfill construction and sand-and-gravel mining, substantial channelization occurred at bridges, more and more of which were built across the Santa Cruz River in Tucson and elsewhere. Bridges with bank protection pinned the channel to a place, reducing the potential of meander belts to widen the channel; this was one of the greatest problems during the 1914–1915 floods (see chapter 6). Patchwork bank protection was insufficient to protect against lateral channel change during large floods, however, and many bridges would fall victim to the renewed meandering.

The Demise of the Great Mesquite Forest

By the 1940s, perennial flow had ceased in the Santa Cruz River,[34] and the velvet mesquites that once were sixty to seventy feet tall and with diameters of four feet and more had been cut for firewood and replaced by secondary growth and resprouting of mesquites that attained, at best, "a height of some twenty or twenty-five feet."[35] We will never know, but the Great Mesquite Forest south of Tucson may have been near its zenith at the time of the first aerial photographs taken in 1936 (see fig. 7.5). Its destruction was documented by ornithologists who conducted research in the bosque in the 1950s and 1960s.[36] Woodcutting and groundwater pumping had decimated the larger mesquites.[37] These tenacious mesquites clung to life as the groundwater literally dropped away from their roots, and the bird populations remained only as long as there were habitat and food offered by the bosque. By the 1960s, the Great Mesquite Forest had been reduced to a fraction of its original size by agricultural fields. During the 1970s, construction of Interstate 19 bisected the remains of what had been the most celebrated mesquite bosque in the American Southwest.[38]

The Last Bird Observations

Joe T. Marshall conducted ornithological research in the Great Mesquite Forest from 1955 to 1963, and his student, R. Roy Johnson, worked there from 1958 to 1960. Marshall kept extensive field notes and field trip lists during several dozen trips to the bosque, largely in conjunction with his work on Brown Towhees.[39] In addition to assisting Marshall with his fieldwork,[40] Johnson made numerous trips to the Great Mesquite Forest and published information on its destruction.[41] Beginning in 1958, and continuing into the early 1960s, Marshall, several of his graduate students, and others regularly visited the bosque. Other studies included Marshall's studies of owls, especially the Western Screech-Owl;[42] Patrick J. Gould's studies of the Northern Cardinal and Pyrrhuloxia;[43] Richard S. Crossin's work on breeding Song Sparrows;[44] and William G. George's work on a number of species.[45]

Numerous bird species that now have greatly reduced populations or have disappeared from the Tucson Basin remained in the Great Mesquite Forest in the 1950s. The Rufous-winged Sparrow is of special interest because several species of sparrow had either disappeared from the bosque or their populations had been greatly reduced by the 1950s, when Marshall and his coworkers began their work. They found this species and three others—the Inca Dove, the European Starling, and the Great-tailed Grackle—that had not been observed previously. The Inca Dove had gradually moved northward into southern Arizona and is now abundant around agricultural areas and in suburban Tucson and Phoenix, and the Great-tailed Grackle similarly had moved northward from Mexico into the Southwest.[46] The European Starling, an introduced species from Europe, first appeared in Arizona in the 1940s.[47]

The Rufous-winged Sparrow is a permanent resident in southern Arizona.[48] It was found regularly in southern Arizona from its discovery in 1872 to the early 1880s, especially in the Tucson Basin.[49] The last record for southern Arizona in the 1800s was a Tucson specimen taken by Herbert Brown on 7 February 1886,[50] and no further Arizona records were obtained until 1915, when it was recorded in the Coyote Mountains west of Tucson.[51] For the next fifteen years, few records were obtained, and the species was considered extirpated in Arizona.[52] In the 1930s, the American Ornithologists' Union checklist stated that it had "formerly" occurred in Arizona.[53] In the fall of 1956, a large influx of the sparrow from Mexico increased the southern Arizona population,[54] and today it is so plentiful that it frequently is encountered in the Tucson Basin, with adults and young even feeding in suburban Tucson yards. Although some have postulated that heavy grazing led to the species' earlier demise,[55] its apparent "recovery" occurred without any concerted effort by humans.

The Gray Hawk has presented a taxonomic problem for ornithologists.[56] Scattered individuals have been reported nesting as far north as the Phoenix region, and it certainly has been a breeding bird in southeastern Arizona. The Great Mesquite Forest was one of the favorite nesting areas for Gray Hawk before its destruction in the 1950s (see appendix A). In addition to F. Stephens's observations of this hawk nesting in mesquites south of Tucson in the 1890s, most ornithologists who worked in the Great Mesquite Forest, from Swarth in 1902 to Marshall in 1958, reported this species.[57] Willard found only a single bird and no nests in 1911, although he was there for only two days.[58] The final nesting record for the Great Mesquite Forest is a report of two young in a nest by W. George on 1 July 1959.[59]

Other birds observed in the Great Mesquite Forest are more rare and have not been seen since the demise of the Great Mesquite Forest. The Crested Caracara is another species that ranged into southern Arizona from Mexico, and it was more common during the late 1800s and early 1900s. It was known from Arizona as early as 19 November 1886, when Mearns collected a specimen on the Salt River in central Arizona.[60] Herbert Brown reportedly raised young taken from a Tucson nest "16 miles southwest of town."[61] It was not uncommon before 1920, but there have been only scattered records since that time, and small numbers are permanent residents, largely on Tohono O'odham lands west of Tucson.[62]

Likewise, the Rose-throated Becard is one of the rarest birds in the United States, and it was first found nesting along the Santa Cruz River. The only report of the species from the Great Mesquite Forest was of an unsuccessful nesting attempt observed in 1959, when an immature male built a nest in a cottonwood tree on the bank of the Santa Cruz River.[63] On 1 July 1959, a male was calling, and the bird was again observed on 18 July 1959. Most of the riparian trees along the Santa Cruz River were dead by 1959, and the few remaining cottonwoods were senescent.[64]

Extirpations

The decline in habitat induced by the interacting effects of groundwater withdrawal, woodcutting, agricultural encroachment, and freeway construction forced the extirpation of many species from the Tucson Basin (see appendix H). Perennial vegetation, particularly obligate riparian species, was decimated, with at least seven species lost and two species—Frémont cottonwood and black willow—significantly reduced in numbers and distribution.[65] The species that no longer occur along the Santa Cruz River in the Tucson Basin include Arizona ash, Arizona walnut, coyote willow, Mexican elder, netleaf hackberry, screwbean mesquite, and soapberry. In addition, arrowweed, not mentioned in the early species lists, was known but also is no longer present (see fig. 7.1B). As previously discussed, approximately ten species of birds were extirpated by the 1970s. All six species of fish, eight species of amphibians and reptiles, and at least two mammals no longer occur in the basin. The fish species, in particular, were killed when perennial flow disappeared by 1945.[66]

Observations of Remaining Trees

The ornithologists working in the Great Mesquite Forest were keenly aware of the decline in the forest around them and what it meant to animal populations dependent upon this once massive riparian area. Brandt, for example, wrote:

> In 1945, with Doctor [Harry C.] Oberholser, I returned to what was left of the grand mesquite forest. With sad heart, I found that every noble tree had been hacked away for firewood, leaving only the tall naked stumps as sad monuments to greed and ignorance. Only brush heaps remained where once grew the mighty mesquites. . . . The poor-burning hackberry trees are untouched, and while they are not numerous, yet they give a semblance of woodland to an otherwise cutover region of open, new-forming chaparral, basking unattractively in the sunglare. Again the heedless hand of man destroys an immeasurably old climax growth, ruining forever an exceptional monument of nature. Birds are now much scarcer in the forest, and this contrast alone offers a disheartening example of what habitat destruction does to a wild, primitive paradise. California has saved her monumental trees; in Santa Fe they restored their old, pioneer shrines. It seems as though, in Arizona where it is needed most, the fundamentals of conservation are practiced least.[67]

Aerial photography (fig. 8.8) documents the decline of the Great Mesquite Forest following what might have been its maximum extent in 1936 (see fig. 7.5). By 1967, groundwater levels had dropped well below the rooting depths of riparian trees (see fig. 7.2), reaching a depth of about eighty feet at that time. Mesquites dropped their leaves when large pumps in the vicinity were started to provide irrigation water for adjacent fields in the springs of 1959 and 1960,[68] and the last remnants of a cottonwood-willow forest were dying. Construction of Interstate 19, which began in the early 1960s, was nearing completion, and roads ran east from the freeway corridor into the Santa Cruz River and sand-and-gravel mines built where the forest once prevailed (fig. 8.8A). By 1971 (fig. 8.8B), the freeway construction was complete, the dirt roads had narrowed or were removed, and floods in the Santa Cruz River eliminated the sand-and-gravel mines, but the damage had been done.

The tipping point past which recovery was impossible for these biotic communities rapidly came and went in the early 1970s. The final blow to the forest came with the completion of Interstate 19. An obituary for the Great Mesquite Forest was written by ornithologists R. Roy Johnson and Steven W. Carothers in 1982. They summarized the situation, referring first to Arnold's observations in 1940 and then to conditions in 1982:

> Trees 20–25 feet high are poor substitutes for the original stand with trees exceeding 60 feet in height. . . . Today "The Grand Mesquite Forest" looks like a depauperate thorn scrubland. . . . [T]o the problems caused by woodcutters, "progress" has added (a) a lowered watertable due to excess groundwater pumping for domestic and agricultural use; (b) additional erosion and habitat loss from increasing farming activities; and (c) Interstate 19 constructed longitudinally through the heart of the old forest paralleling the Santa Cruz [fig. 8.8]. The passing of the forest was accompanied by the death of most cottonwoods and other trees along the river. And, as one can easily guess, recreational values are nil while it is difficult to discuss water quality in a river which has ceased to flow. If the sad history of Santa Cruz River were an isolated situation, ecologists would not be waving so many red flags of alarm.[69]

Documenting Changes in the Desert

During the heart of the midcentury drought, Raymond M. Turner arrived in Tucson as an assistant professor at the University of Arizona. His initial work focused on a variety of ecological issues with desert plants, following in the tradition of the Desert Botanical Laboratory west of downtown. In the late 1950s, he was joined by J. Rodney Hastings, a young graduate student and mayor of his hometown of Hayden, Arizona. Together they began a long-term project that persists into the twenty-first century. Their tools were archives, old photographs, and permanent plots, and their goal was documenting long-term changes in the landscapes of the Sonoran Desert.

Hastings, in particular, saw the value in old photographs and archives. He believed that a long-term perspective was required to begin to understand changes in the desert, and particularly changes in desert rivers. In 1959, he published a highly influential paper documenting changes in riverine conditions in southern Arizona.[70] He was the first to realize the magnitude of changes that could be gleaned from perusal of archives, and he had the long-term vision to begin his own monitoring program using photographs. For the Santa Cruz River in Tucson, he went to most river crossings, both bridges and fords, and took upstream and downstream images, capturing the conditions present along this river in 1958–1959.

For his doctoral dissertation, Hastings wanted to use perennial vegetation, and changes in its distribution on landscapes, to evaluate bioclimatic conditions and controls in the Sonoran Desert. Using old photographs and matching them with photographs taken with medium-format cameras, Hastings and Turner toured southeastern Arizona and Sonora, Mexico, securing photographic matches through an elevation range from the lower oak woodland to the Sea of Cortez in Mexico. Although the Santa Cruz River was not one of the prominent subjects of their work, at least not in Tucson, they laid the groundwork required to evaluate long-term changes in desert rivers. Many efforts later sprang from the seeds they planted,[71] including this book.

9

Arroyo Management in the Time of Floods, 1976–1995

For Tucsonans, many of whom immigrated to southern Arizona from wetter areas where even the smallest creek flows year-round, the Santa Cruz River in 1976 was little more than a dry trench lined with garbage dumps. To county and city officials entrusted with local floodplain management and aware of the arroyo legacy, the river had become an accident waiting to happen. Encroachment of buildings and infrastructure adjacent to the arroyo through downtown Tucson was substantial, setting up the potential for a major disaster should the dormant river roar to life. On 12 September 1981, the *Arizona Daily Star* reported the conclusions of a preliminary report by the US Army Corps of Engineers that "damage would total at least 321 million dollars if a 100-year flood hit the Tucson area." This warning was prescient.

Tucson's attitudes toward flood hazards had been lulled by low-flow conditions during the midcentury drought. There had been no recent floods like those in 1887, 1890, 1905, and 1915, and the collective memory of the populace was of a dusty, barren ditch occasionally carrying some storm runoff. The flood of December 1914 was not exceeded until August 1961; it was exceeded again in December 1968, October 1977, October 1983, and January 1993 (see fig. 2.7). After the 1977 flood, floodplain managers began installing bank protection in the form of soil cement through the center of Tucson.[1] The peak discharge of the 1983 flood, the flood of record, was 3.5 times greater in magnitude than the 1914 peak and 2.3 times greater than the previously accepted 100-year flood. The flood damage in southern Arizona was a half-billion dollars in 1983;[2] after that experience, bank protection was installed throughout the Tucson metropolitan area and beyond.

Climate in the final three decades of the twentieth century wreaked havoc on rivers in southern Arizona, and the epicenter may well have been the Santa Cruz River. The floods from 1977 through 1993 redefined its flood hazard and forced floodplain managers to constrain its banks. In addition, this period of increased flood frequency influenced the national perception of floodplain management on ephemeral streams in the Southwest, spurring a reevaluation of everything from how flood hazard is regulated to the critical assumption of stationarity that underlies recommended methods of flood-frequency analyses. How design and regulatory floods were determined would be completely changed.[3] Although some have claimed that flood hazards were increasing in response to urbanization, the period of large floods abruptly ended in 1995, suggesting that a brief period of wetter conditions was the real culprit.

The midcentury drought lingered into the 1970s after its low point in the 1950s. Although several wet years in the 1960s seemed to break the drought, more dry years followed. Then, beginning in 1976, changes began worldwide that led to a different climate regime, referred to as the late twentieth-century wet period or the late twentieth-century pluvial.[4] Winter rainfall would increase, as would the intensity of storms, and the tropical cyclones that seemed to miss the Santa Cruz River on both sides scored direct hits on its watershed. Tucson managers tried hard to be prepared for the resulting floods, but they were unprecedented in the history of the region.

Flood Frequency and Floodplain Management

In the 1970s, local authorities were making every effort in responding to federal floodplain legislation. The National Flood Insurance Act, passed by Congress in 1968, offered below-cost property insurance for buildings in flood-prone areas.[5] This national program focused on inundation, not the lateral channel changes that plagued Tucson in the late nineteenth and early twentieth centuries. These insurance subsidies were an inducement for local authorities to restrict new construction in the most flood-prone areas and to compel the use of better siting, design, and construction practices. The gamble was that the subsidies would cost the national treasury less than the disaster-relief grants and reconstruction loans handed out after floods.

In 1973, Congress raised the stakes by passing the Flood Disaster Protection Act. Communities not participating in the National Flood Insurance Program became ineligible for federal disaster relief. Federally insured lenders had to require borrowers to purchase flood insurance for loans secured by floodplain property. The Federal Emergency Management Agency (FEMA) was formed to administer the federal flood insurance programs and oversee floodplain insurance studies by each community. Each study would identify the extent of flood hazards; determine flood-flow frequencies for discharges having 10-, 50-, 100-, and 500-year recurrence intervals; develop water-surface elevation profiles for each of those discharges; and produce a map of the floodway that would contain the 100-year discharge. Again, that map depicted inundation assuming a stable channel geometry and ignored any lateral changes associated with a shifting channel.

By the mid-1970s, methods of flood-frequency analysis had evolved with the advent of statistical procedures.[6] In 1967, a federal interagency work group recommended a log-Pearson type III distribution as a uniform technique for determining flood frequencies from a series of annual flood peaks.[7] This mathematical function describes a three-parameter distribution based on the mean, variance, and skew coefficient of the annual flood series (also known as the *moments*) and was considered by the working group to be superior to other distributions. The tail of the log-Pearson type III distribution—where all the interesting low-frequency floods were modeled—could bend upward or downward to catch the extreme events and incorporate them into one seamless distribution of all annual floods experienced on a river. In other words, the common floods could be used to estimate the magnitude of the extreme events.

Other assumptions were at play as well. The method assumes that the hydrological characteristics of the stream have not changed over the period of flow measurements; in other words, the moments of the distribu-

tion are *stationary*, or time invariant. This assumption is invalid where significant watershed or climatic changes produce trends—either increases or decreases—in the annual flood series. If, for example, a major forest fire occurred that increased runoff, the use of the log-Pearson type III or any other parametric distribution would be inappropriate because the past data would not reflect future conditions. Because climate was assumed to be unchanging (time invariant) with respect to generation of floods, the assumption of stationarity seemed appropriate when this method was implemented. However, if the mean and variance of the flood-frequency statistics change with time, the time series is considered to be *nonstationary*.[8]

Despite these assumptions and several others, log-Pearson type III remains the flood-frequency distribution used by all federal agencies in the early twenty-first century and was the technique used to assess flood frequency in southern Arizona at the start of the late twentieth-century wet period.[9] The method is codified in a publication known as Bulletin 17B,[10] and it has been implemented in several computer programs used by hydrologists to estimate flood hazard. Bulletin 17B allows for use of other techniques if the stationarity assumption can be shown to be invalid. Those other techniques are only vaguely described, but the potential of an alternative method to a log-Pearson type III distribution for assessing flood frequency, particularly in light of demonstrable nonstationarity in flooding, was applied to the Santa Cruz River.

Once a regulatory flood was established, whether a 100-year event or an alternative one because of nonstationarity concerns, hydrologists had to estimate how much area would be inundated by the flood. The standard flood-hazard assessments use hydraulic models that rely on rigid banks and beds in the channel and essentially calculate the overbank area that a large event would flood. While this works for many rivers in many places, most of the arroyos, deeply incised in their floodplains, shift laterally during floods, eroding meander bends. Floods also carry large sediment loads, which can raise a channel bed or banks and alter flow characteristics. In southern Arizona, as well as in most deserts worldwide, the hazard is not overbank inundation: it is lateral channel change. Tucsonans had vividly observed this phenomenon during the 1914–1915 floods at Congress Street, but that was sixty years before modern floodplain management began. It was forgotten from the collective memory, and the few hydrologists in Tucson who knew what happened could not accommodate that history in their hydraulic models.

In 1973, the Arizona legislature passed House Bill 2010, naming the governing body of each city, town, and county the floodplains board for their jurisdiction. The bill required each board to adopt floodplain regulations and delineate floodplain properties. Pima County entered the Emergency Phase of the National Flood Insurance Program and enacted its first floodplain management ordinance in 1974. What was required was a regulatory flood, which FEMA designates as the 100-year flood, so that the county would know what area of the floodplain to regulate and what, if any, bank protection would be required.

In the 1930s, F. C. Knapp arrived at a value of 12,500 ft^3/s for the 100-year flood at Congress Street based on twenty years of record.[11] Some forty-five years later, FEMA adopted a 100-year flood discharge of 30,000 ft^3/s for the purpose of floodplain mapping,[12] despite objections by some local hydrologists that even this value might seriously underestimate actual flood hazards. After the 1977 flood and before October 1983, there had been much controversy about the techniques for computing a 100-year flood and the values derived from them (table 9.1). Using various techniques, different hydrologists recommended 100-year flood values

Table 9.1. **Estimates of the 100-year flood on the Santa Cruz River at Tucson, Arizona**

Reference	Years of Record	Method or Probability Distribution[a]	100-Year Flood (ft^3/s)
US Army Corps of Engineers (1972)	—	(1)	45,200
Roeske (1978)	1915–1975	(2)	20,300
	1915–1975	(3)	22,600
Malvick (1980)	1915–1978	(3)	63,900
Federal Emergency Management Agency (1982)	1915–1978	(2)	30,000
Boughton and Renard (1984)	1915–1979	(4)	20,200
	1915–1979	(2)	23,500
	1915–1979	(5)	77,000
Michael Zeller (Simons and Li Associates, written commun., 1984)	—	(6)	50,100
Eychaner (1984)	1915–1981	(2)	22,100
	1915–1981	(3)	23,200
Reich (1984)	1960–1984	(2)	54,000
	1960–1984	(7)	96,400
	1962–1984	(2)	50,100
	1962–1984	(7)	98,200
Ponce et al. (1985)	—	(8) 24-hr	58,600
		(8) 48-hr	67,100
		(8) 96-hr	47,000
Hirschboeck (1985)	1950–1980	(2)	26,000

Source: From Webb and Betancourt (1992: 5).

Note

Estimates made by previous investigators after 1970.

[a] Methodologies and probability distributions are as follows:

(1) Curve, comparison with floods in other watersheds in southern Arizona.

(2) Log-Pearson Type III distribution, method of moments fitting.

(3) Log-Pearson Type III distribution plus regression analysis.

(4) Log-Pearson Type III distribution plus envelope curve.

(5) Log-Boughton distribution, method-of-moments fitting.

(6) Rain, estimated from 100-year rainfall.

(7) Log-Extreme Value distribution, method-of-moments fitting.

(8) Model, estimated from rainfall-runoff model with 100-year, 24-, 48-, and 96-hour duration storms. This value is currently being used in Pima County for compliance with Federal Emergency Management Agency regulations.

ranging from 20,200 to nearly 100,000 ft³/s. In comparison, use of the log-Pearson Type III distribution and annual flood peaks from 1915–1981 yielded a 100-year flood of 22,100 ft³/s,[13] which was less than the peak discharge reached during the October 1977 flood.[14]

Between 1975 and 1980, flood-hazard areas were identified in Arizona communities, and flood-hazard boundary maps were issued for those in Pima County. The Federal Insurance Administration held meetings with local authorities to determine specific areas for flood insurance studies.[15] Pima County adopted a written policy requiring building setback requirements of three hundred feet for new residential units and one hundred feet for rental units along major watercourses.[16] No regulatory consideration was given to the problem of lateral channel change. This regulatory plan would be severely tested in the coming decade.

Hurricane Heather and the October 1977 Flood

On 4 October 1977, a tropical depression in the Pacific Ocean off the west coast of Mexico began to gather strength in its counterclockwise circulation. As noted in chapter 2, tropical cyclones—including tropical storms and hurricanes—commonly develop in the summer and fall in the eastern North Pacific Ocean. Tropical Storm Heather evolved into Hurricane Heather as it coursed parallel to the west coast of Mexico, then up the west coast of Baja California, finally ceasing its existence as an organized storm midday on 7 October. This storm never made landfall; most tropical cyclones in the eastern North Pacific Ocean begin and end at sea. If it were not for a developing cutoff low-pressure system off of the California coast northwest of Heather's track, this hurricane would have just been another statistic in tropical meteorology.

Instead, the advancing cutoff low sheared off the moisture from Hurricane Heather and steered it into the southwestern United States, where it caused widespread rainfall for three days.[17] In southern Arizona and New Mexico, rainfall was heavy between 6 and 10 October, and as much as fourteen inches fell in the highlands surrounding Nogales, causing significant flooding on the Santa Cruz River and its tributaries.[18] Although only two inches of rainfall fell in Tucson, the city endured its first major flood since 1914–1915 and its first taste of flood-hazard mitigation in many decades.

Precipitation was highest in the headwaters of the Santa Cruz River, and the peak discharge attenuated with distance from the maximum source of precipitation.[19] Peak discharges of 31,000 to 26,500 ft³/s were estimated at Nogales and Continental, respectively. At Tucson, the gaging station at Congress Street was destroyed, and the flow record had to be reconstructed from high-water evidence along the banks. The flood peaked in the early morning hours of 10 October at 23,700 ft³/s. Floods were about as large on the San Pedro and Gila Rivers, and about $15 million in damage occurred throughout the region, more than half in agricultural losses. Although floodwaters stayed in the arroyo through Tucson, one bridge was destroyed in town and several others were damaged.[20] Downstream at the town of Rillito, the flood left the channel and spread over a 4,000-foot width of floodplain before recombining into one channel. Downstream from Marana, Greene's Canal captured most of the floodwaters, but even farther downstream, water spread over a three-mile width near Chuichu. The floodwaters reached the Gila River, with a peak discharge of around 2,000 ft³/s arriving on 13 October, making the 1977 flood one of the few historic events to flow continuously from headwaters to terminus at Laveen.[21]

The 1977 flood underscored the flaw in FEMA floodplain regulations aimed at reducing losses caused by inundation. If the low values of the 100-year flood discharges are accepted (table 9.1), the 1977 flood was greater than a 100-year event. The floodwaters did not leave the channel in Tucson; instead, they widened the channel, much like the floods in 1914 and 1915. An average of seventy-five feet of widening occurred in Tucson over a one-mile reach during the 1977 flood, with a maximum increase of two hundred feet at one point.[22] Although no houses were destroyed in Tucson, the threat was very real, and no one with regulatory responsibility for flood hazard in Tucson missed this lesson. Because it was too late to move residences and businesses off the floodplain, it was time to start the expensive task of installing bank protection (figs. 9.1 and 9.2).

In 1976, the City of Tucson undertook plans to revitalize its downtown by constructing a small community of 1,100 homes on the west bank of the Santa Cruz River upstream from St. Mary's Road. The overall project was called Rio Nuevo,[23] a name ironic in the twenty-first century because this redevelopment plan continues and hardly is new anymore. City ordinances required the lowest finished floors of the houses to be one foot above the 100-year water-surface elevations, which remained unchanged despite the 1977 flood. Channel stabilization, in the form of eight-foot-thick

Figure 9.1.

The Santa Cruz River upstream from A Mountain. **A.** (19 August 1958) This upstream view shows the Santa Cruz River south of A Mountain. Near the end of the midcentury drought, aided in this reach by revetment, the Santa Cruz River had a relatively narrow channel with a floodplain between its arroyo walls. Farmland still was the dominant land use in this reach. (J. R. Hastings F IX 9, courtesy of the Desert Laboratory Collection.) **B.** (16 August 2001) Soil cement was installed in this reach in 1982, and the channel was swept bare during the 1983 and 1993 floods. Following the latter event, channel floodplains reestablished with xerophytic riparian vegetation, including mesquites, paloverdes, and native grasses. As can be seen upstream from the 22nd Street (now Starr Pass Road) Bridge, the floodplains can occupy as much as two-thirds of the width between the soil-cemented banks. (D. P. Oldershaw, Stake 2496b, courtesy of the Desert Laboratory Collection.)

Figure 9.2.

The Santa Cruz River at Silverlake Road. **A.** (19 August 1958) This upstream view was taken from the bridge over the Santa Cruz River at Silverlake Road. The channel is relatively narrow and shallow at this time, but a floodplain has formed in the bend at the far center. Revetment, installed in the 1930s in an attempt to control channel widening, appears under the trees on the left. The trees along the banks appear to be a combination of mesquite and nonnative Athel tamarisk. (J. R. Hastings F IX 16, courtesy of the Desert Laboratory Collection.) **B.** (16 August 2001) At the time of this photograph, this was the upstream end of the channelized reach through Tucson. The contrast between the channelized banks (right foreground) and the arroyo walls (right background) is striking, but also notable is the development of the vegetated floodplain on the left and in the center. The channel deepened by perhaps six feet and widened considerably in the intervening years. (D. P. Oldershaw, Stake 2492, courtesy of the Desert Laboratory Collection.)

retaining walls of soil cement to replace the river-banks, was required to contain a design flow of 45,000 ft³/s, and all riparian vegetation was removed from the banks and bed. The performance of the soil cement was problematic, particularly at the upstream and downstream ends where bank protection abutted natural arroyo walls. Bed scour during the 1977 flood was substantial near Nogales (it was not estimated at Tucson), and the threat of undercutting of the soil cement during floods was mitigated by installing concrete sills, also known as grade-control structures, across the channel. Many reaches away from the city center were not altered (fig. 9.3).

Budget cuts, combined with faith in flood-frequency estimates obtained from relatively long gaging records, created another controversy. On 30 September 1981, the US Geological Survey ceased operation of the gaging station on the Congress Street Bridge after sixty-six years of operation. Only a year later, the reach around the bridge was channelized and protected with soil cement (fig. 9.4). At this time, the US Geological Survey was using optimization techniques to design gaging-station networks, and the history of annual floods at specific sites was considered to be irrelevant because floods were assumed to be temporally independent, one of the assumptions of the Bulletin 17B technique. Before October 1983, the contradictory estimates of flood hazard had confused the general public and had fostered some skepticism about the federal program among those charged with enforcing floodplain legislation. In addition, there was no gaging station to measure the coming deluge.

Tropical Storm Octave and the 1983 Flood

Unusual weather conditions developed over southern Arizona in late September 1983. After a typical monsoon season, a weak cold front trailed across the southern Great Basin, part of a long, southwest-to-northeast trough in the upper atmosphere that channeled tropical moisture from the Pacific Ocean into northwestern Mexico and southern Arizona. By 24 September, when it rained at almost every climate station in southeastern Arizona, the month had already been unusually wet. Gaging stations still in operation on the Santa Cruz River registered peak discharges of 5,600 ft³/s at Continental and 2,750 ft³/s at Cortaro Road on 22 September. These relatively small floods created saturated channels through the Tucson reach.

When Tropical Storm Octave developed on 27 September,[24] conditions were ideal for recurvature and persistent advection of large amounts of moisture into southern Arizona. In September, sea-surface temperatures were abnormally high along the west coast of North America, which is typical for the end of an El Niño period. The warm waters favored northwesterly travel of tropical storms from their southern points of origin up to positions off the Baja California peninsula. A deep, low-pressure trough formed off the West Coast that was timed perfectly to guide the oncoming cyclone. Tropical Storm Octave never made landfall, but its moisture was swept into the trough moving across the southwestern United States, and the combined moisture converged on southern Arizona.

For five days beginning on 28 September, rainfall was widespread, persistent, and in some cases intense throughout southern Arizona. Of forty-five official climate stations in southeastern Arizona, 20 percent recorded more than one inch of rainfall on 28 September, 30 percent on 29 September, 60 percent on 30 September, 50 percent on 1 October, and 75 percent on 2 October. In the Santa Cruz watershed, total rainfall for these five days exceeded six inches at most stations; in the higher mountains, on the west slope of the Santa Rita Mountains, and in the Tucson area, the totals exceeded eight inches.[25]

On 30 September, peaks of 5,610 and 10,400 ft³/s were measured at Continental and Cortaro Road, respectively; a daily discharge of 18,000 ft³/s was recorded at Cortaro Road on 1 October.[26] The Santa Cruz continued to rise in the middle of the night, reaching a peak of 45,000 ft³/s just before 3 a.m. at Continental and 65,000 ft³/s at about 6 a.m. at Cortaro. A peak discharge of 52,700 ft³/s had passed under the Congress Street Bridge sometime between 3 and 6 a.m. on 2 October (fig. 9.5); this peak had to be estimated, not measured, because the gaging station had been discontinued in 1981. On the Rillito River, the peak discharge was estimated to be 29,700 ft³/s at another discontinued gaging station, and fortunately for the residents of Marana downstream on the Santa Cruz River, the peak discharges on the Santa Cruz and Rillito Rivers were not coincident.

Four people died during the flood, underscoring the lack of hazard awareness and early warning systems. Damage owing to lateral channel change along the Rillito River, which had little bank protection other than at bridge crossings, was extensive.[27] In total, 154 residential units were destroyed, with many more damaged.[28] A video of a building falling into the raging

Figure 9.3.
The Santa Cruz River at the Drexel Road ford. **A.** (19 August 1958) Near the end of the midcentury drought, the Santa Cruz River was a relatively narrow trench within its former floodplain south of town. An active floodplain that appears on either side is covered with xerophytic riparian vegetation, which appears to be burrobrush; mesquite grows on the top of the arroyo walls. A powerline crosses the arroyo, following the alignment of Drexel Road. (J. R. Hastings F IX 21, courtesy of the Desert Laboratory Collection.) **B.** (16 August 2001) Channel widening has been substantial in this reach, but floodplains are still present on the left. At the time of this photograph, the river was not channelized in this reach. (D. P. Oldershaw, Stake 2491a, courtesy of the Desert Laboratory Collection.)

A   B   C

Figure 9.4.
Aerial photographs of the Santa Cruz River at the Congress Street Bridge. **A.** (22 February 1933) This high-altitude aerial photograph has low resolution at this amount of magnification, but it shows a wide, barren Santa Cruz River (flowing bottom to top) in the reach leading to the Congress Street Bridge in downtown Tucson (top). Congress Street is only two lanes wide. **B.** (23 July 1953) During the midcentury drought, the channel narrowed even as Congress Street was widened and paved, as seen in this view. Riparian vegetation is now apparent within the arroyo walls. **C.** (13 October 1960) In-stream mining of sand and gravel is impacting this reach. **D.** (19 August 1967) In-stream sand-and-gravel mining

floodwaters remains an enduring reminder of this flood; it is still shown on the anniversary of this flood. The intense rainfall also triggered record flows elsewhere in the Gila River basin, wreaking havoc throughout southern Arizona and accruing about a quarter of a billion dollars in damages.[29]

Flood damage along the Santa Cruz River resulted from overbank inundation, with some channel widening in the reach from Continental to the southern end of the San Xavier Indian Reservation and downstream from Cortaro Road to the confluence with the Gila River. In the area once occupied by the Great Mesquite Forest, channel widening in the arroyo was extreme just upstream from Martinez Hill, contributing in part to the failure of the Interstate 19 southbound bridge (see fig. 8.8E). In the Tucson metropolitan area, only

D **E** **F**

Figure 9.4. (*continued*)
has apparently ceased, but the channel has not fully recovered in the lower part of this aerial photograph. However, the reach just upstream from the bridge appears to be lined with riparian vegetation. E. (8 April 1970) Floodplain development has removed riparian vegetation on the right side of this view. F. (17 April 1976) The channel is relatively narrow and lined with riparian vegetation in this view, taken a year and a half before the

one bridge—the Congress Street Bridge—withstood the flood, while all others failed or required extensive repairs. The worst damage occurred because of lateral erosion that eliminated one or both abutments or vertical erosion that undercut bridge piers. As in 1977, flood flows did not leave the channel through metropolitan Tucson, where channel cross sections were stabilized with soil cement or had become large enough to accommodate the discharge. Instead, most of the damage in the Tucson area was caused by widening of meander bends with no bank protection, a repeat of the stories in 1915 and 1977.[30]

Bank protection, such as the soil cement installed only a year before between St. Mary's Road and Congress Street, performed exceedingly well except at locations where soil cement abutted unprotected banks.[31] Wherever bank protection was piecemeal or nonexistent, lateral erosion resulted in collapse of homes into the river and destruction of bridge approaches. In places, lateral erosion moved the thalweg beyond the

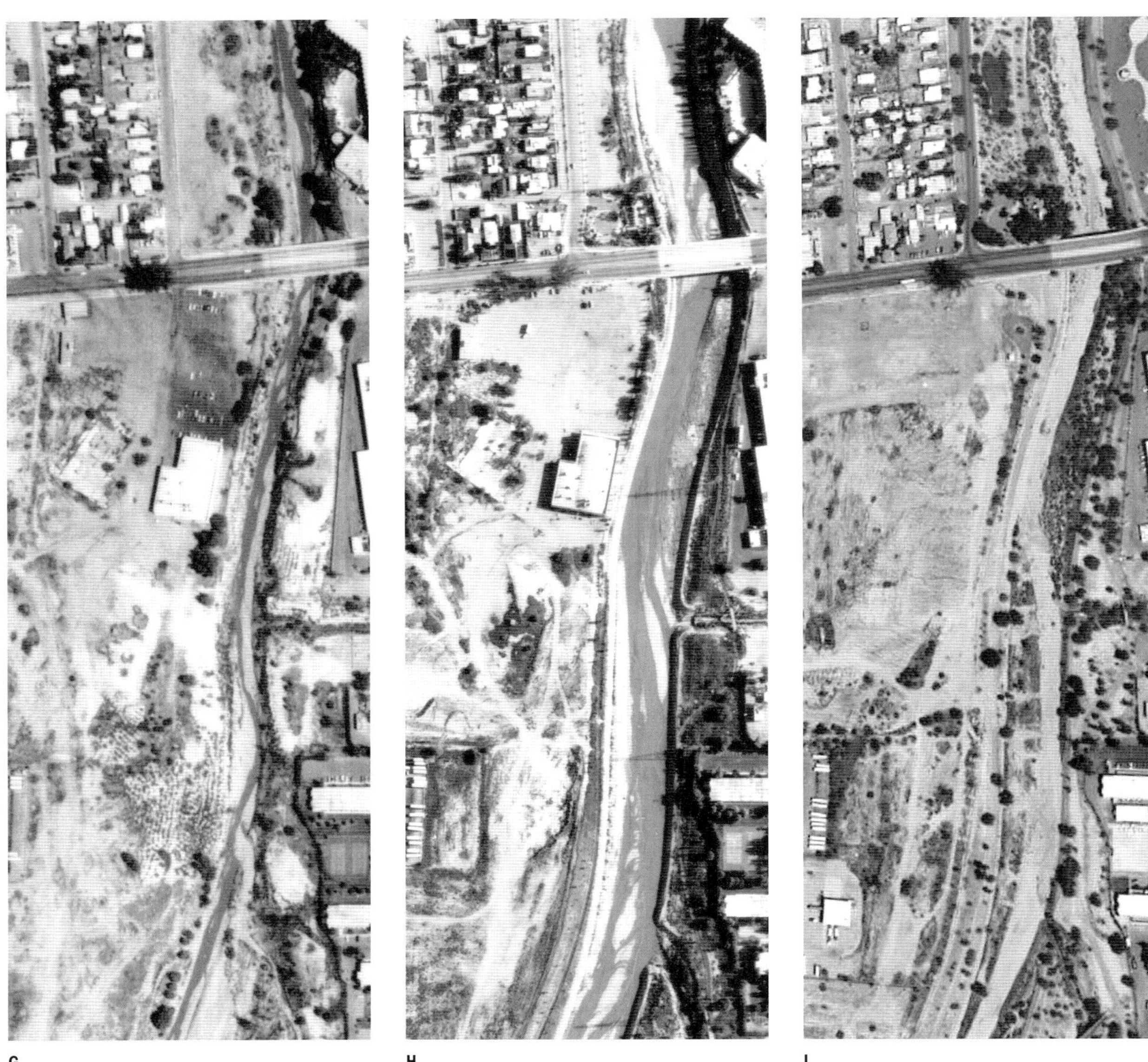

G   H   I

Figure 9.4. *(continued)*
1977 flood. **G.** (8 September 1978) The 1977 flood has widened the channel considerably, removing much of the riparian vegetation that had become established. **H.** (11 July 1984) Two factors—the installation of soil cement and the 1983 flood—created a channel that is much wider and devoid of riparian vegetation. **I.** (2005) Within its soil-cemented banks, the Santa Cruz River has narrowed by as much as half as sedimentation on floodplains and establishment of riparian vegetation has decreased flow conveyance. (Imagery courtesy of Google Earth.)

boundaries of the 100-year flood and even outside of the 500-year floodways.[32] Along other reaches, however, the channel contained the peak discharge of 2 October despite predictions that discharges greater than 30,000 ft³/s would overtop the banks. The discrepancy was thought to be due either to errors in hydraulic modeling or in mapping of surface-water elevations, or to an increase in channel conveyance during the flood as a result of widening and (or) deepening.[33] In reality, the flood probably increased its flow area in channels by scouring the bed, which then backfilled before the floodwaters receded.

One of the more impressive erosional features from the 1983 flood resulted from upstream migration of the headcut at Greene's Canal near the boundary between Pinal and Pima Counties (fig 9.6; see fig. 1.1). From 1915, when the headcut first formed, to 1983, the Soil Conservation Service and local farmers had managed to restrict

Figure 9.5.
(2 October 1983) Photograph of the 1983 flood near its peak discharge of 52,700 ft^3/s at the St. Mary's Road Bridge.
(J. L. Betancourt, Stake 1320, courtesy of the Desert Laboratory Collection.)

further erosion to a relatively small area north of the county line. However, the 1983 flood, which approached a width of four miles along the lower Santa Cruz River, either overtopped or went around protective measures and cascaded into the headcut, which eroded upstream along what previously was a broad, shallow river channel. The headcut moved upstream and around most of the structures installed to restrict upstream migration. Other headcuts formed upstream near Marana.[34]

Once again, the Santa Cruz River flowed continuously from headwaters to terminus. The magnitude of this flood and its effects on the hydrology of southern Arizona were unprecedented in the recorded history of the region. The 1983 flood and the ones that preceded it recharged the alluvial aquifers along the Santa Cruz River and its major tributaries in Santa Cruz County, temporarily reversing the declines that had occurred during the midcentury drought, and water-level rises of as much as forty-seven feet were measured in the severely depleted Tucson Basin (see fig. 7.4). The major tributaries in Santa Cruz County were perennial for about eighteen months after October 1983, something that had not occurred since 1939.[35]

In the wake of the 1983 flood, there was little question in the minds of any hydrologist familiar with this river that the 100-year flood discharge of 30,000 ft^3/s adopted by FEMA was inappropriate. Using the alternative methods clause of Bulletin 17B, and relying on a somewhat controversial rainfall-runoff model,[36] both the City of Tucson and Pima County opted to designate a "regulatory flood" and a "design flood" with discharges of 60,000 ft^3/s and 70,000 ft^3/s, respectively. These regulations, adopted by January 1985, were implemented for the Santa Cruz River between San Xavier and the Rillito River confluence and were needed to initiate infrastructure repair and planning for additional bank protection. The regulatory flood was designated to map floodway boundaries required by the National Flood Insurance Program, and the design flood was used to engineer bank protection and bridge piers.

Pleased with the stabilized channel's performance during the 1983 flood, while also alarmed at the damage at the beginnings and ends of bank-protected reaches, Pima County accelerated channel stabilization with soil cement throughout the Tucson Basin in the late 1980s. Installation of bank protection reportedly cost between one

Figure 9.6.
(14 June 1988) Upstream aerial view of the active headcut of the Greene's Canal arroyo, about three miles upstream of the canal's
diversion point from the Santa Cruz River, which runs vertically through the photograph. During the October 1983 flood, the overflow,
one to three miles wide in this area, cascaded into the headcut, at right, promoting headward as well as lateral erosion. With continued
upstream migration of this headcut during future floods, sediment eroded from the Santa Cruz arroyo–Tucson Basin reach and
deposited in lower reaches between 1890 and 1990 was later transported farther downstream. (J. L. Betancourt.)

and three million dollars per mile.[37] The US Geological
Survey, prompted by its own hydrologists, reestablished
the gaging station at the Congress Street Bridge on 1 October 1987, this time at a site along a soil-cemented bank
where the gaging station might survive the battering effects of another major flood. Annual flood peaks for the
years without a gaging record were estimated from direct observations or indirect evidence. Those responsible
for flood hazard mitigation in Tucson and Pima County
were determined that there would be no repeat of 1983,
and they were severely tested again only a decade later.

The January 1993 Floods

El Niño conditions prevailed in the eastern North Pacific
Ocean in the fall and winter of 1992–1993, setting up
ideal conditions for above-average rainfall. Winter

began with high precipitation in December, when three
times the average monthly precipitation occurred during relatively cold weather. Then, in early January and
proceeding into late February, a series of warm storms
passed through the southwestern United States.[38] In
Arizona, rainfall on the snowpack built up during the
December storms generated extremely large floods
statewide, rivaling the last-known flood of such regional extent that occurred in 1891. Floods of record
were established at thirty-seven gaging stations in the
state.[39]

Although the drainage basin, and particularly the
contributing area to the Rillito River, was saturated by
early January, the Santa Cruz River itself was impacted
by only one of four major storm periods in January and
February 1993: the storms of 13–19 January.[40] Runoff
from about six inches of rain just north of Nogales produced a peak discharge of 37,400 ft³/s at the Congress

Figure 9.7.
(18 January 1993) Upstream view of the 1993 flood at approximately its peak discharge of 37,400 ft^3/s. The Congress Street Bridge appears in the background. (J. T. C. Parker, courtesy of the US Geological Survey.)

Street Bridge (fig. 9.7). Although it was the second-highest peak discharge known for the Santa Cruz River, no lives were lost, no bridges were damaged or destroyed, and no bank failures were reported within the city limits. Inundation again was a problem downstream from Tucson, primarily damaging agricultural lands, and the river once again flowed for its entire length to the Gila River.

In fifteen years, Tucson's Santa Cruz River had three floods, all of which exceeded the 100-year flood that was estimated before 1977. Flood-mitigation efforts, piecemeal or hinted about before 1977, came together for the 1993 floods. No lives were lost, and emergency responders finally were on the same page: following disorganization in 1983, Tucson's emergency management agency merged into the Pima County Sheriff's Department, allowing disaster coordination between city and county, as well as providing an effective disaster-response structure within local government.[41] All the focus was on floods, bank protection, and disaster response, and only a few thought about what might have been. If the massive bosque still remained upstream from Tucson, would the flood peaks between 1977 and 1993 have been less?

After the Deluge

Nonstationarity in Flood Frequency

Imagine, if you will, the plight of a poor floodplain manager, a fictional character who was hired in 1970 to estimate flood hazard for the Santa Cruz River through downtown Tucson. Just consider what this person would have faced: from 1970 through 1993, the 100-year flood kept rising the longer he or she stayed on the job (fig. 9.8), increasing from 20,870 ft^3/s to 38,180 ft^3/s in twenty-three short years, essentially the length of most careers. This 83 percent rise in the magnitude of the 100-year flood was outside of the 95 percent confidence limits in the estimated 1970 discharge. How could this happen if the annual flood series for the Santa Cruz River were stationary? How can any individual, group, or agency possibly assess the appropriate magnitude of flood hazard that requires mitigation?

Floods in the Santa Cruz River from 1977 to 1993 did not occur in isolation from the hydrology of the rest of the region. In central and southern Arizona between 1977 and 1995, a total of six discrete events—and this

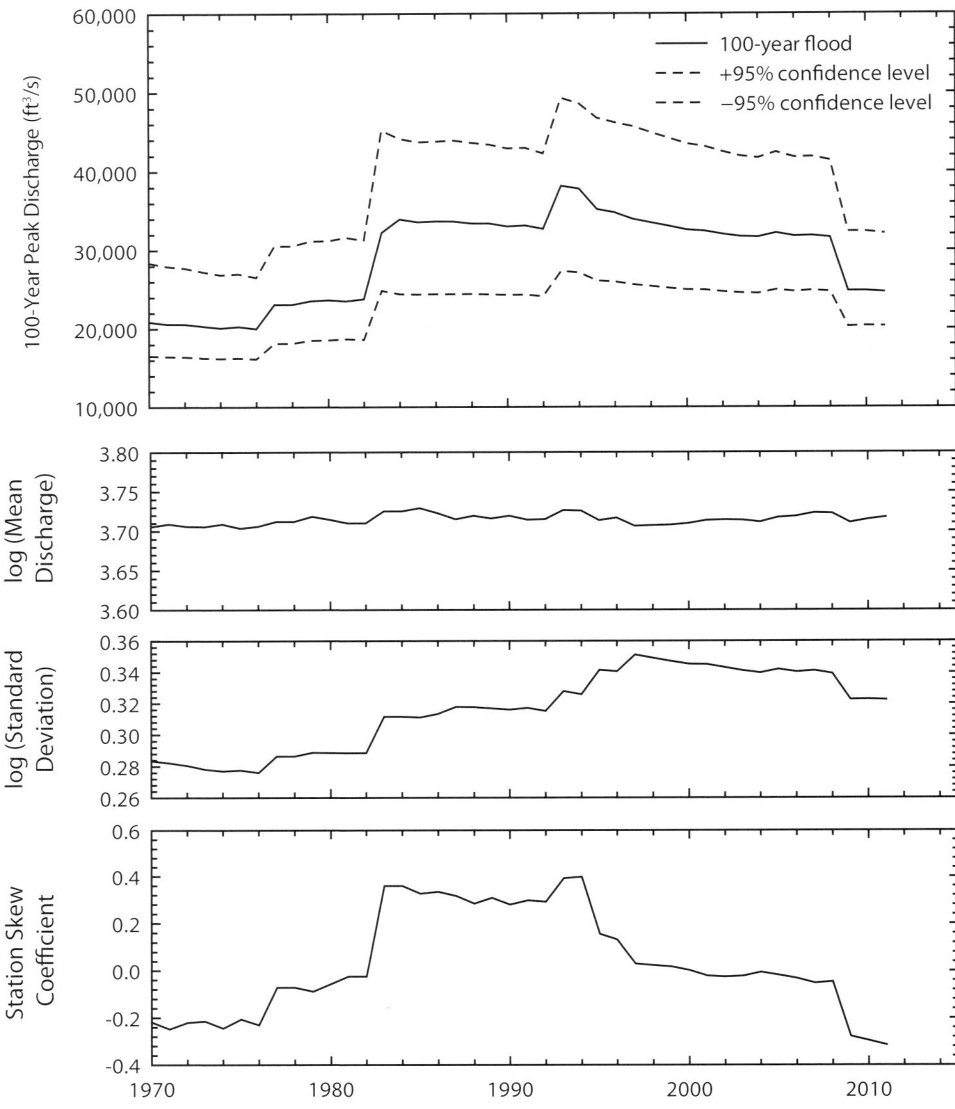

Figure 9.8.
Top: The 100-year flood estimated for each year from 1970 through 2011 for the Santa Cruz River at Tucson (see fig. 2.7). The statistical confidence limits of +95% and −95% are shown as dashed lines. The lower three graphs are the record moments used to calculate the 100-year flood by year. Flood frequency was estimated using a log-Pearson type III distribution using standard techniques (US Water Resources Council 1981) and incrementally adding each year's peak discharge to the record to recalculate the 100-year flood.

includes the four storm periods of 1993 as one event— led to peak discharges exceeding previously estimated 100-year floods.[42] Every climatic indicator that has been evaluated, including El Niño occurrence, other measures of the ocean-atmosphere system affecting the West Coast of the United States,[43] the frequency of tropical cyclones (see fig. 2.6), and cool season rainfall in Tucson (see fig. 2.5C), supports the hypothesis that an unusual climatic episode of increased flood-generating storms occurred during this period.

Other factors could have conspired to increase peak discharges. Changes in channel topography, such as those that happened as the Santa Cruz arroyo developed into the modern channel, are known to alter conveyance of flow waves.[44] The Santa Cruz River was unincised at the southern end of the San Xavier Indian

Reservation at the beginning of the gaging record at Congress Street, and flow had to pass through the Great Mesquite Forest before it could get to Tucson. Through the reservation, the channel deepened ten to fifteen feet between 1915 and the late 1930s and another six to ten feet afterward.[45] All those trees in the bosque would have held floods back, with an expected effect of decreased peak discharges but longer flow duration.

The zero-flow elevation at Congress Street dropped ten to fifteen feet between 1946 and 1984,[46] mostly because of encroachment of the channel by landfills and highway construction. Hypothetically, the flood in the winter of 1915, which lasted several months and produced a peak discharge of 15,000 ft^3/s at Congress Street, might have produced a much higher peak if routed through the modern channel. Conversely, the 1983 peak might have been much less if routed through the smaller 1915 channel and the dense riparian vegetation in the Great Mesquite Forest. There has been little effort to reconcile the effects of changing channel geometry on flood discharges, although some researchers have maintained that channel changes—not unusual climate variation—are responsible for the increase in annual peaks.[47] However, record precipitation suggests a climate reason for the floods, with channel conveyance issues being of secondary importance.

The Santa Cruz flood series shows a lack of uniformity in the seasonality of flood peaks (see fig. 2.7) that may partly account for the increase in annual peak discharges since 1960.[48] In the periods of 1915–1930 and 1960–1964, almost half of the annual flood peaks occurred in early fall (September–October) or winter (November–February). In the intervening period of 1931–1959, which is representative of the midcentury drought, 93 percent of the peaks occurred in July or August. Seven of the eight largest peaks in the flood series were produced by fall or winter storms, and five of these occurred after 1960. The seasonal pattern is not peculiar to the Santa Cruz River; it is repeated in other annual flood series from southern and central Arizona (such as those for the San Francisco, Gila, Rillito, and San Pedro Rivers).

Whether related to climatic or watershed changes, the results of flood-frequency analysis change with the period of measurement or with the type of storm.[49] Analyses using the partial duration series from 1950 to 1980 suggest that the 100-year flood for winter frontal storms and fall tropical storm-cutoff lows is double that of summer monsoonal storms, which are the most common source of flood peaks.[50] Mixed-population analysis indicates that flood hazard is driven by the fall and winter storms and that monsoonal-caused floods are stationary (time invariant).[51]

Bank Protection and Flood-Hazard Mitigation

Installation of bank protection along the Santa Cruz River was an unqualified success, if the only goal were flood-hazard mitigation. Soil-cemented banks withstood both the 1983 and the 1993 peak discharges preventing lateral channel change.[52] Problems of flow over the soil-cemented banks, which eroded the material behind them, were relatively minor. This solution to lateral channel change was exported throughout the Southwest and now is used, with local variation, to stabilize river channels in California, Nevada, New Mexico, Utah, and elsewhere in Arizona.

From the perspective of 1995, both watershed and climatic changes appeared to have contributed to larger flood peaks along the Santa Cruz River. Most hydrologists familiar with flood frequency and floodplain management expected this trend to continue, and many responsible for flood-hazard mitigation were complacent knowing that the 1993 flood had been contained within the bank protection prompted by the 1983 flood. Then everything changed again, with climatic conditions reverting back to the relatively dry conditions of the middle of the twentieth century or even to the extremes that marked the end of the nineteenth century. El Niño conditions, once thought to guarantee large floods on this river—and they did from 1977 to 1995—came and went with insignificant peak discharges. As its history into the twenty-first century shows, the behavior of the Santa Cruz River is not as predictable as might be preferred for a floodway that runs through a major metropolitan area.

Is the Santa Cruz River Through Tucson Still an Arroyo?

Bank-protection and grade-control structures have eliminated the ability of the Santa Cruz River to shift its channel through the Tucson metropolitan area. Large floods no longer can erode channel bends, widen the channel, or downcut into the valley fill; local scour and fill, however, typically at bridge crossings, remains a possibility. Channelization increased the gradient, which purportedly would make the river self-cleaning and minimize long-term deposition by small floods. All these engineered changes make for an unnatural river channel. So does this necessarily mean that the Santa

Cruz River through Tucson has been tamed and is no longer subject to the same forces as more-natural channels elsewhere in the region?

Regional generalizations of arroyo cutting-and-filling, particularly during the nineteenth and twentieth centuries, are based on a series of channel changes that occurred in discrete time intervals (see fig. 3.1).[53] First, there was the late nineteenth-century period of arroyo downcutting, which typically occurred during one or more large floods. This was followed in the early twentieth century by a period of frequent smaller floods, which widened and deepened the arroyos. Gradually, and especially during the midcentury drought, channels began to narrow and floodplains were deposited between arroyo walls. Large floods occurred again, causing widening and removal of floodplains. This is the stage at which floodplain managers stabilized the channel of the Santa Cruz River.

Floodplain managers would like to think that this river channel is unchanging and can convey a large design discharge of the magnitude of the 1983 flood. Events of the coming decades would show that this view is optimistic at best and dangerous at worst. Regionally, the early twenty-first-century drought has caused fewer large floods, and channels have responded by narrowing and depositing floodplains; riparian vegetation has flourished in this setting.[54] Although the soil-cemented banks of the Santa Cruz River are unnatural, regional climate variation would change the rules, and this unnatural river through Tucson would again behave like all those natural regions throughout the Southwest: it would begin to aggrade.

The Shaky Entry of CAP Water

Arizona's total water use, which is driven by agricultural production in low-desert fields, peaked in the mid-1980s, driven largely by a post-1980s decline in agriculture.[55] In 1985, Tucson used 93,000 acre-feet of water, all of which was pumped out of the ground (fig. 9.9), and Pima County used 260,000 acre-feet of groundwater. Even though it seemed like water was everywhere, getting the drops to drink was making water managers in Pima County extremely nervous. The passage of the landmark groundwater act by the Arizona legislature in 1980 created the Active Management Areas, including in the Tucson Basin. The need to regulate groundwater use, long overdue, was prompted by the related trends of declining groundwater levels and subsidence. In Tucson, as much as four inches of subsidence was associated with a

water-level decline of forty-five feet between 1989 and 2005.[56] Another source was desperately needed, and the only possible source was the Colorado River. The problem was that lots of entities already had claims on that water, and Tucson literally was at the end of the line.

What appeared to be the solution to Tucson's water problems had its origins on 24 November 1922, when delegations from the seven states of California, Arizona, Nevada, Colorado, Wyoming, Utah, and New Mexico agreed to a large-scale division of the waters of the Colorado River. The Colorado River Compact was signed pending ratification by the state legislatures.[57] The geopolitical boundary between the upper basin states and the lower basin states of California, Arizona, and Nevada was a theoretical place in Arizona called Compact Point, which was on the Colorado River a short distance downstream from the Utah-Arizona border and at the head of Grand Canyon. The lower basin states were entitled to 7.5 million acre-feet of Colorado River water, and Arizona was given 2.8 million acre-feet and an additional 1 million acre-feet from the Gila River.

In a dispute with California, the Arizona legislature refused to ratify the compact, which meant that Congress could not ratify it either. The political wrangling would take years to resolve, and armed conflict was threatened. In the mid-1930s, by order of Arizona governor Benjamin B. Moeur, the Arizona National Guard defended the east bank of the river near Parker against any attempt, by California or the Bureau of Reclamation, to build any type of water-diversion structure in the riverbed.[58] Despite the threat of troops, the handwriting was on the wall for Arizona. Control—specifically, federal control—of the Colorado River was inevitable. The Arizona legislature finally signed the compact in 1944, allowing Congress to ratify it that year; a new treaty included 1.5 million acre-feet for Mexico.[59] Climate again reared its controlling head, as hydrologists had overestimated the amount of water produced long-term in the Colorado River basin; they had no knowledge of the early twentieth-century pluvial, which inflated water yield.[60]

It was all well and good that Arizona had legal access to Colorado River water, and agriculture benefited along the lower Colorado and Gila Rivers, but the need for water was greatest in the cities and farmlands of central and southern Arizona, including the Tucson Basin. Getting the water to where it was needed most would require money and another act of Congress. The Central Arizona Project Association was created in 1946 to lobby for federal government support to solve the financial problem of building a canal from Colorado

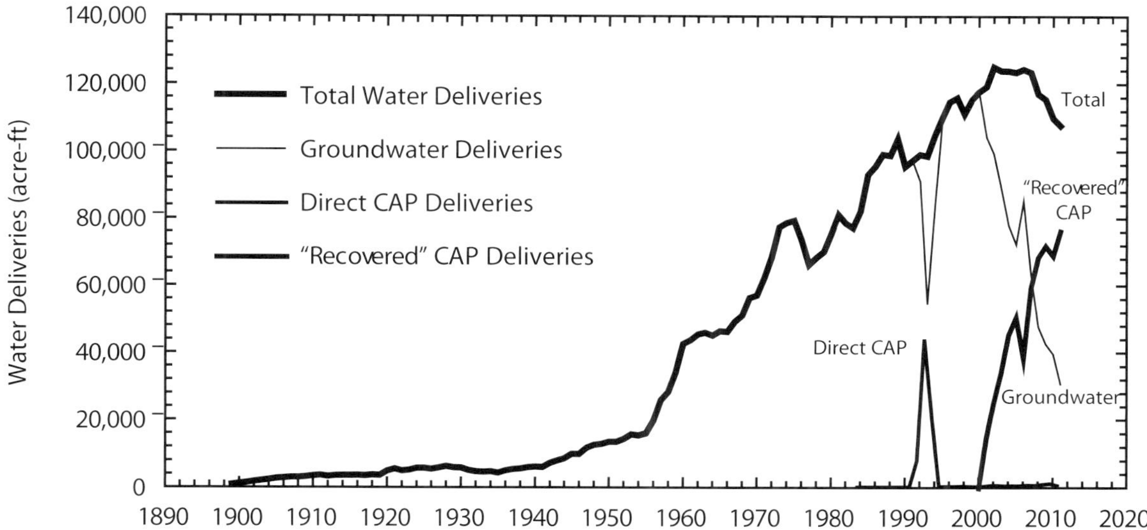

Figure 9.9.
Water deliveries by Tucson Water, 1899–2011. Between 1899 and 1983, all water deliveries were groundwater. The short-lived direct delivery of Central Arizona Project (CAP) water in the 1990s is shown as a spike mirrored by a downward drop in groundwater usage. "Recovered" CAP water generally comes from blended groundwater from the Avra Valley west of Tucson. (Data courtesy of Tucson Water.)

River reservoirs in western Arizona to the agricultural fields and metropolitan areas of central and southern Arizona. In 1968, President Lyndon B. Johnson signed a bill approving construction of the Central Arizona Project (CAP) that provided funding to the Bureau of Reclamation to complete the task, but there was a catch: Arizona had to repay the federal government for part of the costs of construction after completion.[61]

In 1971, the Central Arizona Water Conservation District was given the authority to do the financial heavy lifting required to sell water and repay the construction costs. In addition, this agency manages and operates CAP. Construction began at Lake Havasu on the Colorado River in 1973. The numerous engineering challenges of the CAP canal system included a complex set of pumping stations to move water over mountains and siphons to move water beneath roads, freeways, and other infrastructure. In addition, the canal route was designed to skirt areas of land subsidence in central Arizona caused by groundwater extraction.

Although the Phoenix metropolitan area was eager to get any additional water to supplement its surface-water supplies from the Salt and Verde River systems and its own rapidly declining groundwater resources, Tucson was reluctant to receive CAP water.[62] Tucson residents were embroiled in a debate about controlled growth, and importation of more water meant that

development could continue without limits other than the availability of land. After years of delay, the canal to Tucson was approved and paid for. The Central Arizona Project reached the Phoenix metropolitan area in 1983 and was extended to the west side of Tucson in 1993.[63] Direct deliveries in Tucson had an immediate effect on reducing groundwater pumping (fig. 9.9).

The initial reception to Colorado River water in southern Arizona was thumbs-down, at least at first.[64] Farmers found the water to be high priced and of lower quality than the groundwater beneath their fields.[65] Residents blamed CAP water for ruptured pipes in houses and brown water pouring from taps; the chemistry of CAP water was different from native groundwater, and old pipes shed years of built-up compounds that literally glued them together. This became a public relations nightmare, and flow of CAP water for domestic delivery was quickly terminated (fig. 9.9). With the expensive canal and the new water-treatment plant idle, many proposals of what to do with CAP water were circulated, some practical and others fanciful. For a while, a proposal was made to recharge groundwater in the Tucson Basin by discharging CAP water into the Santa Cruz River near Martinez Hill, but this option was never taken seriously. Even though this could have been a viable means for "restoration" of the Santa Cruz

River, practical matters, ranging from algal sealing of the bed that would minimize recharge to losses from increased evapotranspiration to possible contaminant entrainment from those channel-bank landfills, eliminated this option.

Another plan was to pump CAP water across the City of Tucson to the Rillito-Pantano system, which had far fewer landfills in its banks and a wide, sandy channel.[66] The cost of the distribution plumbing eliminated this plan. Direct injection into the Tucson Basin aquifer was also out of the question. CAP water eventually was recharged into the Avra Valley aquifer using a large network of shallow spreading galleries so that it could be blended with groundwater before it was "recovered" and piped into the Tucson water system.

Another way to reduce groundwater pumping was to reduce the use of potable water for landscaping. In 1984, the City of Tucson built the Tucson Reclaimed Water Treatment Plant (RWTP) to produce reclaimed water from the wastewater stream. Part of this reclaimed water was pumped into a new distribution system to be used for irrigation of parks, golf courses, and other public places. By 2006, reclaimed water use provided more than 10 percent of the total deliveries by Tucson Water.[67]

CAP water deliveries were only one part of a complicated change in Tucson water use that has led to rises in water levels beneath the city. The importation of Colorado River water allowed reduction of groundwater extraction from the Tucson Basin. The constant drumbeat of water conservation, particularly during summer months, slowed water use (fig. 9.9). Treated effluent is used to water some golf courses and most city parks. Optimization of well pumping has spread the amount of groundwater extraction around the basin. Even with the most draconian conservation and water recycling efforts, however, there is little to no chance that water levels will rise sufficiently in the alluvial aquifer to allow natural regeneration of riparian vegetation along the Santa Cruz River. Another source of water would be required to accomplish this, and it would come out of pipes discharging wastewater into the river on the north side of town.

By 1995, approximately 15 percent of the water used in Pima County was surface water—water imported through canals from the Colorado River. The Central Arizona Project, the last major waterworks that came out of the Colorado River Compact of 1922, was finally having an impact on reducing groundwater extraction from the Tucson Basin. This solution to Tucson's water problem was way too late for the Great Mesquite Forest, but it did initiate a regional water conservation plan using reclaimed wastewater and reductions in water use that at least slowed water-use rates, a first small step toward water sustainability. Water conservation would be very appropriate for what was up next for Tucson and the Santa Cruz River: the early twenty-first-century drought was on the way.

10

Channel Filling and River Restoration Efforts, 1996–2012

The 100-year flood has come and gone, so, by all rights, Tucsonans
should enjoy another century of great Southwest weather.
—*press release by Metropolitan Tucson Convention and Visitors Bureau
following the 1983 flood (quoted in Saarinen et al. 1984)*

From 1977 through 1993, floods on the Santa Cruz and Rillito Rivers raged through Tucson, causing extensive damage and forcing installation of expensive bank protection. Then the big floods ceased as long-term drought set in, beginning with the 1996 La Niña. The flood regime on the Santa Cruz returned to low peak discharges reminiscent of the midcentury drought (see fig. 2.7). Once again, new residents streaming into fast-growing Pima County encountered a floodplain more prone to blowing dust than raging water. Another flood, this time from the Rillito River, would remind everyone in Pima County of how much water these dry channels could carry, and why millions of taxpayer dollars were literally poured into cementing their banks. Another stage of arroyo evolution was only beginning; the start of channel filling was perceptible to only a few hydrologists and engineers fixated on what was happening between the soil-cemented banks.

Conservationists in southern Arizona did not care much for the channelized Santa Cruz River, which was usually dry and lined with ugly soil-cemented banks. Under the relentless pressure of pumping, groundwater levels dropped beneath the channel, leading to more dieback in riparian vegetation that had been temporarily rejuvenated by nearly two decades of floods, but new vegetation slowly was becoming established. Calls for river restoration increased through the 1990s and into the 2000s. Few ever considered the obvious question: restore the river to exactly what past condition? Fanciful ideas abounded that the river once was lined with cottonwoods and willows over its entire length, that beaver and fish once lived in its crystalline pure waters, and that humans—particularly their cows—were the sole reason the river had become entrenched.

A more serious approach to management of the Santa Cruz River begins with asking the simple question of what society wants from its ephemeral urban rivers: do we want flood protection or fully functioning riparian ecosystems? Those goals may be mutually exclusive. River restoration, regardless of its goals, requires water to create and maintain ecosystems, and Tucson needs water to keep expanding its human environment. Finding water for riparian vegetation would not be easy, and the riparian vegetation that becomes established—either through restoration efforts or naturally—traps sediment and impedes flow, increasing flood hazards.

Reduced Flood Frequency

Following the 1993 floods, flood discharges in the Santa Cruz River at Tucson decreased in response to the early twenty-first-century drought (see fig. 2.7). Between 1995 and 2011, flood peaks mostly occurred during the summer months (as they did during the midcentury drought) with the exception of four floods in the fall and one in the winter. The largest peak discharge was 16,300 ft^3/s in August 2005, and surprisingly enough, this flood caused a brief scare (see later section). With the exception of 2005–2007, each of which had seven peaks above base discharge, most years had fewer than five floods above base discharge (see fig. 8.9). Precipitation decreased and the arroyo quieted down, producing fewer and smaller floods than in the previous two decades. The channel, by then mostly lined with soil cement, began to respond by aggrading with sediment that sprouted new riparian vegetation, which in turn trapped more sediment and increased aggradation.

While it might seem intuitive that floods would decrease in size and frequency during droughts, flood frequency in semiarid watersheds like the Santa Cruz River is not necessarily related to total annual or seasonal precipitation. Individual storms, possibly embedded in a relatively brief but intense period of moisture influx into the region, can generate floods, even during periods when the tendency is for dry conditions. Near misses occurred, particularly in 1997 when the remnants of Hurricane Nora passed west of Tucson, generating record floods in west-central Arizona. Similarly, winter storms in 2005 passed north of Tucson, generating extremely large floods in the Salt and Verde Rivers, while the Santa Cruz River carried only minor runoff.

One set of storms in 2006 showed how the Santa Cruz River basin responds to its component tributaries in generating floods. In late July, a low-pressure system stalled over northern New Mexico, which is unusual for the North American monsoon. For a week, its counterclockwise spin pushed moisture-laden air southward, from central into southeastern Arizona, generating nightly storms in this part of the state. One recipient of frequent storms was the Santa Catalina Mountains northeast of the Tucson Basin, where the daily storm totals climbed to unprecedented levels. Whereas multiday storm frequency commonly was at or above the 50-year event, some sites had rainfall that exceeded what little we know of the 1,000-year event.[1]

Tributaries draining the southern flanks of this range responded with flash floods that increased in size as the week went on. Sabino Canyon, one of the largest tributaries flowing south from this mountain range, had a 40-year flood on 29 July, followed by a 60-year flood on 30 July, followed by a torrent that exceeded the 100-year event on 31 July.[2] Although the storm in the early morning hours of 31 July was not particularly unusual in intensity or magnitude for the monsoon, it fell on saturated ground.[3] The result was a spate of slope failures and debris flows that added to the flood in Sabino Canyon, pushing large quantities of water and sediment into the Rillito River.

Water poured into the Rillito River from record floods generated in Sabino Creek and its other major tributaries—Tanque Verde Creek, Pantano Wash, and Rincon Creek—resulting in a combined peak discharge of 38,700 ft^3/s that is the largest in stream-gaging history of this watershed.[4] Although soil-cemented banks were overtopped locally, the bank protection held as designed,[5] and the record discharge surged into the Santa Cruz River. The peak discharge at the Cortaro Road gaging station was 40,900 ft^3/s, the second-highest peak at this station after October 1983. Despite all the rain in the Santa Catalina Mountains, as well as in the Rincon Mountains to the east of Tucson, only a modest amount of rain fell upstream in the Santa Cruz River watershed. The peak discharge at the Congress Street Bridge in Tucson was only 7,200 ft^3/s, the annual peak discharge for this station in 2006 (see fig. 2.7).

As the 1983 and 1993 floods showed, near misses can become direct hits, with the probabilities increasing with the varying generation, at the decadal-to-multidecadal timescales of the types of moisture sources required to induce floods. Take tropical cyclones, for example: two of the three largest floods on the Santa Cruz River at Tucson were directly related to an influx (or *advection*) of moisture from dissipating tropical cyclones into southern Arizona.[6] If the number of tropical cyclones generated in the eastern North Pacific Ocean were constant, then the possibility of large floods generated by tropical cyclones on the Santa Cruz River might be expected to be constant as well. As the record shows (see fig. 2.6), however, the number of tropical cyclones, including tropical depressions, tropical storms, and hurricanes, decreased after about 1995. While an average of ten hurricanes and seventeen tropical cyclones occurred between 1976 and 1995, an average of seven hurricanes and fourteen tropical cyclones occurred between 1996 and 2011. Fewer possible chances translate to fewer possible floods.

The Return of Riparian Vegetation

Most observers, such as the one quoted here, expected channelization in the Tucson reaches to spell the end of riparian vegetation along the Santa Cruz and Rillito Rivers: "Soil-cementing the banks sealed the fate of the riparian zones along the Santa Cruz River. Trees and shrubs don't grow through cement."[7]

True enough, trees and shrubs only rarely grow through cement, but the bed of the Santa Cruz River, unlike that of the Los Angeles River, was not lined with cement and remains permeable, except at the widely spaced concrete grade-control structures installed in the 1980s and early 1990s that were designed to keep the channel from downcutting beneath the soil cement and bridge piers during floods. These sills were not designed to stop the channel from filling with sediment; instead, the channel gradient was increased with the idea that this, in combination with a fixed-channel cross section, would make the channel self-cleaning. If it were not for the reestablishment of mostly facultative riparian and xeroriparian vegetation within the channel and the decrease in flood magnitude, the self-cleaning design actually might have worked.

Instead, a floodplain within the trapezoidal channel began forming soon after the 1993 flood (figs. 10.1, 10.2). The wide, barren channel became colonized with ruderal vegetation, although its initial stages are difficult to discern in aerial photographs (see fig. 9.4). Native and nonnative perennial grasses and forbs were the first arrivals, followed by native and nonnative trees and shrubs. A prominent black willow grew on the upstream east side of the Congress Street Bridge for many years (fig. 10.1D), with its canopy reaching above the bridge deck before a flood destroyed it. More commonly, mesquites and Athel tamarisks became established between the walls of soil cement.

The weeds, shrubs, and trees grew on a low terrace that only occasionally was overtopped during flash floods. When floods occurred, sediment carried in the turbulent waters dropped out in the dense vegetation, which increased roughness and slowed flow velocities. A positive feedback mechanism developed: as long as flood discharges and durations remained low and sediment concentrations were high, the floodplains grew in height as sediment trapped in the dense vegetation promoted more vegetation growth. Eventually, the extent of floodplains spread toward the center of the channel, leading to a narrowing of the active channel width between the soil cement (fig. 10.3). Urban debris, from shopping carts to trash cans, also trapped sediment, enhancing the cycle of sedimentation and riparian plant growth.

No one would ever mistake the nascent riparian assemblages through downtown Tucson as a restored version of what was lost in the most verdant reaches through the Tucson Basin. The lone black willow at the Congress Street Bridge aside, few if any obligate riparian species grew in this reach. The splash of summer greenery against the drab gray soil cement suggests possibilities driven solely by processes of natural vegetation establishment and reversible only by another large flood or by floodplain managers who recognize the threat this vegetation poses to flood control and fight it with vegetation removal and (or) dredging. Another possibility, simultaneously more ominous and interesting, lies in the posi-

tive feedback between riparian vegetation and sediment trapping. The Santa Cruz River through the Tucson Basin may be refilling its entrenched channel.

Is the Channel Filling?

Following the 1993 flood, the Santa Cruz River through Tucson had a trapezoidal cross section within its soil-cemented banks. The series of relatively small floods that followed and their interaction with the increased riparian vegetation created a "channel within a channel" consisting of a barren, low-flow conduit constrained within vegetated floodplains within the stabilized banks. Even following the 1983 flood, the channel upstream from Martinez Hill (fig. 8.8), with its natural arroyo walls, narrowed in response to lower flood peaks.[8] At the beginning of the twenty-first century, the channel at Congress Street was narrowing and starting to aggrade (fig. 10.4).

Channels do not fill only by *vertical aggradation*, during which all elevations on the bed raise simultaneously. In fact, the lowest point on the bed of the Santa Cruz River, known as the *thalweg*, remains at its 1982 level (figs. 10.4 and 10.5). Instead, a growing body of evidence indicates that channels first narrow, probably in response to reduced flood discharges, and deposit floodplains, generally on *point bars*, which typically form on the inside of channel bends, or *lateral bars* on either margin of straight reaches. Other floodplains are deposited on one or the other side of the channel, seldom simultaneously on both sides. This process is known as *lateral aggradation*, and it is commonly observed in arroyo systems in the region.[9]

Vegetation begins to grow on the aggrading floodplains, increasing the roughness faced by overtopping floodwaters, which slows flow velocity and allows sediment to drop out. Near the Congress Street Bridge, the plants established between the soil-cemented banks include Athel tamarisk, paloverde, mesquite, and burrobrush, although numerous species of native and nonnative forbs and grasses—as well as a few nonnative palm trees—have become established. As long as flood discharges remain low, the floodplain creates a positive feedback mechanism of vegetation growth followed by sediment deposition and more vegetation growth. Deposition can also occur on what is termed the *streamside margin*, or the edge of the floodplain next to the active channel, again spurred by riparian vegetation that traps sediment from flows that do not overtop the floodplain.

Figure 10.1.
The Santa Cruz River at the Congress Street Bridge. **A.** (23 December 1914) Following repeated floods and channel change in the early twentieth century, bridges replaced fords across the Santa Cruz River near downtown Tucson. This steel-and-wood bridge was built after the floods of 1904 and 1905. The short approach structure shown in the center of this view failed soon after this photograph was taken. (Godfrey G. Sykes, Sykes Family Collection, PC 240, Box 7, f.125, Arizona Historical Society.) **B.** (November 1926) The replacement bridge, a reinforced arched concrete span, lasted through much of the first half of the twentieth century. Taken during a relatively small flood, this image shows a relatively wide channel with some laid-over riparian vegetation in the foreground. (Photographer

Figure 10.1. (*continued*)
unknown, 28765, Arizona Historical Society.) **C.** (12 September 1983) The arch bridge was replaced in the 1970s by a simple reinforced concrete span that accommodated four lanes of traffic and a center lane. The flood of 1977 caused significant channel change in Tucson, prompting the installation of bank protection (right midground) to create a trapezoidal cross section. (R. M. Turner.) **D.** (17 June 2008) Channel aggradation following the 1993 flood reduced channel area and increased floodplains that support dense riparian vegetation, both native and nonnative species. The bridge withstood both the 1983 and 1993 floods, the largest in the 97-year record. (R. H. Webb, Stake 1084, courtesy of the Desert Laboratory Collection.)

Figure 10.2.
The Santa Cruz River at Congress Street. **A.** (1902) This downstream view shows the Congress Street Bridge at the start of the twentieth century. The deep arroyo that eroded in 1890 and 1891 made river crossings more difficult. By 1902, this Pratt Truss steel bridge had been erected to span the river. The young stand of willows and cottonwoods was probably established after the 1890 flood. (A. Hadsell, 26698, courtesy of the Arizona Historical Society.) **B.** (20 December 2003) This view shows a much deeper channel with soil-cemented banks. Although some sediment deposition and encroachment of riparian vegetation is evident on the right side, the channel is essentially trapezoidal in shape. (D. P. Oldershaw.)

Figure 10.2. (*continued*)
C. (17 June 2008) In the intervening 106 years between the match and the original, the bridge has been replaced twice.
The City of Tucson installed soil cement and narrowed the channel within the channelized arroyo. (R. H. Webb, Stake 4688,
courtesy of the Desert Laboratory Collection.)

This mechanism of channel aggradation has been observed in many ephemeral channels in the southwestern United States.[10] The principal differences between the Santa Cruz River and other channels in more-natural settings lies in the influence of the urban environment in increasing local runoff and the lowered groundwater levels beneath the channel. Measurements at low flow are difficult because the channel shifts around at low stage, although the number of small floods generated from the upstream drainage within Tucson and its suburbs is certainly to have increased with urbanization. More small floods would be expected to increase the rate of lateral aggradation. Lowered groundwater levels would decrease the potential for obligate riparian vegetation to become established through the center of Tucson; instead, facultative riparian vegetation, such as mesquite and nonnative tamarisk, is more likely to become established on the floodplains.

Channel aggradation has consequences for flood control, the reason why the soil cement was installed beginning in 1982. The channelized river through downtown Tucson had a design capacity to convey a bankfull discharge of 60,000 ft³/s; additional water above the design discharge would overtop the banks. As channel area decreases owing to floodplain aggradation, the amount of water that can be conveyed through the channel decreases. Add riparian vegetation and increased flow resistance, and even less water can be conveyed. Floods with discharges smaller than the design flood can overtop the soil-cemented banks, increasing the hazard to infrastructure and buildings on the landscape beyond the river.

The influence of channel aggradation and riparian vegetation is captured in the records for gaging stations. Gaging stations do not directly measure discharge; instead, automatic recorders measure the elevation of the water surface in the channel. That elevation, known as *stage*, is converted to discharge using a *stage-discharge relation*. Hydrotechs who maintain the gaging station periodically create a new stage-discharge relation by manually taking discharge measurements made over a range of stages. Discharge measurements seldom are made for the largest floods, and stage-rating curves are extended using hydraulic models or by linear projection. Usually, a new stage-discharge relation is established to respond to the effects of channel change on discharge estimation; for example, if a channel erodes, the relation between stage and discharge will change,

Figure 10.3.
The Santa Cruz River from the Congress Street Bridge. **A.** (November 1907) This downstream view shows the Santa Cruz River's narrow channel north of the Congress Street Bridge. Cottonwood trees line the relatively narrow channel. (W. T. Hornaday, 11669, courtesy of the Arizona Historical Society.) **B.** (29 July 1916) This photograph, from a vantage point similar to the 1907 image, was taken after the 1915 flood widened the river channel. Many cottonwood trees adjacent to the river were eroded away, but others farther away from the channel were not affected. (Photographer unknown, University of Arizona Special Collections Library.)

Figure 10.3. (*continued*)
C. (22 December 2003) The banks of the Santa Cruz River are now covered with soil cement. A large amount of riparian vegetation, including nonnative saltcedar, lines the channel. Paved trails, part of the Santa Cruz River Park, are present behind guardrails seen along both sides of the river. (D. P. Oldershaw, Stake 3744, courtesy of the Desert Laboratory Collection.) D. (27 July 2008) In the five years after the previous photograph was taken, some of the shrubs growing in the river bed were removed, presumably by flood flows, while much of the vegetation is now larger in stature. (R. H. Webb, Stake 3744, courtesy of the Desert Laboratory Collection.)

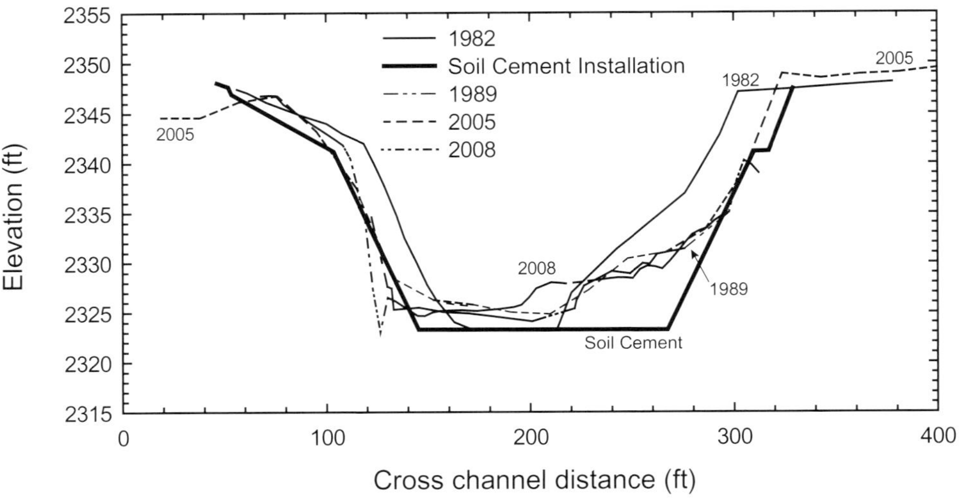

Figure 10.4.

This graph shows channel cross section 4, downstream from the Congress Street Bridge by about three hundred feet, at various times before and after installation of bank protection in 1982. Comparison of the cross sections from 1982 and the soil-cement design shows the amount of channel widening during construction to convey the design discharge of 60,000 ft³/s. Channel narrowing and floodplain aggradation increased after the 1993 flood.

Figure 10.5.

(7 August 2008) This view across the Santa Cruz River shows the concrete grade-control structure downstream from the Congress Street Bridge. The river flows from right to left, and a scour line has developed downstream from the sill. Note the channel narrowing shown by the distance to the far (east side) soil-cemented bank. (R. H. Webb.)

and if the channel aggrades, a different stage-discharge relation will apply. Stage-discharge relations, therefore, reflect the history of channel change as well as channel roughness: riparian vegetation can greatly increase roughness and decrease flow velocity, increasing the stage for a given discharge.[11]

The stage-discharge relations for the Santa Cruz River at Tucson changed significantly after the 1993 flood (fig. 10.6), indicating that the soil-cemented channel cannot convey the same amount of water that it was designed to convey in 1982. In 1992, a stage of ten feet corresponded to a discharge of 25,000 ft³/s; in 2011, that same stage corresponded to a discharge of 9,780 ft³/s (fig. 10.6). These stage-rating relations are calibrated using discharge measurements made over a range of flows. Because no floods larger than about 17,000 ft³/s occurred (see below), the stage-discharge relation was projected to higher discharges.

The soil-cemented channel was designed for a capacity of 60,000 ft³/s to be conveyed without overtopping of the channel banks, which have an approximate height of nineteen feet on the stage-rating curves. Channel aggradation (fig. 10.4) has reduced that capacity by both decreasing the area within the soil-cemented banks and increasing channel roughness. As a result, the channel can now convey about 30,000 ft³/s of bankfull discharge, which had a recurrence interval of less than ten years at the time that the channel stabilization was designed and built. However, flood frequency de-

creased significantly after 1993, which lowered the estimates for all flood *quantiles* (the inverse of recurrence intervals) (see fig. 9.8). Ironically, this means that a channel designed to convey a flood thought to represent the 100-year event in the 1980s can now still convey a flood with a 100-year recurrence interval in the early twenty-first century.

The changing stage-discharge relation had consequences. On 23 August 2005, Tucson residents awoke to startling media reports claiming that the largest flood since 1983 was occurring in the Santa Cruz River through downtown Tucson. First responders were stationed at bridges, waiting for orders to close them to traffic in the event of a threat to their piers or bridge decks. The peak discharge of this flood was a respectable 16,300 ft³/s, but it was far smaller than the 52,700 ft³/s flood in October 1983. Despite the actual discharge, the stage registered high on the stage-discharge relation then in effect (rating curve 11, table 10.1), and floodwaters appeared to be close to the bottom of the bridge deck at Congress Street. Needless to say, this flood had no impact on the city of Tucson or its infrastructure across the Santa Cruz River. We now know that the stage-rating curve had changed in the low-flow years following the 1993 flood, giving anyone viewing the real-time record a false sense of the magnitude of the August 2005 flood. Hydrotechs maintaining this gaging station changed the stage-discharge rating three times after 2006 to the current 2011 rating (table 10.1; fig. 10.6).

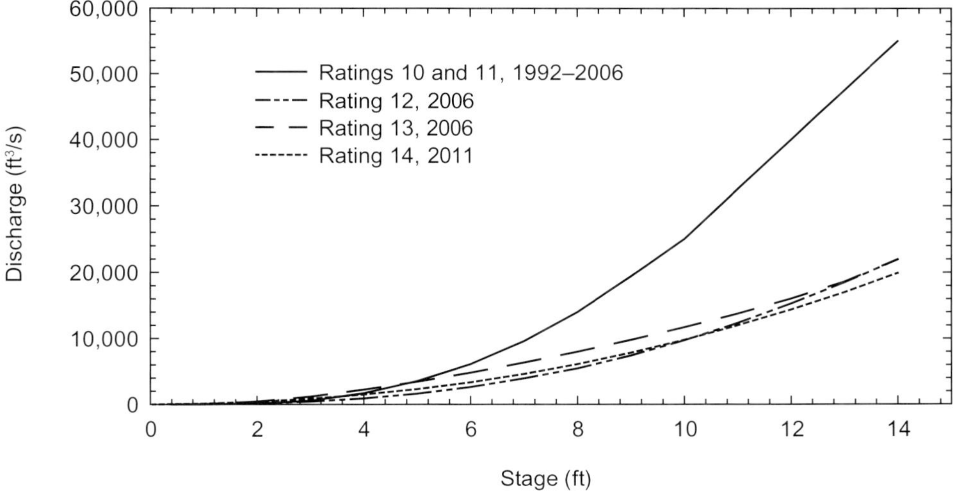

Figure 10.6.
Five stage–discharge relations for three time intervals for the Santa Cruz River at Tucson, Arizona. Curves with higher slope indicate that the channel can convey higher discharges. Ratings 10 and 11, in effect from 1992 through 2006, overlap except at very low discharges.

Table 10.1. Dates of stage-rating curves for the
Santa Cruz River at Tucson After 1992

Rating Curve Number	Starting Date	Ending Date
10	1 October 1992	1 October 1999
11	1 October 1999	9 February 2006
12	9 February 2006	28 December 2006
13	28 December 2006	15 September 2011
14	15 September 2011	Present

The Santa Cruz Flows Again: River Restoration Projects in the Tucson Basin

The loss of perennial surface water in the Tucson Basin did not impede rapid population growth in the greater Tucson metropolitan area in the 1990s and early 2000s. As the demand for potable water increased, the amount of wastewater increased, and the treatment and disposal of effluent became a problem seeking a novel solution. The Santa Cruz River had dried up in response to repeated irrigation projects that mined groundwater. In contrast, the Salt and Gila Rivers in central Arizona had been altered from perennial to ephemeral streams after multiple flood-control and water-supply dams were constructed in the early 1900s. What all these rivers have is new surface-water sources in wastewater effluent, and that new water was all that riparian vegetation needed to regrow, albeit with many nonnative species mixed in with a few native ones.

In the late twentieth century, the densely populated Tucson and Phoenix metropolitan areas faced the problem of increased demand for potable water in addition to disposal of treated wastewater, and both decided that the most convenient means of disposal of effluent was in the nearby dry river channels. By the end of the twentieth century, treated wastewater flowed perennially downstream from urban areas and away from the cities, especially along the Santa Cruz River downstream from Tucson (fig. 10.7).

Clearly, any serious attempt at river restoration for the Santa Cruz River, particularly one that uses some semblance of natural processes instead of irrigating trees in a riverside park, requires water.[12] With water levels more than one hundred feet below the channel, except of course after periodic floods, there was insufficient groundwater to sustain the once-abundant ri-

parian vegetation through the Tucson reach.[13] A proposal was made to discharge Central Arizona Project (CAP) water into normally dry channels upstream from Tucson to artificially recharge the alluvial aquifer, but this proposal was defeated, in no small part because of possible interaction with and contamination from the numerous abandoned landfills in the banks of the river. The only means left for creating any significant riparian ecosystem would be to turn lemons into lemonade: discharge treated wastewater into the channel or create wetlands where no or few wetlands previously occurred using imported water.

Historically, wastewater was a problem requiring an expensive solution. When the arroyo downcut, sanitation in Tucson consisted of outhouses covering shallow pits excavated in backyards. Development of septic tanks was an improvement, but the innumerable leach fields resulted in the potential for groundwater pollution as the wastewater percolated downward into the alluvial aquifer. In 1900, the first sewer was installed in Tucson, extending on a gradient from houses and businesses in town downstream toward the river. It dumped untreated waste onto an agricultural field, and unpleasant smells wafted back into town.[14] Sewage was pushed farther downstream to a farm at the end of Roger Road, but still the putrid odors were unacceptable. In 1928, the first sewage treatment plant was constructed at the end of Roger Road to remove solids from the waste stream; this plant was expanded in the 1940s but still dumped its wastewater onto fields.[15] In 1951, the Roger Road Wastewater Reclamation Facility was completed with the capacity to treat 41 million gallons per day of sewage generated in Tucson south of the Rillito River.[16] It currently treats 64 million gallons per day, and most of the treated wastewater produced by this plant is discharged into the Santa Cruz River.

Growth on the northwest side, including the towns of Marana and Oro Valley, prompted construction of the Ina Road Wastewater Reclamation Facility in 1977. This facility, located south of Ina Road on the west side of the Santa Cruz River, has a capacity of 25 million gallons per day, but the facility currently is being expanded to treat 37.5 million gallons a day. In 1984, the Roger Road plant became part of the Tucson Reclaimed Water Treatment Plant (RWTP). This wastewater is discharged into the Santa Cruz River, which is the reason why, in combination with Roger Road discharges, base flow at the Cortaro Road gaging station is about 40–60 ft^3/s for much of the year. The combined discharge from the Ina and Roger Road plants amounts to 5,355 acre-feet per

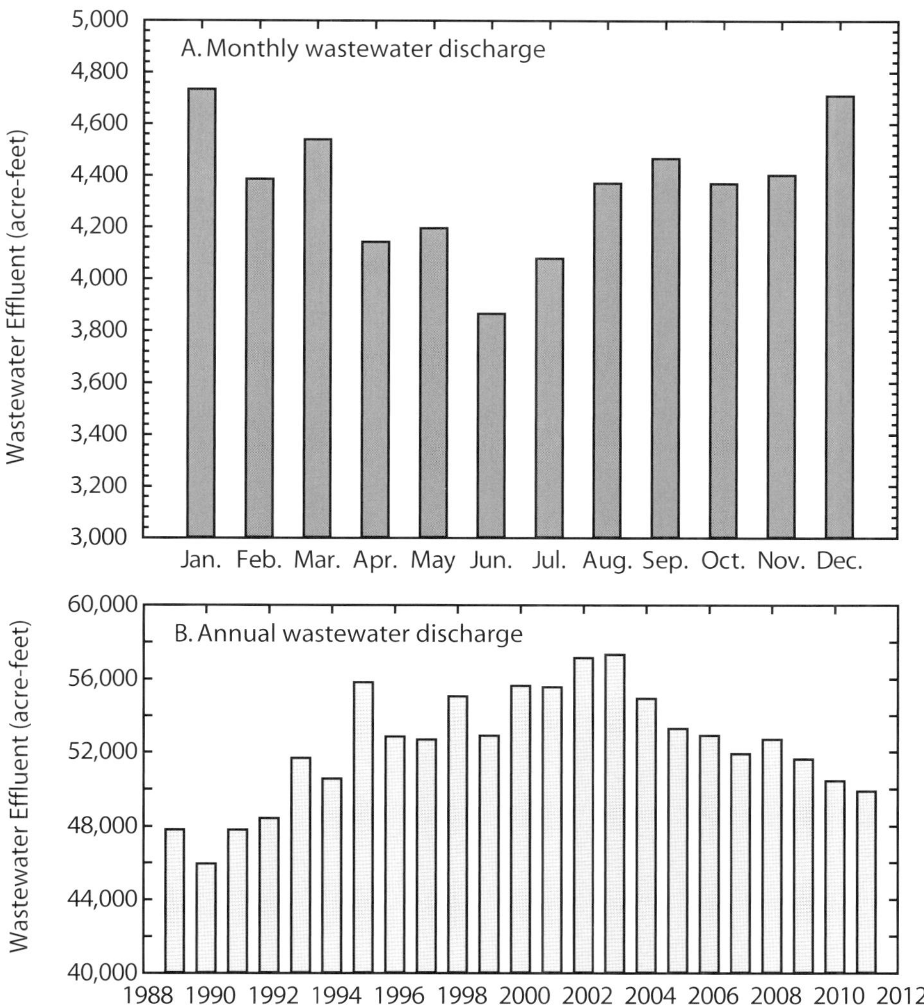

Figure 10.7.
Combined effluent discharge into the Santa Cruz River from the Roger Road and Ina Road Water Reclamation Facilities north of Tucson. **A.** Monthly combined discharges. **B.** Annual combined discharges.

year from 1989 to 2011 (fig. 10.7B). Peak effluent discharge is in December and January, presumably because of decreased evaporation and increased storm runoff, although there is little variation through the year (Fig. 10.7A).

Wastewater and the Sweetwater Wetlands

Discharge of reclaimed water from the RWTP promoted growth of riparian vegetation along the channel of the Santa Cruz River as far downstream as the town of Marana. Growth was sufficient to suggest the potential of a new wetland, partly in a reach (the confluence of the Santa Cruz and Rillito Rivers and the Cañada del Oro) that historically sustained dense riparian vegetation. Some of these riparian ecosystems could help to solve yet another problem: how to improve the water quality of the wastewater stream without additional (and expensive) treatment.[17]

A lawsuit filed by the Arizona Department of Environmental Quality against the City of Tucson over pollution in the effluent further encouraged creation of an artificial, off-channel wetland. Pools sustaining riparian vegetation would help clean up the wastewater. In 1996,

Figure 10.8.
The Santa Cruz River at its confluence with the Rillito River. **A.** (1939) In this downstream (north) view, the Santa Cruz River has a shallow channel lined with scattered cottonwoods, velvet ash trees, mesquites, and other shrubs. (From Willis 1939: plate XV A).
B. (9 November 1983) The river has downcut, and soil cement has been installed as bank protection. Several factors, including groundwater withdrawals and large floods (especially in October 1983), have led to an overall decrease in vegetation in the forty-four years since the original photograph was taken. The trees are mostly mesquites. (R. M. Turner, Stake 1102, courtesy of the Desert Laboratory

Figure 10.8. (*continued*)
Collection.) **C.** (21 December 2003) Thick vegetation, including mesquite, black willow, saltcedar, burrobrush, and blue paloverde, now blocks the view. This riparian ecosystem gets its perennial water from the Tucson wastewater treatment effluent stream. A soil-cemented bank is visible at right center. (D. P. Oldershaw, Stake 1102, courtesy of the Desert Laboratory Collection.)

Tucson Water began construction of a new facility in abandoned agricultural fields adjacent to the RWTP and the river channel. In 1998, the Sweetwater Wetlands was formally established after years of study and observation.[18] With a total area of 17.3 acres mostly consisting of cattails and bulrushes, this site became the largest managed riparian area along the Santa Cruz River in Pima County. The Sweetwater Wetlands contains a flow channel, ponds approximately four feet deep, and bordering shallow-water areas one to two feet deep.[19] From the ponds and shallow water, effluent flows into settling ponds, recharge basins, and the river. Ultimately, this water percolates into the groundwater along the river for future use as potable or irrigation water.

The area of unmanaged riparian habitat extends down the Santa Cruz River, past the north end of the Tucson Mountains and into the south edge of the Santa Cruz Flats before disappearing into the channel bed. In this reach, a narrow strip of vegetation consisting of a mixture of native (cottonwood, willow) and nonnative (Athel tamarisk, or saltcedar) riparian species lines the narrow channel conveying effluent from the Roger Road and Ina Road wastewater treatment plants (fig. 10.8). These trees and shrubs are sustained by the groundwa-

ter mound created as water infiltrates into the alluvial aquifer, although the width of the mound is narrow as it spreads out and passes downward toward the water table. If the effluent, which has channelized itself between floodplains, were spread out over the wide area between the soil-cemented banks, more obligate riparian species might become established. Several species of native trees and shrubs that historically grew along the Santa Cruz River have reestablished on the banks of the various water structures; these include riparian gallery trees, such as Frémont cottonwood and black willow, and smaller trees, such as velvet mesquite (the species of the Great Mesquite Forest) and native shrubs.[20]

Two problems are created by effluent discharge. First, if the effluent infiltrates into the shallow alluvial aquifer, it counts as a positive toward groundwater management in the Tucson Basin.[21] In 1993, 90 percent of the effluent (50,620 acre-feet of discharge) infiltrated into the channel.[22] The nutrient-rich, clear-water effluent promotes algal growth in the riverbed, which can clog the porosity of bed sediments and reduce infiltration unless floods are large enough to scour the bed.[23] By 2011, eighteen years after the scouring 1993 flood and five years after the larger 2006 event, only 45 percent of 49,500 acre-feet of effluent recharged into the bed. Bed clogging means that the potential for groundwater recharge is reduced and that perennial flows may extend out into Pinal County, where any claim over the water is then lost by Pima County and its municipalities.

The second problem is that the clear-water effluent carved a deep narrow channel into the channel alluvium. Near the Sweetwater Wetlands, the downcutting rate is nearly 2 feet per year in the channel and 0.7 feet per year on the floodplain.[24] These rates generally decrease downstream until the effluent is totally infiltrated into the channel near Marana. The Santa Cruz River in downtown Tucson, dry most of the year, has aggrading floodplains and an unchanging thalweg elevation, while the same river downstream with perennial flow is eroding its bed.

Wastewater discharge completely changed the public perception of this once-dry reach of the Santa Cruz River. It has become a favorite destination for birders, who literally flock there to enjoy riparian and other wetland birds otherwise rare in the Tucson Basin and southern Arizona (see appendix M). The nutrient-rich effluent promoted rapid growth of many riparian and wetland plants that had grown along the Santa Cruz River more than a century earlier—only much farther upstream. Bulrushes and cattails provide habitat for Common Gallinule, Marsh Wren, and Common Yellowthroat, which otherwise would be uncommon in the Tucson Basin (see appendix C). Open water in the ponds attracts a large number of wintering and migrating ducks as well as breeding Mallard, Ruddy Duck, and Pied-billed Grebe. Mud flats around areas such as settling ponds are ideal for Killdeer and migrating shorebirds. Long-legged wading birds, including Great-blue Heron, Green Heron, and Black-necked Stilt, are attracted to the deeper water of the Sweetwater Wetlands in their hunt for prey.

Other Restoration Projects

Importation of CAP water and wastewater effluent has led to several other small-scale efforts to create riparian habitat along the Santa Cruz River. Perhaps the most ambitious is along the Santa Cruz River near Mission San Xavier, using part of the CAP water allocated to the Tohono O'odham and imported from the Colorado River to build off-channel wetlands. Known as the San Xavier Indian Reservation Riparian Restoration Project,[25] its goal is to create small riparian areas of 12.5- and 5-acre size with cottonwoods, willows, and mesquites.

These sites are on what once was the northern edge of the Great Mesquite Forest (see fig. 1.1).

Other projects in the planning stages are ambitious but follow the same pattern of substituting effluent or imported water for what once occurred naturally in the Tucson Basin. A project called the Paseo de las Iglesias was planned for the Santa Cruz River floodplain upstream from downtown Tucson, reportedly involving 1,100 acres to be irrigated using water harvesting.[26] The proposal called for planting mesquite and other native riparian species in the floodplain and small groves of cottonwood and willow at tributary junctions. Similarly, the El Rio Medio Project is proposed for 2,675 acres along 4.5 miles of the Santa Cruz River north of downtown; it would use a combination of reclaimed water and water harvesting. Both projects are funded by the US Army Corps of Engineers.

These projects reflect the increasing interest in the Tucson region for additional recreational opportunities, including birding. With the development of the Sweetwater Wetlands, bird-watching along the Santa Cruz River is enjoyed by a large local birding population, supported by the well-organized Tucson Audubon Society. Birding in the wastewater-effluent reaches has contributed to a growing ecotourism business that draws increasing numbers of visitors from outside of the basin, which is important to the local economy. Ecotourism is enhanced by the fact that southeastern Arizona, including the Santa Cruz Valley and the surrounding "sky island" mountains, is home to a large number of species (see appendix D) that birders from throughout the United States and other countries wish to observe in the wild.

The Increasing Value of Reclaimed Water

Reclaimed wastewater is valuable for many nonpotable uses. For example, mining companies in Pima County want to switch to reclaimed water to lower their dependence on groundwater.[27] Landscaping in parks and other recreation areas can use reclaimed water, and power plants and other industries could use it for cooling and cleaning. One of the most valuable uses of wastewater is to recharge the alluvial aquifer of the Santa Cruz River, thereby offsetting the high rates of groundwater withdrawal in the Tucson Basin. As previously discussed, an estimated 84 percent of the reclaimed wastewater is discharged into the channel,[28] and most of it seeps into the alluvial aquifer upstream from the Pinal

County line. Once the effluent began to be discharged into the channel, riparian vegetation became established, and reclaimed wastewater became the key ingredient in small-scale river restoration projects.

Needless to say, 17.3 acres of new Sweetwater Wetlands hardly restores 7 square miles (4,480 acres) of destroyed Great Mesquite Forest. Likewise, the narrow strip of perennial riparian vegetation that lines the channel fed with wastewater effluent does not compare to the wide bosque that once thrived upstream. The politics of water dictates that restoration on the scale of the Great Mesquite Forest is impossible. If the barren, eroding landscape upstream from Martinez Hill somehow was converted to wetlands or a bosque, and if Tucson and Pima County residents actually had the will to propose to restore the iconic riparian area of the Santa Cruz River, the political opposition would kill the project. In a world obsessed with disease eradication, enhancement of mosquito habitat—and the potential for greater incidence of West Nile virus and, eventually, dengue fever—would have difficulty getting past the planning boards. Already, the Sweetwater Wetlands is treated for mosquito abatement, further shifting this artificial wetland away from any semblance of naturalness. Against the specter of additional mosquito habitat, the loss of water to evaporation and transpiration would be minor.

As Tucson moves into the second decade of the twenty-first century, drought generally prevails (see fig. 2.4). Water conservation is working; water deliveries by Tucson Water peaked at 125,000 acre-feet in 2002, one of the most severe drought years, and afterward declined (see fig. 9.9). Conversion of groundwater and CAP water moving through the Tucson waste stream and into the Santa Cruz River has blunted the impact of lowered flood frequency and below-average rainfall. At the time of Tucson's settlement, ancient groundwater discharging at the surface sustained a mission and a large wetland that became the Great Mesquite Forest. That water source was buffered from short-term climate change, which attracted the Tohono O'odham and the Spanish priests, as well as the trees. Now, water again is perennial in the Santa Cruz River, albeit in an entirely different location and geometric configuration. As before, and for completely different reasons, this water once again is buffered from short-term changes in climate and now sustains a riparian ecosystem.

11

Summary of the Past and Some Possible Futures

The Santa Cruz River in the early twenty-first century bears little if any resemblance to the river that attracted the Piman peoples who lived along the river in the seventeenth century and the Spanish colonists and Jesuit priests who wanted to develop the fertile floodplains. While fanciful descriptions invoked a land of milk and honey, with a continuous free-flowing river full of beaver and fish, the reality of the presettlement Santa Cruz River was far more complex, both hydrologically and ecologically. This river seldom flowed continuously from headwater to terminus and was—and still is—better defined by subsurface water than by surface flow. As the groundwater levels changed through the centuries, influenced by either arroyo downcutting or water extraction, so changed the conditions of the riverine ecosystems and the ability of people to use its water, either for domestic supplies or for irrigation.

This river continues to change, despite its banks of soil cement and the lowered water levels beneath its bed. Almost in defiance of hydraulic engineers, who encased its margins in cement, the river refused to retain its trapezoidal shape, instead depositing new floodplains here, downcutting a new narrow channel there, growing new riparian vegetation within its channelized course, and even exhibiting a new tendency for channel filling. This has turned wastewater into wetlands, creating riparian ecosystems that scarcely resemble those that the river once sustained. The change represents a trend toward viable riverine ecosystems, albeit not in the same places where these ecosystems once thrived, and not sustaining the same species as before. Nonetheless, birds and tourists increasingly are attracted to the new habitat. Although local restoration is possible, large-scale restoration is impossible, given the hydrologic reality of increas-

ing demand for water and diminished groundwater supplies.

The Development of Arroyos in the Tucson Basin

The Santa Cruz River experienced a complex alluvial history culminating in a major cut-and-fill cycle between 500 and 300 years ago, with renewed downcutting historically (see fig. 2.3). Short entrenched reaches existed six to twelve miles upstream of Tucson as early as 1849, suggesting that the Santa Cruz River was a discontinuous arroyo through the Tucson Basin. In most reaches, particularly near present-day downtown Tucson, the stream flowed at or near the valley surface over a broad floodplain. Perennial reaches were present upstream from present-day Continental, near San Xavier, at Tucson, and, less certainly, near the confluence of the Santa Cruz and Rillito Rivers. These reaches, constituting only 20 percent of the forty-mile course through the Tucson Basin, supported extensive cienegas near San Xavier and localized riparian areas at Tucson and the confluence of the Rillito and Santa Cruz Rivers.

In the San Xavier reach, a discontinuous arroyo deepened ten feet between 1872 and 1882, and a new arroyo entered the cienega from downstream, eroded by floods in the summer of 1887. In the Tucson reach, four major floods in the summer of 1890 initiated headcut migration that started at the unprotected heading of an intercept ditch. The arroyo extended upstream into the San Xavier reach during winter floods in 1904–1905 and 1914–1915. Ultimately, a continuous arroyo was created by diversion of water in an artificial ditch designed to protect San Xavier Mission by redirecting flow from the West Branch to the east side of the valley. The artificial channel entering the Great Mesquite Forest upstream from Martinez Hill was widened considerably through the twentieth century, especially by floods from dissipating tropical storms in 1977 and 1983. Downstream from the Tucson Basin, the 1914–1915 flood initiated a headcut along a diversion canal (Greene's Canal) near Red Rock, and sediment excavated from the Tucson Basin was deposited farther downstream on the Santa Cruz Flats. Despite protective measures, the headcut at Greene's Canal eroded again during the 1983 flood, but subsequent floods in 1993 and 2006 had little impact here.

Arroyo initiation and extension occurred during relatively wet decades marked by heavy rains and large floods in southern Arizona. During the late nineteenth century, all but one of the years registered more than three days with rain exceeding one inch per day in Tucson. Although many of these wet years were associated with El Niño events, at least one year when large, channel-eroding floods occurred (1890) was a La Niña year. In the twentieth century, the greatest channel changes on the Santa Cruz River were caused by floods in 1905, 1915, 1977, and 1983; all four floods were associated with El Niño episodes.

Arroyo initiation and coalescence into continuous channels occurred during relatively wet decades characterized by high rainfall intensities. The scant observations suggest that most of this moisture may have originated in the Pacific Ocean and not from the Gulf of Mexico, the origin of most of the moisture in the North American monsoon.[1] The late nineteenth-century climate was different from that of the twentieth century and may have differed from climatic regimes in previous centuries. Changing flood probabilities with low-frequency (decadal) climatic fluctuations may partly explain arroyo initiation in the late nineteenth and early twentieth century, as well as periods of floods that occurred in the late twentieth century. Although the increase in flooding in both periods has been attributed to human influences—especially the introduction of cattle and construction of floodplain canals—the evidence suggests that increased flood magnitudes owe more to unusual climate than to land-use practices. Land-use practices, particularly floodplain irrigation systems and roads, clearly affected the amount and location of the erosion induced by flooding.

In summary, the arroyo of the Santa Cruz River formed when climatic conditions heightened the probabilities for occurrence of large floods in southern Arizona. Livestock grazing in Pima County peaked well after arroyos had already downcut (see fig. 8.1b). The climate of southern Arizona is known to have varied historically, but groundwater extraction, in particular, greatly altered riparian ecosystems that could have attenuated peak discharges through the Tucson Basin. Also, far too many roads, canals, and ditches on floodplains downcut into arroyos for us to ignore land-use effects. First, however, we examine the climate record to assess what might have changed to increase the probability of flooding on the Santa Cruz River.

Climatic Variability in Southern Arizona

Previous work has established that large-scale climatic phenomena affect the hydroclimatology of southern Arizona and the watershed of the Santa Cruz River.[2] Precipitation in certain parts of the world are *teleconnected*, or related over long distances,[3] by global-scale phenomena such as the El Niño–Southern Oscillation (ENSO) phenomenon. The suggestion that a teleconnection exists between ENSO and North American weather dates to pioneering studies in the 1920s and 1930s,[4] and the link with seasonal precipitation along the West Coast and in the Southwest has been well studied.[5] For example, the southwestern United States usually has abundant precipitation at the same time that the northwestern United States undergoes drought. Similarly, the southwestern United States has drought when much of Australia is wet. Propagation of teleconnections worldwide suggests that the same climatic process may control concurrent flooding in Arizona and Florida or India and Australia. Teleconnections provide a network for studying the worldwide propagation of low-frequency climatic fluctuations;[6] they also show the strength of connection of local-scale climate to the global-scale climatic systems.

Tucson boasts one of the longest precipitation records (1868–2009) in Arizona (see fig. 2.4), though like other long-lived southwestern stations, it has a complicated station history.[7] That record, and particularly the graphs of seasonal precipitation anomalies (see fig. 2.5), show that precipitation has never been constant but instead has high interannual variability and periods of high or low annual or seasonal precipitation. Certain periods of time identified in regional analyses of low-frequency climate are apparent in this record as well (see chapter 2).[8] Among these are the general droughts at the end of the nineteenth century, in the middle of the twentieth century, and at the start of the twenty-first century; and pluvial (wet) periods in the late nineteenth century and early and late twentieth century. The Santa Cruz River responded to each of these periods of climate, and an understanding of why those periods occurred is essential to understanding both past and future responses.

Frequency of El Niño–Southern Oscillation Conditions in the Twentieth Century

As discussed in chapter 2, ENSO affects various meteorological and oceanographic conditions worldwide, and teleconnections are particularly pronounced during ENSO conditions.[9] ENSO conditions affect the hydroclimatology of the southwestern United States, particularly Arizona.[10] Areas closely teleconnected with the equatorial Pacific Ocean, such as the southwestern United States, have increased variability of seasonal and annual precipitation.[11] Winter frontal storms are more numerous and intense during certain ENSO years because of an intensified Aleutian Low and southerly displacement of storm tracks.[12] The probabilities for incursions of tropical cyclones change during ENSO conditions, but the advection of moisture needed to fuel monsoonal storms is reduced.[13] Hypothetically, ENSO conditions could reduce the number of monsoonal storms but increase the number of frontal systems and tropical cyclones that affect Arizona. El Niño conditions prevailed in more than half the years since 1965, a period with a high incidence of such conditions since AD 1525.[14] This indicates a global climate signal associated with floods on the Santa Cruz River between 1977 and 1993 (see fig. 11.1).

ENSO affects the variability of tropical-cyclone generation. On average, and after 1965, fewer tropical cyclones were generated under El Niño conditions (14.7 tropical cyclones per year) than non-ENSO conditions (17.4 tropical cyclones per year). Even with fewer tropical cyclones possible, the incidence of tropical cyclones dissipating over Arizona increases during El Niño conditions. The largest numbers of dissipating tropical storms per year that affected the southwestern United States occurred in the ENSO years of 1925–1926, 1939, 1957–1958, 1976–1977, and 1982–1983.[15] In September, the peak month for cyclone recurvature, 3.4 tropical cyclones per year were generated during El Niño conditions compared with 2.3 tropical cyclones per year during other years. The annual number of tropical cyclones generated increased from 13.7 per year for

1965–1970 to 16.4 per year for 1983–1988,[16] but then decreased to 15.5 per year for 1989–2011 and 14.8 per year during the first eleven years of the twenty-first century (see fig. 2.6).

Climate in southern Arizona is strongly related to ENSO conditions. For example, correlations between the Southern Oscillation Index (SOI), the most commonly used time series of ENSO, and seasonal precipitation for many Arizona stations are statistically significant.[17] During ENSO conditions, precipitation in Arizona and western New Mexico is enhanced in the normally dry spring and fall because warm sea-surface temperatures off the west coasts of Mexico and California provide the necessary energy for the development of strong West Coast troughs; weaken the tradewind inversion and thus allow moist air to penetrate into the Southwest; and cause more tropical cyclones to recurve to the northeast than usual.[18] As a result, many of Arizona's largest floods have occurred during El Niño years.[19]

Tucson precipitation is seasonally affected by ENSO-related phenomena in some subtle ways.[20] Significant positive correlations are obtained between indices of ENSO and divisional precipitation in the southwestern United States during October, November, and the following February and March,[21] and southern Arizona yields the best correlations for each of these months.[22] Because precipitation is enhanced in spring and fall, seasonality of precipitation during El Niño conditions is blurred.[23] This regional pattern extends to most of Mexico, which experiences summer drought during El Niño conditions.[24] The southeastern United States and northern Mexico have above-normal precipitation during El Niño conditions in 80 percent of the cases studied for the October to March season.[25] In the Great Basin, above-normal precipitation was also found in 80 percent of the cases for the April through October period during El Niño conditions.

Heavy Versus Light Rains

One of the more popular explanations for arroyo downcutting in the Southwest is a shift to a climatic regime favoring heavy rains at the expense of light ones.[26] Though such shifts have been documented in the northern Rio Grande Valley,[27] on the Colorado Plateau,[28] in southern Arizona,[29] and in California,[30] they are generally asynchronous. In the Santa Fe record, the average annual intensity of rainfall was highest in the period 1850–1880 and lowest between 1880 and 1925.[31] On the

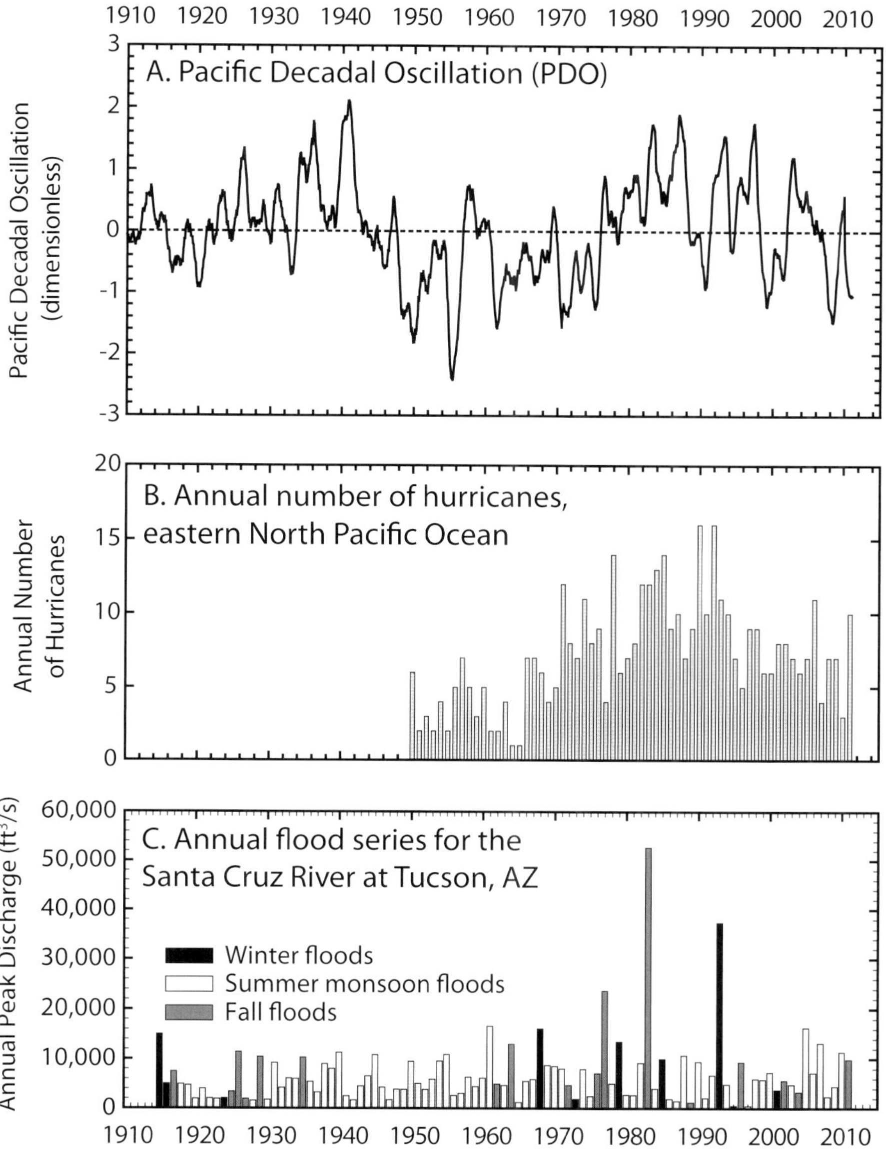

Figure 11.1.

Comparison of climate and hurricane frequency in the Northeast Pacific Ocean with annual peak discharge on the Santa Cruz: **A.** Time series of annual Pacific Decadal Oscillation (PDO), a measure of the climatology of the North Pacific Ocean and the atmosphere over it. Periods of positive PDO correlate with wet conditions in the southwestern United States; periods of negative PDO tend to be dry. **B.** Annual number of hurricanes in the eastern North Pacific Ocean (repeated from fig. 2.6B). **C.** Annual flood series for the Santa Cruz River at Tucson (repeated from fig. 2.7).

Colorado Plateau, rainfall intensities in summer were highest between 1909 and 1932, when many arroyos developed in the region.[32] In southern Arizona, summer rainfall was unusually heavy in the period 1868–1890, and the contrast between heavy and light rain appears to have been quite dramatic.[33] California exhibits the same trends in heavy rainfalls as southern Arizona, with 1884–1890 being the wettest period of record.[34]

Between 1868 and 1893, ten years exceeded 7 inches for July and August, compared with only six between 1890 and 2008. Comparisons of the seasonal (June–September) frequency of heavy (>1 inch) versus light rains (<0.5 inch) rains show a high frequency of heavy rains and a low frequency of light rains from 1868 to 1890 and a return to a high frequency of heavy rains after the 1950s, albeit without a concomitant decrease in light rains.[35] Perhaps only coincidentally, ENSO activity was unusually strong and frequent during the period 1864–1891.[36] Between 1868 and 1890, all but one of the years registering more than three days with >1 inch of rain were during El Niño conditions (1868, 1871, 1874, 1878, 1880, 1884, 1887, 1889).[37] The one exception was the summer of 1890, an important year for arroyo downcutting along the Santa Cruz River. This relation did not persist into the twentieth century, when August rainfall typically was lower under El Niño conditions.[38] Increased summer rainfall, which was associated with arroyo downcutting and widening in the late nineteenth and early twentieth centuries, does not appear to be related to global-scale climate in the early twenty-first century.

Rainfall during the period of 1868–1890 appears to be unique compared with the instrumental record in the twentieth century and precipitation reconstructed from tree rings in the past few centuries. As previously discussed, recent reconstructions of ENSO suggest that the frequency of ENSO conditions was higher in the past 150 years than the previous five centuries.[39] Intensified ENSO activity during the period 1864–1891 may be symptomatic of long-term climatic change, perhaps associated with a regional response to the end of the Little Ice Age.[40]

The Floods of 1890

The floods in the summer of 1890, which were the most important in coalescing the arroyo segments in the Tucson Basin, were preceded by notable floods in 1886 and 1887. The 1890 floods received a great deal of attention in the newspapers, comparable perhaps to coverage of the floods in 1915, 1977, and 1983. The season was unusual in the amount of precipitation, the anomalously cool temperatures, and the number of runoff events. The first flood swept through the valley during the five-day period of 27 July–1 August, producing overflows more than two thousand feet wide and up to twelve feet deep in the Tucson area. The next round of floods between 4 and 7 August downcut the Sam Hughes's Ditch and eroded the headcut upstream from Congress Street. These were singular events in the history of arroyo development in the Tucson Basin.

At the same time, flooding on the Rillito River caused downcutting. One astute observer noted that the deepened channel of the Rillito River could now carry a third more water without reaching the flood stage of a few days before. The Santa Cruz River rose again on 13 August, washing out the dam at Silver Lake. Flooding resumed on 23–25 August, reportedly reaching the highest flood stage of the summer. Arroyo initiation, then, was produced by four major floods spaced three to ten days apart in the summer of 1890. With the exception of August 1955, the number and persistence of summer floods in 1890 has not been repeated in the past 120 years.

Rainfall in the summer of 1890 was unusual. Low surface pressure and low-level winds from the south-southwest prevailed over most of July, intensifying in August.[41] The primary source of moisture was the Pacific Ocean rather than the Gulf of Mexico.[42] A total of 11.81 inches of rainfall was recorded for the two months at Camp Lowell (in Tucson) and other stations throughout the upper Santa Cruz and San Pedro valleys. Cooler-than-normal high temperatures occurred over several days during the periods of sustained rainfall. During July and August at Tucson, the probability of observing maximum daily temperatures equal to or less than 85°F, which occurred over four days in 1890, is about 1 percent.[43] These lower-than-normal high temperatures can be explained only by advection of cooler air carried in a cold front or in a tropical air mass from the Pacific Ocean.

Unlike most twentieth-century floods on the Santa Cruz River, the 1890 floods occurred during La Niña conditions characterized by a relatively cool Pacific Ocean.[44] When El Niño conditions returned in 1891, large winter floods occurred statewide, and this scenario was repeated only in 1993. With the exception of August 1961, all of the annual peaks at Congress Street above 15,000 ft^3/s have been associated with El Niño conditions and winter frontal storms or dissipating tropical cyclones. Perhaps it is significant that 1961 experienced strong North Pacific atmospheric circulation (an intensified Aleutian Low), which is more characteristic of ENSO events.[45] By every measure, climate and flooding in the summer of 1890 was unique, and the exact climatic drivers defy simple explanation.

Late Twentieth-Century Floods and Channel Change

Following the severe floods and channel change of the late nineteenth and early twentieth century, the Santa Cruz River settled down in response to primarily drought conditions in the mid-twentieth century. The start of the period of increased flood magnitude is uncertain; it may have begun in the 1960s,[46] or it could have begun during what has been called the step-change of 1976–1977.[47] By the late 1970s, climate had shifted into a period of large storms and repeated floods that wreaked havoc on southern Arizona. Extensive channel changes and property losses forced floodplain managers to install bank protection on the Santa Cruz River. Almost as fast as the period of flooding started, a low-flow period settled in, persisting into the second decade of the twenty-first century. This unusual history of low to high to low flood frequency has prompted the conclusion that flood frequency on the Santa Cruz River is nonstationary,[48] which challenges traditional methods for assessing flood hazard that are based on assumptions of time-invariant annual floods with no low-frequency climatic influence.

Numerous indices (too numerous to describe here) have been used to describe decadal-scale climate variability in North America. Previous work has used direct measures of ENSO—whether the Southern Oscillation Index or indices of sea-surface temperature—or has used other measures that reflect the combined effect of sea-surface temperatures and sea-level pressures. One such index used previously is the Pacific Decadal Oscillation (PDO).[49] The PDO reflects a persistent, El Niño–like pattern of climate variability that directly affects the southwestern United States and has been used to explain streamflow in the region.[50] The PDO has positive and negative values that are associated with persistent wet and dry periods, respectively, of twenty to thirty years' duration, reflecting the persistence of sea-surface temperatures in the Pacific Ocean. Twentieth-century fluctuations in the PDO show periodicities of fifteen to twenty-five and fifty to seventy years.[51]

The PDO helps to explain two phenomena about the Santa Cruz River and potentially elucidates the reasons for late twentieth-century flooding and subsequent drought (fig. 11.1). Extending from the 1970s through the mid-1990s, the PDO is persistently positive through the period of large floods on the Santa Cruz River. The first part of this period has been previously identified using only persistent ENSO conditions,[52] and this relation clearly shows a change occurring in about 1996 (fig. 11.1A). Even more interesting is the relation with the record of hurricanes in the eastern North Pacific Ocean (fig. 11.1B): the decrease in hurricanes (also all tropical cyclones [see fig. 2.6]) tracks both the PDO trend into the negative and the reduction in the annual flood series of the Santa Cruz River. This relation clearly shows

the close relation between flooding and global climate indices and reinforces the conclusion that flood frequency responds to climate, not land-use practices or urbanization.

Human Impacts on the Santa Cruz River Floodplain

Human activities on the floodplain affected flow conditions and channel change along the Santa Cruz River beginning in the 1880s. On one hand, it may have been more surprising if arroyo downcutting had not happened, given the number of irrigation ditches and livestock roaming the riparian areas. On the other hand, water erodes floodplains, and without significant floods, it is not clear that the arroyo would have downcut. The general pattern of land use, although more intense at the time of arroyo downcutting, had been in place for a couple of centuries. Entrenched channel segments were documented in the vicinity of Martinez Hill and San Xavier as early as 1849, and at least one of these could be associated with irrigation works. Even though the Santa Cruz River surely had large floods in the mid-nineteenth century (for example, the late 1850s and 1860s), the reach near Tucson was not continuously entrenched until Sam Hughes's Ditch eroded in 1890.

Before the river downcut in the Tucson Basin, floods traveled as broad sheets of water that would gather wherever the floodplain narrowed, as it does at the base of Sentinel Peak. In the 1880s, sufficient flow coming in from upstream of Tucson entered Silver and Warner's Lakes, and their dams were not engineered with large enough floodgates to pass large discharges; they stored and released water for operation of mills and irrigation. Despite floods in 1886 and 1887, which damaged the dams, an arroyo did not downcut in the reach downstream.

The irrigation technology at that time included intercept ditches,[53] which, depending on the fall of the river, would be excavated at a lower gradient than the river to intercept the near-surface groundwater at the upstream end. Water flowed from the start of the intercept ditch to agricultural fields, where the bottom of the ditch reached the level of the floodplain surface. Sam Hughes built an intercept ditch in 1888 but failed to protect its heading properly, instead relying on floods to excavate the heading, saving him the expense of labor. If Hughes had not started his intercept ditch, the arroyo may not have downcut when it did.

Once downcutting began on 4 August 1890, subsequent floods extended the headcut upstream to Silver Lake. The headcut continued to migrate upstream during floods in the 1890s, 1900s, and 1910s, eventually forming a continuous arroyo well into the San Xavier Reservation. Arroyo cutting may have not happened when it did if several decades of low flow conditions, such as occurred between 1930 and 1960, had prevailed in the 1890s.

The arroyo at Greene's Canal offers a similar challenge. The diversion canal constructed in 1910 was designed to widen during a single flood, which occurred in 1915. Had Greene's Canal not been built, the arroyo likely would not have developed in this reach during the 1915 flood. No other arroyos have developed in the lower Santa Cruz since the downcutting of Greene's Canal in 1915, and Greene's Canal, which once was thought to threaten the Tucson Basin,[54] has not moved upstream despite large floods in 1993 and 2006. Arroyo downcutting, whether at Tucson in 1890 or at Greene's Canal in 1915, was the result of unusually large floods eroding canal headings soon after these ditches were constructed.

Despite two centuries of land use in the Tucson Basin—even though the floodplain in the area of San Xavier and Tucson had been heavily cultivated and grazed—downcutting capable of coalescing discontinuous arroyos did not occur until 1890. On the Santa Cruz River, poorly engineered dams and ditches concentrated flows, which were apparently of large magnitude and certainly of long duration, to focus the erosive power of floods and downcut arroyos through wide floodplains. The history of the Santa Cruz River supports the hypothesis that increased precipitation and flooding initiated arroyo downcutting, but land use on floodplains influenced the result, perhaps even increasing the potential for erosion and certainly guiding the course of the consequent channel.

In summary, the arroyo of the Santa Cruz River formed when climatic conditions heightened the probabilities for occurrence of large floods in southern Arizona. Construction of ditches that were intended to erode as part of their design resulted in abrupt changes in the longitudinal profile of the stream, which further augmented probabilities that any additional floods would initiate an arroyo at the head of these ditches. Large floods in the late twentieth century forced installation of bank protection, but reduction in flood frequency, again in response to climate and despite increasing urbanization, has allowed floodplain deposition within the stabilized channel and establishment of riparian vegetation that increasingly traps sediment. In the future, changing flood probabilities with low-frequency climatic fluctuations and decreased flow conveyance owing to the initial stages of channel filling will further complicate management of the arroyo in an increasingly urbanized floodplain of the Santa Cruz River.

Past the Tipping Point: Riparian Ecosystems in the Tucson Basin

The Great Mesquite Forest, under pressure from woodcutting and agricultural clearing, was destroyed by groundwater overdraft (fig. 11.2) well after the arroyo had downcut.[55] Agricultural clearing dated back to the days before Father Kino arrived to establish Mission San Xavier, and woodcutting was relentless for several centuries. Woodcutting in the Great Mesquite Forest during the early 1900s resulted in the loss of the bosque's towering old-growth trees, although woodcutters kept cutting and the trees regrew; woodcutting did not kill the trees but altered the structure and function of the bosque. Agricultural encroachment and roads decreased the acreage and overall extent of the bosque.[56] The bosque withstood arroyo downcutting, clearing, and woodcutting, but this unique ecosystem could not withstand those stresses combined with the lowering of the underlying water tables.

The Great Mesquite Forest occupied at least 7.8 square miles (see chapter 7), or about 5,000 acres. Now, the area once occupied by dense mesquite resembles a mudflat with scattered xerophytic mesquite, creosotebush, and ruderal shrubs (fig. 11.3). In 2008, the Santa Cruz River from the Mexican border to the Santa Cruz–Pima County border (see fig. 1.1) sustained 5,710 acres of riparian woodland in 409 discrete patches.[57] Agricultural clearing, in large part, is responsible for this present-day habitat fragmentation, creating discontinuous riparian woodlands along the upper Santa Cruz River that are not ecologically equivalent to a large contiguous bosque.

The tipping point past which recovery was impossible for the Great Mesquite Forest occurred in the early 1970s, when groundwater levels plunged below the depth generally thought to be the limit for obligate riparian species (see figs. 7.3 and 11.2). In 1982, two wetland experts summarized the demise with these words: "Once one of the finest mesquite bosques in the Southwest, ground water pumping has now virtually destroyed this interesting community."[58] In the mid-1980s, all that remained of the Great Mesquite Forest were

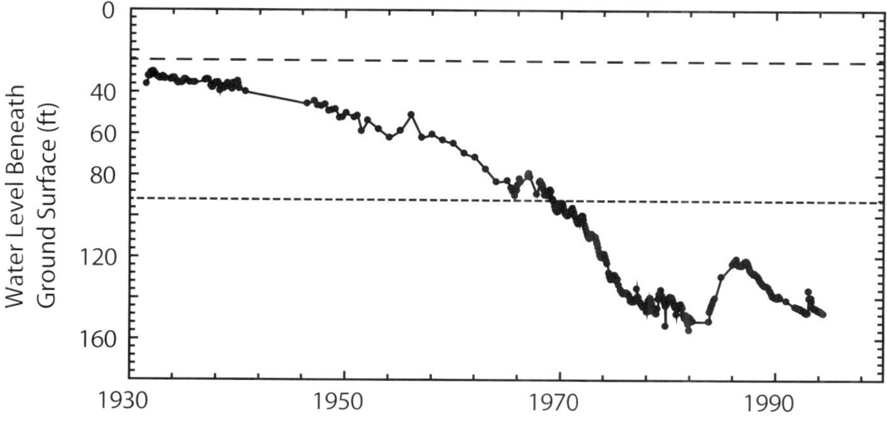

Figure 11.2.

Groundwater levels for well D-15-13 11CBA near the Santa Cruz River between Valencia and Drexel Roads in Tucson, Arizona. The dashed line represents the bottom of the Santa Cruz River channel, and the dotted line represents the generally accepted maximum rooting depth of ninety feet for most riparian trees.

Figure 11.3

Oblique aerial photograph of the Santa Cruz River upstream from Martinez Hill (middle distance). This downstream view shows the approximate area once occupied by the Great Mesquite Forest, which is sparsely vegetated and light gray in color. The Tucson Mountains appear on the left horizon; on the right, Tumamoc Hill appears in front of the Santa Catalina Mountains.

stumps of what once had been large trees, which were replaced by xerophytic mesquite, desert shrubs, and, in some areas, nonnative Athel tamarisk.

By the middle nineteenth century, Anglo-Americans had brought concepts and technology to the arid American Southwest from more mesic eastern landscapes.[59] At first, irrigation "greened the desert," but water diversion and groundwater withdrawal drastically lowered water tables, and both natural and cultural resources suffered. Groundwater demand for irrigation and domestic supplies for the growing Tucson metropolitan region accelerated under the assumption that subsurface water was a sustainable resource. By the time the rosy projections of the 1930s were proven false, the decline of this unique forest was irreversible, and the water table had been lowered well below the roots of the mesquites, hackberries, cottonwoods, and willows.

Perennial reaches were eliminated from the Tucson Basin, and the Santa Cruz–Rillito River system was converted into a network of ephemeral channels. Where people once picnicked under giant mesquite and cottonwood trees, swam and fished in the river, and hunted ducks, doves, and quail, the Santa Cruz River now flows only after it rains, and the band of riparian vegetation along its channel is characteristic of ephemeral streams. Nowhere in the Southwest has water development been more detrimental to riparian ecosystems and avifauna than along the Santa Cruz River.[60] Beyond the statuesque mesquites it once sustained, the Great Mesquite Forest supported some of the densest populations of certain birds (for example, Lucy's Warbler)[61] and provided habitat for many species at the northern extent of their ranges. The ebb and flow of populations of birds at the periphery of their ranges is well known,[62] but the loss of an entire avifauna is far beyond the natural stresses that control expansions and contractions of species ranges.

Quantifying losses in the mesquite bosques along the Santa Cruz River, as well as other lowland rivers of the Southwest, is impossible. Robert Webb documented changes in riparian vegetation in the southwestern United States, mostly using repeat photography.[63] Most increases in riparian vegetation involve linear bands along watercourses, changes in species composition, and changes along higher-elevation perennial streams. The largest decreases are caused by groundwater withdrawal and construction of large reservoirs.[64] Actual measurements that document changes in extent of bosques or canopy of riparian gallery forests are nonexistent; other than bird lists and some vertebrate col-

lections, we do not know precisely what animal species were in these bosques, or even their general extent.

We will never know precisely what was lost from the Great Mesquite Forest. Methods for quantifying plant and animal populations were not developed until the middle of the twentieth century, well after the decline in mesquite bosques began. Whereas information on changes in plant communities has been lost, avifaunal data from the Santa Cruz River, although incomplete, is exceptional compared to that of most southwestern riverine ecosystems.[65] Three decades would pass before quantification of avian populations would become an important undertaking for ornithologists,[66] and five decades would pass before avian populations along a southwestern river system would be quantified.[67]

The avifauna of the Santa Cruz River before Anglo-American colonization is unknown, although reports from ornithologists beginning shortly afterward, in the mid-1800s,[68] provide a glimpse of spectacular bird populations. Most of the avian species that nested historically in the Great Mesquite Forest were obligate or preferential riparian nesting species.[69] Species that were considered accidental visitors during the late 1800s and early 1900s may have earlier been regular nesting species along the Santa Cruz, especially waterbirds such as the Black-bellied Whistling-Duck.[70]

We used historic records to document the changing suite of bird species that lived in the Tucson Basin between about 1872 and the present (see appendixes A and B). Approximately 85 species of summer birds were recorded historically from the Great Mesquite Forest, with more than 75 species recorded as nesting there before its demise (see appendix A), and there are indications that several other species may have nested here as well.[71] This compares to 80 species of summer birds for the Rillito River, with approximately 65–70 species nesting in its now-destroyed riparian areas (see appendix B). Most of those not nesting either foraged in these riparian ecosystems or flew overhead. This compares to approximately 65 species occurring along the Santa Cruz River today, and several of those species are recent additions because of the Sweetwater Wetlands, sustained by wastewater effluent (see appendix C).

Comparisons of the Rillito and San Pedro Rivers indicate that approximately 60 species of breeding birds are common to both watercourses but that each riparian system had 6 unique species.[72] The 65–70 avian species that nested historically along the Rillito River were in riparian vegetation that is similar to what once occurred along the Santa Cruz River. Mesquite, cottonwood-willow, and adjacent Sonoran desertscrub ecosystems were less extensive along the Rillito River, owing to the geography of the eastern Tucson Basin. The Santa Cruz River is in a broader valley with deeper alluvial soils that permitted more extensive growth of riparian ecosystems.

Bird Species No Longer Present in the Tucson Basin

We know that numerous breeding species no longer use or are present in the Tucson Basin; several species recorded historically either have been extirpated or their habits have drastically changed since settlement of the Tucson region by Anglo-Americans (see appendix H). Several species of birds that had been recorded earlier along the Rillito River had apparently been lost from the Great Mesquite Forest by the start of the twentieth century. These were largely riparian or wetland species, including Killdeer, Black Phoebe, Common Yellowthroat, and Song Sparrow. Other birds, such as the Gray Hawk and Willow Flycatcher, disappeared from both rivers (see appendixes A and B). In some cases, such as Wild Turkey and Ferruginous Pygmy-Owl, these losses are well documented. In other cases, we must extrapolate from historical writings or by comparing the current avifauna with earlier records; for example, although the Pied-billed Grebe and the Green Heron were not observed at the start of the twentieth century, strong evidence suggests that both were breeding in the Tucson Basin before and afterward.[73] Several large species—including the Common Black-Hawk, Zone-tailed Hawk, Crested Caracara, and Spotted Owl—disappeared from the Tucson Basin; other smaller species no longer here include the Yellow-billed Cuckoo and the Song Sparrow (see appendixes A and B).

Some of these species have recolonized the Santa Cruz River,[74] usually because of newly created habitat from increased surface water, especially effluent from wastewater treatment plants. Such areas include the Sweetwater Wetlands, the narrow riparian zones sustained by wastewater effluent from the Roger and Ina Road treatment plants, and the lush riparian woodland along the river upstream from Tubac, fed by the Nogales International Wastewater Treatment Plant.[75] The missing habitat elements for many of these previously extirpated species are largely bodies of water, cattail marshes, cienegas, and woodlands, such as

cottonwood-willow gallery forests and mesquite bosques.

At least three large species that originally occurred along the Santa Cruz River now occur as higher-elevation montane species. The first, Wild Turkey, now generally an inhabitant of montane forests and meadows,[76] was noted by early explorers during the last half of the nineteenth century along lowland rivers of southeastern and south-central Arizona, including the Santa Cruz and San Pedro Rivers.[77] Wild Turkeys were collected along the San Pedro River in the 1880s,[78] recorded "about the valley of the Santa Cruz,"[79] and found on the Santa Cruz at Tubac in 1849 and 1864.[80] In 1849 near Tubac, "thousands of wild turkeys came here to the river to drink";[81] there they also would have found mesquite beans, fruit from hackberry and elderberry trees, and grasshoppers, a favorite insect food.[82] We do not know when Wild Turkey disappeared from the Santa Cruz River, other than it was sometime between the mid-1800s and when it had been "shot out" in southern Arizona by 1907.[83]

In contrast, we have a 135-year record tracing the loss of the Ferruginous Pygmy-Owl, not merely from the Rillito–Santa Cruz River system but from the entire Tucson Basin. This is an example of a bird whose population changed during historic times from being a relatively common, widespread riparian nesting species in southern and south-central Arizona to an uncommon or rare and, finally, an extirpated one.[84] This diurnal bird is abundant and widely distributed in other regions, and it is probably the most common owl in the American tropics.[85] The Ferruginous Pygmy-Owl was discovered in the United States on the Rillito River in 1872,[86] but there is only one record for the Santa Cruz River, collected in 1884.[87] However, because of the similarity of habitat and resident bird populations between the Santa Cruz and Rillito Rivers, we surmise that this owl had earlier been extirpated from the Santa Cruz River (see appendixes A and B).

Although the Ferruginous Pygmy-Owl is best known in the Tucson Basin, this small, resident owl was later found on the Gila, Salt, and New Rivers and Cave Creek in central Arizona. A Tucson population of the Ferruginous Pygmy-Owl adapted to Sonoran desertscrub, using saguaro cacti for nesting, a situation that was apparently an *ecological sink.*[88] Near Phoenix, the species was recorded in vegetation along banks of irrigation canals as far as ten miles from the Salt River,[89] perhaps another ecological sink. The Ferruginous Pygmy-Owl in the Phoenix region continued to be an obligate riparian nesting species until last recorded in

May 1971.[90] The Arizona population was listed as a federally endangered species in 1997 and was delisted in 2006; by 2008, all known Ferruginous Pygmy-Owl breeding populations from the Tucson region northward had been extirpated.[91]

Other species were surprising finds in the Tucson Basin. In 1872, a Mexican Spotted Owl nest was found in a cottonwood tree ten miles northwest of Tucson.[92] Also in 1872, a Zone-tailed Hawk nest was found on the Rillito River.[93] Like the Wild Turkey, the Spotted Owl and Zone-tailed Hawk now inhabit the montane forests in southeastern Arizona;[94] they appear to be partial to montane canyons, often with riparian ecosystems.[95] Both species now are unknown on the desert floor. The Gray Hawk nests along lowland streams,[96] as does the Common Black-Hawk (unlike the Wild Turkey, Spotted Owl, and Zone-tailed Hawk).[97] Both species nested in the Tucson Basin in the Santa Cruz–Rillito riparian ecosystems during the first half of the 1900s but were extirpated by the mid-1900s.[98] Another hawk, the Crested Caracara, was earlier reported from both the Santa Cruz and the Rillito River as a breeding bird; it was last recorded in 1917 (see appendix A).[99] Almost like the final gasp of the disappearing avifauna, nests of one of the rarest breeding birds in the United States, a Rose-throated Becard, were found in 1958 and 1959,[100] and on 1 July 1959, a male was found calling in a dying cottonwood.[101]

Loss of Vertebrates from the Tucson Basin

The elimination of perennial flow in the Santa Cruz River meant the loss of most associated hydroriparian and mesoriparian ecosystems, including large bosques and smaller and more isolated riparian areas.[102] Cienegas and pools that supported cattails and other aquatic plants disappeared following surface-water diversions and groundwater extraction. Some species of plants and animals associated with these ecosystems disappeared entirely, while populations of many others were greatly reduced. In some cases, local extirpation of entire species of fish, amphibians, reptiles, birds, and perhaps mammals occurred (see appendix H). Many of these depended on moist soil for burrowing and (or) egg laying.[103] Others, such as the Sonoran spiny lizard and ornate tree lizard, are basically tree dwellers.[104] Barren channel banks and depauperate floodplains presented unsuitable habitat for most vertebrates. Gone were the healthy, water-based environments that had

provided prime habitat in the arid Tucson Basin for a large percentage of insects that were the food supply for many of these vertebrates.[105]

Of the 175–180 vertebrates historically recorded in the Santa Cruz–Rillito River system, some have been even more severely impacted than birds.[106] None of the five species of fish once known from the middle reaches of the Santa Cruz River remain, and one species, the Santa Cruz (Monkey Spring) pupfish, has been completely extirpated (see appendix H). The California floater, a large edible unionid clam found in the Santa Cruz River in June 1880,[107] was dependent on fish (probably the Gila chub) and is no longer present in the watershed. The known riparian herpetofauna consisted of forty-four species: ten amphibians, sixteen lizards, and eighteen snakes, including the lowland leopard frog, Sonora mud turtle, and northern Mexican gartersnake, all extirpated during the 1900s (see appendixes F and H).[108] Today, we know of only seventeen species of reptiles and amphibians occurring along the river (see appendix F), with some others surviving in the surrounding uplands.

We do not know how many of the thirty-eight mammals recorded historically in the Great Mesquite Forest still occur along the river (see appendix G) or how many others may have occurred before records were kept. It would seem, for example, that muskrat should have occurred in the river adjacent to the forest, based on its distributional range and habitat preference.[109] The only record we have are remains recovered from a Classic Hohokam site (AD 1100–1450) at the former site of the bosque (see appendix H). Because of a lack of adequate records, we will never know whether muskrats were present in the Tucson Basin at the time of the earliest Anglo-American settlement in the mid-1800s or even when the Spanish arrived here in the 1600s.

Socioeconomic Losses: The Value of Mesquite Bosques

The loss of riparian birds, fish, and other riparian and aquatic plants and animals from the Tucson Basin is lamentable, but the most serious impact was the elimination of the aquatic/riparian ecosystems on both the Santa Cruz and the Rillito River without any chance of short-term recovery. On one level, this loss is roughly equivalent to the burning and permanent loss of forests in the Santa Catalina Mountains. Not only has there been an irretrievable loss of plants and animals, but recreational, scenic, and aesthetic values also were compromised.[110] Early accounts tell of boating, fishing, and swimming in Silver Lake and other localities along the Santa Cruz River, as well as picnics in the shade of the Great Mesquite Forest. One of the largest losses, however, might have been ecotourists coming to bird-watch in exceptional surroundings. In a recent study, sites in southeastern Arizona were named the leading birding destinations in the United States.[111] As a major industry, in 2001, birding and related wildlife recreation in Pima County alone was valued at more than $326 million and created in excess of 3,100 jobs.[112]

In the minds of some ecologists, mesquite bosques have been relegated to second place in importance as avian habitat in the Southwest, behind cottonwood-willow gallery forests. Conventional wisdom is that cottonwood-willow forests have higher population densities and, at least by inference, possibly a higher avian biodiversity (or species richness) than mesquite bosques.[113] This would seem to be true, based on recent avian studies in southwestern mesquite bosques,[114] as compared to studies in cottonwood-willow forests.[115] However, based on the sixty-year record from the Great Mesquite Forest, there are several reasons for suggesting that mature mesquite bosques support the greatest number of species and, possibly, the highest population densities of any North American riparian ecosystem or any ecosystem in the United States.[116]

Reasons for undervaluing the avian habitat importance of mature mesquite bosques compared with cottonwood-willow forests largely come down to history. No early studies of avian density were conducted in virgin mesquite bosques before they were decimated by woodcutters and replaced by secondary growth mesquites. Research has shown that bird species diversity is positively related to foliage height diversity, a measurement of density of foliage structure.[117] Comparisons of foliage height diversity between mature cottonwood-willow stands and mesquite bosques, such as the Great Mesquite Forest, were not made before the mature bosques were destroyed. Therefore, the premise of higher avian species biodiversity and population densities of cottonwood-willow forests has been advanced with half the information missing: mature cottonwood-willow forests are being compared to secondary growth of mesquite bosques. Along an estimated five miles of the Santa Cruz River, approximately eighty-five species of breeding birds were recorded historically in the Great Mesquite Forest and adjacent ecosystems (see appendix A), which is the same number of breeding species that was recorded along a

length of more than two hundred miles of the lower Colorado River valley.[118]

The loss of water-based recreation not only affects local residents but also reduces the socioeconomic health of the region. Ironically, mesquites are now widely planted in the Tucson Basin as xerophytic, low-water-use shade trees. Several species are widely used as ornamental plantings around homes and offices, along roadways, in parks and parking lots, and for greenery. A proposal has been advanced by planners to plant 500,000 mesquite trees in Tucson.[119] The following values (year 2000 dollars) have been applied to this planting proposal: (1) the estimated value of each tree in reducing cooling costs to buildings is $20.75 per tree annually, a total savings of $10,375,000 annually; (2) trees annually remove 6,500 tons of particulate matter from the air that would prevent the necessity of an alternative dust control program that costs $1,500,000 annually; and (3) vegetation reduces runoff that would otherwise require $90,000 in construction of detention structures, perhaps more in the future. These trees, spread through a metropolitan area, would not have offset the loss of a dense bosque such as the Great Mesquite Forest.

New Riparian Habitat: Introduced Saltcedar

While the gigantic mesquites of the Tucson Basin were being destroyed during the late 1800s and early 1900s, a new plant was introduced into the Southwest. A group of shrubs—eight species of saltcedar (tamarisk)—was imported from eastern Europe and the southern Mediterranean area and planted as ornamentals, as windbreaks, and for stream-bank stabilization. Saltcedar soon escaped from cultivation and the erosion-control sites, and by the 1920s at least 10,000 acres of riparian habitat in the southwestern United States consisted of various species of the genus. By 1970 this had increased to 1.3 million acres.[120]

In the 1970s, efforts to determine water-use savings by saltcedar eradication were replaced with a concerted effort to determine the impacts of nonnative species—particularly saltcedar—on native ecosystems.[121] This issue is complicated because saltcedar has both beneficial and negative impacts on riparian ecosystems. Initially, saltcedar was considered undesirable compared to most native riparian species, although some birds, particularly doves, were attracted to the dense, monospecific stands for cover and food. Stands of saltcedar-clogged

Figure 11.4.

Photographs of the Santa Cruz River showing three scenarios of channel management. **A.** (10 December 2011) View upstream from the Valencia Road Bridge showing Martinez Hill in the distance. The relatively natural arroyo banks are not protected with soil cement except on the far right side. A floodplain clad with woody riparian vegetation has developed on the left, and several cottonwood trees grow in an area that once was a bosque. These trees persist despite overall declines in groundwater

Figure 11.4. (*continued*)
levels in this reach (see fig. 11.2) and may be tapped into perched groundwater directly beneath the channel. (R. H. Webb.)
B. (9 December 2011) View downstream from the walkway beneath the Irvington Road Bridge showing soil-cemented banks
prominently on the left and behind vegetation on the right. This channel originally had a trapezoidal cross section, but
deposition of a floodplain has fostered establishment of facultative riparian vegetation. (R. H. Webb.) **C.** (9 December 2011) View
upstream from the Irvington Road Bridge showing a greatly widened channel with a broadly trapezoidal cross section and
low-flow channel soil cement. This reach is designed to contain floods of any conceivable size and with established woody
riparian vegetation increasing channel roughness. (R. H. Webb.)

stream channels slowed the flow of water while also increasing flood hazards, and saltcedar's perceived high water use meant loss of water resources through evapotranspiration. Although the goal of erosion control was achieved, saltcedar stands meant decreased water delivery downstream for domestic, industrial, and agricultural uses.[122] Once saltcedar became established, it outcompeted native cottonwoods, preventing them from becoming reestablished.[123] The word *phreatophyte* took on negative connotations when applied to saltcedar.[124] Large governmental and private sector efforts were undertaken to control and eradicate saltcedar along major watercourses in the Southwest, including the Colorado, Gila, and Salt Rivers and the Rio Grande.[125]

Saltcedar was considered to have low value for wildlife, especially for riparian birds.[126] This was largely because of the lower stature and structure of saltcedar, which diminished nesting opportunities compared to native trees such as cottonwoods and willows. Some birds, such as most woodpeckers and other cavity-nesting species, are unable to nest in saltcedar because it lacks large enough branches to support excavation of nesting and roosting cavities. In addition, saltcedar leaves possess salt glands that produce a saline exudate that is toxic to many native plants and presumably animals, including birds.[127] This has not been borne out by the research, however.[128] Additionally, saltcedar has been found to provide suitable habitat for numerous species of birds (see appendix L),[129] and it rivals mesquite bosques as nesting habitat for White-winged and Mourning Doves.[130]

A review of the literature documents approximately forty species of nesting riparian birds of the southwestern lowlands that use saltcedar (see appendix L). This is half of the approximately eighty species of land birds recorded historically as regular riparian nesting species of the Santa Cruz–Rillito River system (see appendixes A and B). Although saltcedar stands do not replace avian values of cottonwood-willow riparian gallery forests, and perhaps other riparian ecosystem values,[131] they are an improvement over barren stream banks and floodplains. As a result, some biologists have resisted vilifying this newcomer to riparian ecosystems. In considering impacts of saltcedar establishment within native riparian ecosystems, one researcher stated early on that "to the extent that saltcedar supplies the needs of the organisms from the previous biotic communities [replaced by saltcedar], the habitat disruptions will have little impact."[132] As late as 1987, a saltcedar expert wrote that "although tamarisk has undesirable effects in many water courses, it is apparent that this species may be an ecologically neutral or even beneficial addition to some riparian ecosystems."[133]

Where saltcedar intermingles with native riparian species in narrow galleries along the margins of watercourses, such as the Sweetwater Wetlands in the middle Santa Cruz River and the Colorado River in the Grand Canyon, the impact to the native ecosystem may be minimal.[134] In the Grand Canyon, there has been an increase in distribution and numbers of several species of native plants as well as riparian breeding birds consequent to the spread of saltcedar, and native riparian vegetation has become established along the dam-controlled river managed for flood control.[135] The most severe impacts occur where large stands of essentially monospecific saltcedar are established along major watercourses, such as reaches of the Gila and Colorado Rivers where natural or reservoir deltas occur. Both the biological components and the ecological processes of many of these ecosystems have been so altered by large-scale water projects that the modified ecosystems have been called a "reclamation disclimax," a plant community that would be structurally different without the modification.[136]

Introduction of a nonnative insect, the tamarisk beetle,[137] was designed to control saltcedar. Among other places, these beetles have been recently found along the Colorado River in Grand Canyon.[138] The tamarisk beetle controls saltcedar through defoliation, raising the question about how defoliation of large expanses of saltcedar might impact riparian birds. Because saltcedar is winter-deciduous, most impacts would be during the summer months, which is the critical time for breeding birds. This loss of foliage would reduce shelter for nesting birds and leaf availability for insects that some insectivorous breeding birds use for food. Some investigators have proposed that "the rate of regeneration and/or restoration of native cottonwoods and willows relative to the rate of tamarisk loss will be critical in determining the long-term effect of this large-scale ecological experiment" on the avifauna.[139] This is still an experiment in progress, but questions remain as to whether the trade of saltcedar for native species is a net ecosystem benefit or a zero-sum game in terms of habitat improvement.

The Great Mesquite Forest as an Outdoor Laboratory: A Lost Opportunity

The University of Arizona of the twenty-first century has an international reputation for its work in astronomy, optical sciences, and planetary sciences. Much of this success is owed to excellent year-round visibility for telescopes due to the dry, pollution-free air, few cloudy or foggy days, and low interference from atmospheric moisture and particulates. Also, the Sky Islands reach elevations 5,000 to 10,000 feet above the desert floor, allowing for the placement of telescopes in thinner air above interference from urban lights.

At the beginning of the twentieth century, the Santa Cruz–Rillito River system, with its Great Mesquite Forest and more scattered riparian areas so close to the University of Arizona, presented the opportunity to showcase an unusual natural ecosystem. Included were some of the rarest birds in the United States. The Great Mesquite Forest offered unequaled opportunities for biological field research in a poorly known North American ecosystem. Individuals from more than twenty-five scientific institutions and agencies conducted studies in the Great Mesquite Forest (see appendix I), and several of these individuals were directors or members of some of the country's leading biological institutions and agencies.

These ornithologists relied on their scientific knowledge to conduct studies in the forest; in turn, they acquired new information and published papers from these studies that were important to the development of the science of ornithology. Scientists also collected specimens of birds, nests, and eggs—often rare—which were archived in many of the country's leading museums. Thus, there was a feedback loop between ornithologists and the Great Mesquite Forest, with the forest playing a pivotal role in the development of ornithology as an international science. Continuation of riparian studies in the Tucson Basin ceased with the degradation of the Santa Cruz–Rillito River system, especially with the loss of the Great Mesquite Forest.

Some Glimpses of the Future of the Santa Cruz River

Prospects for meaningful ecological restoration for the Santa Cruz River are bleak, despite the worthy attempts exemplified by the Sweetwater Wetlands and other projects that have been proposed or are in progress. Before groundwater withdrawals eliminated the possibility of natural bosques, the riverine ecosystems along the Santa Cruz River were changing in response to both climatic fluctuations and channel change. Early human exploitation of these ecosystems for food and fuel drove ecological changes, but it was the mining of both shallow and deep groundwater that had the most profound impacts. When turbine water pumps were added, the riparian zones were doomed. Given the considerable depth to the water table and current water demands, the riparian habitat of the late nineteenth and early twentieth century no longer can be restored. The best-case scenario is already happening: the rate of water demand has reversed, stabilizing groundwater levels or allowing slow rises in certain areas where recharge occurs. Constructed greenbelt parks irrigated from reclaimed water are welcome additions but do not replace or serve the same ecological or hydrological functions as cienegas, bosques, and riparian plant communities.

Politicians and managers still remember the hazard posed by the now-ephemeral Santa Cruz and Rillito Rivers, and development immediately adjacent to their channels has slowed. In the 1980s, when flooding caused hundreds of millions of dollars of damages, the impetus was to build channels capable of conveying a maximum discharge. Floodplain managers believed that large floods resulted from urbanization and widespread impervious surfaces, not short-term climate variation. We now know that short-term periods of wet climate recur in southern Arizona but that most of the time, the major rivers have to convey small discharges, not catastrophic floods. In the 1950s, Tucsonans pushed landfills out into the dusty channels; now, a different ethic calls for rethinking ephemeral channel design to see whether these channels can be adapted for multiple purposes beyond just large-flood conveyance and urban wastewater drainage.

The Sweetwater Wetlands, albeit an off-channel cienega, and the establishment of native riparian trees downstream from the wastewater treatment plants without clearing to maintain channel conveyance are a good start to developing multipurpose channels in the Tucson Basin. These are a pale shadow of what once was, but they are an improvement over barren flood conduits. New attempts at increasing infiltration into the channelized bed, with the goal of increasing groundwater levels beneath the river, could enhance the establishment of riparian vegetation by expending some of the valuable water resources on evapotranspiration while allowing much of the reclaimed wastewater to sink deeper into the basin aquifers. One looming problem is the clogging of bed sediments with fines, precipitates, and (or) algal biofilms, which decreases infiltration into the alluvial aquifer; this artificial infiltration barrier, created by effluent releases, is periodically broken by floods that erode channel sediments.[140] Periodic sealing of bed sediments, with consequent reductions in infiltration, could partially explain the narrow strip of riparian vegetation that develops along the Santa Cruz River downstream of the effluent-discharge points: infiltration is reduced, and shallow groundwater levels accessible to plants drop quickly away from the center of the channel.

Multipurpose channels, even if unintentional, have major benefits for ecosystem services and aesthetics, but a single large flood that overtops banks because of decreased channel conveyance could reverse the trend toward greener flow regulation. After the 2006 floods on the Rillito River, floodplain deposition resulted in an unacceptable loss of conveyance, and some reaches of the channel were dredged. With reduced flood frequencies, this type of channel and floodplain disruption may not occur frequently unless channel aggradation continues to the point of frequent overtopping of the soil-cemented banks.

Some new design for flood control may be taking hold, at least on a local scale. Near the south edge of Tucson, segments of arroyo remain without significant channelization, spliced between channelized sections (fig. 11.4A). These reaches are rare because they still pose a flood-control threat in the event that valuable housing or infrastructure is built nearby; the potential for lateral channel change during floods is ever-present, despite a mostly quiescent watershed in the early twenty-first century. In the flood-ravaged 1980s and 1990s, these reaches, if housing or critical infrastructure were nearby, would be channelized with soil cement and an imposed trapezoidal cross section, fixed width, and widely spaced sills to thwart downcutting (fig. 11.4B). These are the reaches through most of Tucson that now have low floodplains, xerophytic riparian vegetation, and reduced channel conveyance.

What if environmentalists and flood-control managers could both have their ways? What if channels could be engineered to accommodate floodflows without eroding laterally and still be able to support riparian vegetation? This experiment in new thinking for engineered flood control has been put into practice in a short reach of the Santa Cruz River south of Irvington Road in Tucson (fig. 11.4C). The engineered channel is overly wide and constrained within low, soil-cemented banks; instead of steep-angled soil cement extending up to the surrounding land surface, gentler slopes of natural materials provide substrate for desert vegetation, planted or naturally established. The conveyance of this overly wide cross section is so high that no conceivable flood on the Santa Cruz River could overtop the banks, regardless of how much vegetation grows along the channel.

A Requiem for the Santa Cruz

The title of our book, *Requiem for the Santa Cruz*, refers to the irreversible changes that have happened in the Santa Cruz River's reach through the Tucson Basin, principally as a result of urbanization and groundwater overdraft. The word *requiem* normally evokes a religious ritual or musical composition associated with death and mourning, but its meaning has evolved to encompass remembrances for victims of wartime and environmental disasters. Our environmental history of the Santa Cruz commemorates the many victims of unsustainable water development. Principal among these are the loss of groundwater levels high enough to discharge along the valley floor and support local cienegas, the destruction of a unique mesquite bosque rich with birdlife, and the conversion of an unstable arroyo into a cemented ditch that serves little ecological function but conveys floods and protects the public.

For most of the Holocene, and really until the 1940s, the high groundwater table gave the Santa Cruz River its resilience and enabled newly entrenched arroyos to rapidly revegetate and refill. Most of the groundwater lost to overdraft in the past fifty years accumulated during wetter times, particularly during Pleistocene climates. It cannot be recharged to past levels under present-day climates and current or additional water demands. On floodplain management timescales, whether a decade or a century, the Santa Cruz River of old can be neither restored nor revived. It is time for local residents and authorities to acknowledge the river's demise and begin to engineer a new and sustainable course that minimizes flood risk while maximizing desirable ecosystem services. This will require both an acknowledgment of history and fresh perspectives on how to manage rivers and floodplains in urban areas of the Southwest.

Appendixes

The following appendixes list all vertebrate species from the Great Mesquite Forest (GMF) that were recorded before its destruction in the middle of the twentieth century. This was one of the few named mesquite bosques (woodlands) in the Southwest; although known by other names, it was often referred to as the San Xavier bosque because of its proximity to Mission San Xavier del Bac. Those species known to occur historically along the middle Santa Cruz River in the Tucson Basin are included as well. Historic records are most complete for birds, and all historically known species of riparian breeding birds for the Tucson Basin are listed in appendixes A, B, and C. Additional species that may have also been part of the breeding avifauna are listed in appendix H.

We have generally followed taxonomic systems and listed scientific names in use at the beginning of 2000. In several cases, newly proposed taxonomy and scientific names have been proposed, often based largely on DNA analyses. We have included many of these recent proposed changes in classification and (or) nomenclature, as asterisked alternative scientific names, along with older, more widely known names, since these newer names may or may not be accepted over time.

Appendixes C and L are of special interest because they compare the current avifauna of the Santa Cruz River with the past. Perennial water in the Santa Cruz River is provided by Tucson and Marana sewage effluent that then flows downstream past Marana. The riparian and aquatic ecosystems responding to this supplemental water attract a large number of birds, many of which had been largely or entirely extirpated. Riparian vegetation along much of this stretch of the river is largely nonnative tamarisk (*Tamarix* spp.), introduced shrubs and trees that have generally replaced native riparian vegetation along much of the Santa Cruz River, as well as many other rivers throughout the Southwest. Two species occur in this region: the large shrub, saltcedar; and a larger tree, Athel tamarisk.

Appendix A

Summer birds recorded from the Great Mesquite Forest (GMF) and the adjacent Santa Cruz River by ornithologists before its destruction in the middle of the twentieth century. Habitats include mesquite bosque (forest), wet riparian ecosystem along the Santa Cruz River adjacent to the bosque (for example, cottonwood-willow gallery forests), and Sonoran desertscrub immediately adjacent to the bosque on the more arid, upland side (for example, paloverde, saguaro–mixed cacti, and so forth).

A=Arnold (1940).
O=Eggs and (or) nest recorded and (or) collected.

X?=It is not clear whether this species was seen by this person or he was referring to past records.
*Species probably not breeding in the immediate vicinity but flying over, foraging in, or passing through the area during the breeding season.
**Extirpated in the Tucson Basin during the late 1800s or during the 1900s (see appendix H).
***Recent changes in classification or nomenclature that may or may not be accepted over time (American Ornithologists' Union 2012).

		Published Accounts						
		Swarth[1] 1905	Willard[2] 1912	Dawson[3] 1921	Bent/ Willard[4] 1919–1968	Brandt[5] 1951	Arnold[6] 1940	Marshall[7] 1960, 1964, 1968
Years of Work		1902–1903	1911	1917	1922	1935–1945	1938–1940	1951–1963
Common Name	Scientific Name[8]							
Gambel's Quail	*Callipepla gambelii*	X	X	XO	X	X	A	X
Pied-billed Grebe[9]	*Podylimbus podiceps*							
*Great Blue Heron	*Ardea herodias*			X				
Green Heron[10]	*Butorides virescens*	X		X			A	
*Black-crowned Night-Heron[11]	*Nycticorax nycticorax*	X		X			A	
*Black Vulture**[12]	*Coragyps atratus*					X	A	X
*Turkey Vulture	*Cathartes aura*	X		X	X		A	X
Cooper's Hawk	*Accipiter cooperii*	X	X	XO		X	A	X
Common Black-Hawk**	*Buteogallus anthracinus*				X	X	A	
*Harris's Hawk	*Parabuteo unicinctus*	X		X			A	
Gray Hawk**	*Asturina nitida*[13]***	X	X	XO	X	X	A	X
*Swainson's Hawk	*Buteo swainsoni*	X		XO			A	
*Zone-tailed Hawk**	*B. albonotatus*			X				X
Red-tailed Hawk	*B. jamaicensis*	X		XO		X	A	X
*Crested Caracara**	*Caracara cheriway*	X		X	X			
American Kestrel	*Falco sparverius*	X		XO		X?	A	X
*Peregrine Falcon[14]	*F. peregrinus*							X
*Prairie Falcon	*F. mexicanus*			X			A	X
Killdeer	*Charadrius vociferus*			X		X	A	X
White-winged Dove[15]	*Zenaida asiatica*	X	X	XO	X	X	A	X
Mourning Dove	*Z. macroura*	X	X	XO	X	X	A	X
Inca Dove	*Columbina inca*						A	X
Common Ground-Dove**	*C. passerina*	X	X	XO	X	X	A	X
Yellow-billed Cuckoo**[16]	*Coccyzus americanus*	X		X			A	X
Greater Roadrunner	*Geococcyx californianus*	X		XO		X	A	X
Barn Owl	*Tyto alba*			X		X	A	
Western Screech-Owl	*Otus kennicottii*[17]***	X	X	XO	X?	X	A	X
Great Horned Owl	*Bubo virginianus*	X		X		X	A	X
Ferruginous Pygmy-Owl**[18]	*Glaucidium brasilianum*							
Elf Owl	*Micrathene whitneyi*	X	X	XO		X	A	X
Burrowing Owl[19]	*Athene cunicularia*			X				
Lesser Nighthawk	*Chordeiles acutipennis*	X		XO			A?[20]	X
Common Poorwill	*Phalaenoptilus nuttallii*	X					A	X

(continued)

Common Name	Scientific Name	Swarth 1905	Willard 1912	Dawson 1921	Bent/ Willard 1919–1968	Brandt 1951	Arnold 1940	Marshall 1960, 1964, 1968
*White-throated Swift	*Aeronautes saxatilis*	X		X			A	
Black-chinned Hummingbird	*Archilochus alexandri*	X		XO			A	X
Gila Woodpecker	*Melanerpes uropygialis*	X	X	XO	X	X	A	X
Ladder-backed Woodpecker	*Picoides scalaris*	X		XO			A	X
Gilded Flicker	*Colaptes chrysoides*	X	X	XO		X?	A	X
Northern Beardless-Tyrannulet	*Camptostoma imberbe*	X			X?	X	A	X
Willow Flycatcher**[21]	*Empidonax traillii*	X		X				
Black Phoebe	*Sayornis nigricans*			X		X	A	X
Say's Phoebe	*S. saya*			X			A	X
Vermilion Flycatcher	*Pyrocephalus rubinus*	X	X	XO	X	X	A	X
Ash-throated Flycatcher	*Myiarchus cinerascens*	X	X		X?	X	A	X
Brown-crested Flycatcher	*M. tyrannulus*	X	X	X		X?	A	X
Tropical Kingbird**[22]	*Tyrannus melancholicus*					X?		X
Cassin's Kingbird	*T. vociferans*	X?[23]		X		X	A	X
Western Kingbird	*T. verticalis*	X		XO		X	A	X
Rose-throated Becard[24]	*Pachyramphus aglaiae*							X
Loggerhead Shrike	*Lanius ludovicianus*	X		XO			A	X
Bell's Vireo	*Vireo bellii*	X	X	XO		X	A	X
*Chihuahuan Raven	*Corvus cryptoleucus*	X		X	X		A	X
*Common Raven	*C. corax*	X		X			A	X
Purple Martin	*Progne subis*	X		X		X?	A	X
Northern Rough-winged Swallow	*Stelgidopteryx serripennis*	X?[25]		XO		X	A	X
Verdin	*Auriparus flaviceps*	X	X	XO		X	A	X
Cactus Wren	*Campylorhynchus brunneicapillus*	X		XO	X?[26]		A	X
Rock Wren	*Salpinctes obsoletus*					X	A	X
Canyon Wren	*Catherpes mexicanus*	X		X	X?	X	A	X
Bewick's Wren	*Thryomanes bewickii*	X	X	XO	X?	X	A	X
Black-tailed Gnatcatcher	*Polioptila melanura*	X		X			A?[27]	X
Northern Mockingbird	*Mimus polyglottos*	X	X	XO	X	X	A	X
Bendire's Thrasher	*Toxostoma bendirei*	X		XO		X?[28]	A	X
Curve-billed Thrasher	*T. curvirostre*	X		XO		X?	A	X
Crissal Thrasher	*T. crissale*	X		XO		X?	A	X
European Starling[29]	*Sturnus vulgaris*							X
Phainopepla	*Phainopepla nitens*	X	X	XO	X	X	A	X
Lucy's Warbler	*Vermivora luciae*[30]***	X	X	XO	X	X?	A	X
Yellow Warbler	*Dendroica petechia*[31]***	X		X		X	A	X
Common Yellowthroat	*Geothlypis trichas*			X				X

Yellow-breasted Chat	*Icteria virens*		X	X	XO		X		X
Canyon Towhee	*Pipilo fuscus*[32]***		X		XO	X?[33]	X?	A	X
Abert's Towhee	*P. aberti*[34]***		X	X	XO	X?	X	A	X
Rufous-winged Sparrow	*Aimophila carpalis*[35]***								X
Black-throated Sparrow	*Amphispiza bilineata*		X		X		X	A	X
Song Sparrow**[36]	*Melospiza melodia*							A	X
Summer Tanager	*Piranga rubra*		X	X	X		X	A	X
Northern Cardinal	*Cardinalis cardinalis*		X		XO	X	X	A	X
Pyrrhuloxia	*C. sinuatus*		X	X	XO		X	A	X
Blue Grosbeak	*Passerina caerulea*		X		X			A	X
Red-winged Blackbird	*Agelaius phoeniceus*		X		XO			A	X
*Great-tailed Grackle[37]	*Quiscalus mexicanus*								X
Bronzed Cowbird[38]	*Molothrus aeneus*				XO			A	X
Brown-headed Cowbird	*M. ater*		X		XO			A	X
Hooded Oriole	*Icterus cucullatus*		X	X	XO	X?[39]	X	A	X
Bullock's Oriole	*I. bullockii*		X	X	XO	X?	X?	A	X
House Finch	*Carpodacus mexicanus*[40]***				XO			A	X
Lesser Goldfinch	*Carduelis psaltria*[41]***							A	X
House Sparrow[42]	*Passer domesticus*						X	A	X
Number of Species									
—Nesting Species[43]		73±	55±		63			66	68±
—Total Species[44]		85±	63±	25±	72	?[45]	?[46]	74	80±

Notes

1. Harry S. Swarth, a professional ornithologist, worked 17–23 May 1902 with O. W. Howard, an amateur Arizona ornithologist, and 1–14 June 1903 with Frank Stephens; see appendix I.

2. Frank Willard, an amateur Arizona ornithologist, spent one day working around the edge of the GMF, 24 May 1911, and one day in the mesquite bosque, 25 May 1911; see appendix I.

3. William L. Dawson, a professional ornithologist, collected eggs 7–26 May and 2–11 June 1917 with a crew of four and a cook from the Museum of Comparative Oology, Santa Barbara, California. In twenty-two days, 572 nests were recorded and 165 sets of eggs were taken from forty-one species; see appendix I.

4. A. C. Bent visited Arizona in the spring and early summer of 1922, with F. C. Willard as his guide. Details of Bent's trip were not published, but species listed are mentioned in various volumes of Bent's *Life Histories of North American Birds* (Bent 1919–1968); see appendix I.

5. In 1935, Brandt, an amateur Ohio ornithologist, began studies in southeastern Arizona that lasted through eight seasons, scattered until 1948, but spent most of that time in other locations besides the GMF. His last work in the forest was in 1945; see appendix I.

6. Lee W. Arnold (1940) was working on a master's thesis at the University of Arizona; see appendix I.

7. From field trip cards of J. T. Marshall (former ornithologist at the University of Arizona), including observations made by Marshall, students, and associates, during his work in the GMF, especially towhee studies (Marshall 1960, 1964; Marshall and Johnson 1968); see appendix I.

8. After American Ornithologists' Union (1998, 2012).

(continued)

9. Vorhies saw Pied-billed Grebes at Indian Dam in the GMF frequently in the 1930s, including three calling on open water on 15 April 1934 (Vorhies et al. 1935). They may have been nesting here, because the species begins nesting in December and January in Arizona lowlands (Corman and Wise-Gervais 2005); see appendix H.

10. Vorhies saw Green Heron nests at Indian Dam in the GMF for "three years in a row" in the mid-1930s (Vorhies et al. 1935: 244).

11. Swarth (1914) and Phillips et al. (1964) discounted nesting of Black-crowned Night Heron along streams of southern Arizona, but newly fledged young have been found along the Salt/Gila River system south of Phoenix in a situation similar to that of the GMF (Rea 1983, Johnson and Simpson MS a). Others, including Herbert Brown (Swarth 1914) and Scott (1886), also reported the species from the Santa Cruz River in spring and summer; see appendix H.

12. "At least a dozen" Black Vultures were seen "in the bottom of a ditch" on 7 May 1922, along the Santa Cruz River, twelve miles south of Tucson (Kimball 1923: 109); see appendix H.

13. From *Asturina nitida* to *Buteo nitidus*.

14. Peregrine Falcon was listed as an endangered species in 1970 and delisted in 1999 (US Fish and Wildlife Service 2012).

15. The GMF was probably the most noted site in Arizona for hunting White-winged Doves in the early 1900s (Arnold 1941, 1943; Brown 1989).

16. Yellow-billed Cuckoo, a candidate for listing as a threatened or endangered species (US Fish and Wildlife Service 2012), is the last breeding bird to arrive annually, generally in early June (Phillips et al. 1964, Johnson and Simpson MS b); see appendix H.

17. From *Otus kennicottii* to *Megascops kennicottii*.

18. Herbert Brown took a Ferruginous Pygmy-Owl south of Tucson in 1884, the only Santa Cruz River record (Johnson et al. 2003). The Arizona population was listed as federally endangered in 1997 and delisted in 2007; it was considered extirpated in the Tucson Basin by 2009 (RRJ); see appendix H.

19. Western Burrowing Owl (*Athene cunicularia hypugea*) is considered "a species of concern" (US Fish and Wildlife Service 2012); see appendix H.

20. Lesser Nighthawk was misidentified as Common Nighthawk (*C. minor*) by Arnold.

21. Southwestern Willow Flycatcher (*Empidonax traillii extimus*) is a federally endangered species (US Fish and Wildlife Service 2012) that was extirpated earlier; see appendix H.

22. Tropical Kingbird was first reported as a breeding species for the United States along the Santa Cruz River between Tucson and the GMF (Phillips 1940, Phillips et al. 1964). Brandt thought he saw one in the GMF but was uncertain. The species was first recorded in Arizona (and the United States) by a specimen taken by H. H. Kimball near Fort Lowell on 12 May 1905 (Peters 1936, Phillips et al. 1964). Whether it was a solitary bird or a member of a breeding pair or colony is unknown. See appendix H.

23. Swarth thought he saw Cassin's Kingbird but was uncertain.

24. A Rose-throated Becard nest was found in 1958, and a calling male was recorded in 1959 (Phillips et al. 1964; see chapter 7); see appendix H.

25. Swarth listed Bank Swallow (*Riparia riparia*), which resembles Northern Rough-winged Swallow but does not breed in Arizona (Phillips et al. 1964, Corman and Wise-Gervais 2005).

26. Bent (1953: 130) used the term "wrens" without specifying species.

27. Black-tailed Gnatcatcher misidentified as Blue-gray Gnatcatcher (*P. caerulea*) by Arnold.

28. Brandt used the term "thrasher" without saying which species, or all.

29. European Starling was introduced into the United States from Europe; it was first reported for Arizona in 1946 (Phillips et al. 1964, Monson and Phillips 1981).

30. From *Vermivora luciae* to *Oreothlypis luciae*.

31. From *Dendroica petechia* to *Setophaga petechia*.

32. From *Pipilo fuscus* to *Melozone fusca*.

33. Bent (1953: 130) used the term "towhees" without saying which species, or both.

34. From *Pipilo aberti* to *Melozone aberti*.

35. From *Aimophila carpalis* to *Peucaea carpalis*.

36. Song Sparrow, formerly a common riparian nesting species in the Tucson region (Bendire 1872a, 1872b), was extirpated from the region shortly after 1895 (Phillips et al. 1964). Records by Arnold and Marshall were probably migrating individuals; see appendix H.

37. Great-tailed Grackle was first reported arriving in Arizona from Mexico in 1936 (Phillips et al. 1964, Monson and Phillips 1981).

38. Bronzed Cowbird was first reported arriving in Arizona from Mexico in 1909 (Visher 1909, Gilman 1914, Phillips et al. 1964).

39. Bent (1953: 130) used the term "orioles" without saying which species, or both.

40. From *Carpodacus mexicanus* to *Haemorhous mexicanus*.

41. From *Carduelis psaltria* to *Spinus psaltria*.

42. House Sparrow was introduced into the United States from Europe; it was first reported in Tucson in 1902–1903 (Howard 1906, Phillips et al. 1964).

43. "Nesting species" includes only species considered as nesting in or adjacent to the GMF.

44. "Total species" includes the total number of species observed breeding, foraging, and (or) flying over the area during the breeding season from 1902 to the 1960s. It also includes species, especially those nesting, not recorded by listed ornithologists but for which other records exist (for example, Pied-billed Grebes and nesting Green Herons).

45. No total number of species can be given, because of the difficulty of finding all records in the multiple accounts in Bent's *Life Histories of North American Birds*.

46. No total number of species can be given, because of a lack of lists and difficulty finding all records in Brandt's 725-page book.

Appendix B

Summer birds recorded from the Rillito River and its floodplain, mostly near Fort Lowell, by ornithologists during the late 1800s and early 1900s. Species with no X occurred in the Great Mesquite Forest (GMF, see appendix A) and are included for comparison but were not recorded for "Rillito Creek." Ecosystems along the Rillito River include cottonwood-willow forest, cottonwood-mesquite, mesquite bosque, and upland habitats immediately adjacent to the floodplain (for example, paloverde, saguaro–mixed cacti, and so forth).

O=Eggs and (or) nest recorded and (or) collected.
T=Location given as "Tucson" without specific locality, although it may have been here: Bendire did most of his work along Rillito Creek,[1] and Stephens collected extensively along Rillito Creek, including at Fort Lowell.

*Species probably not breeding in the immediate vicinity but flying over, foraging in, or passing through the area during the breeding season.

**Extirpated in the Tucson Basin during the late 1800s or 1900s (see appendix H).

***Recent changes in classification or nomenclature that may or may not be accepted over time (American Ornithologists' Union 2012).

		Published Accounts			
		Bendire[2] 1892, 1895	Stephens[3] 1882–1883	Howell[4] 1916	Brandt[5] 1951
Years of Work		1871–1873	1881	1915–1916	1935–1948
Common Name	Scientific Name[6]				
Gambel's Quail	*Callipepla gambelii*	XO	T	X	X
*Great Blue Heron	*Ardea herodias*	T	T		X
Green Heron	*Butorides virescens*				X
Black-crowned Night-Heron[7]	*Nycticorax nycticorax*				
Black Vulture	*Coragyps atratus*				
*Turkey Vulture	*Cathartes aura*	T	T	X	X
Cooper's Hawk	*Accipiter cooperii*	XO		X	X
Common Black-Hawk**	*Buteogallus anthracinus*	XO			X
Harris's Hawk	*Parabuteo unicinctus*	TO			
Gray Hawk**	*Asturina nitida*[8]***	XO	T		
Swainson's Hawk	*Buteo swainsoni*	TO			X
Zone-tailed Hawk**[9]	*B. albonotatus*	XO	T[10]		
Red-tailed Hawk	*B. jamaicensis*	XO		X	X
Crested Caracara**	*Caracara cheriway*	X	T	X	
American Kestrel	*Falco sparverius*		T	X	
*Prairie Falcon	*F. mexicanus*	T			

Common Name	Scientific Name	Bendire 1892, 1895	Stephens 1882–1883	Howell 1916	Brandt 1951
Killdeer	*Charadrius vociferus*	X	T		X
White-winged Dove	*Zenaida asiatica*	XO	T	X	X
Mourning Dove	*Z. macroura*	XO	T	X	X
Inca Dove	*Columbina inca*	X		X	
Common Ground-Dove**	*C. passerina*	XO	T	X	X
Yellow-billed Cuckoo**	*Coccyzus americanus*	XO	T		X
Greater Roadrunner	*Geococcyx californianus*	XO	X	X	X
Barn Owl	*Tyto alba*	X			
Western Screech-Owl	*Otus kennicottii*[11]***	XO	X	X	X
Great Horned Owl	*Bubo virginianus*	X	X	X	X
Ferruginous Pygmy-Owl**[12]	*Glaucidium brasilianum*	X	X	X	
Elf Owl	*Micrathene whitneyi*	X	X		X
Burrowing Owl	*Athene cunicularia*				
Lesser Nighthawk	*Chordeiles acutipennis*	XO	X		X
Common Poorwill	*Phalaenoptilus nuttallii*	XO			
*White-throated Swift	*Aeronautes saxatalis*	T			
Black-chinned Hummingbird	*Archilochus alexandri*	XO	X		X
Gila Woodpecker	*Melanerpes uropygialis*	XO	X		X
Ladder-backed Woodpecker	*Picoides scalaris*	XO	T		X
Gilded Flicker	*Colaptes chrysoides*	X	X	X	X
Northern Beardless-Tyrannulet	*Camptostoma imberbe*		T		X
Willow Flycatcher**[13]	*Empidonax traillii*	XO	T		
Black Phoebe	*Sayornis nigricans*	XO	T		X
Say's Phoebe	*S. saya*				
Vermilion Flycatcher	*Pyrocephalus rubinus*	XO	X	X	X
Ash-throated Flycatcher	*Myiarchus cinerascens*	XO	X		X
Brown-crested Flycatcher[14]	*M. tyrannulus*		X		X
Tropical Kingbird[15]	*Tyrannus melancholicus*				
Cassin's Kingbird	*T. vociferans*	XO	T		
Western Kingbird	*T. verticalis*	XO	X	X	X
Rose-throated Becard**	*Pachyramphus aglaiae*				
Loggerhead Shrike	*Lanius ludovicianus*	XO	T		X
Bell's Vireo	*Vireo bellii*	XO	X	X	X
Chihuahuan Raven	*Corvus cryptoleucus*	XO	T	X	
*Common Raven	*C. corax*	X	T		X
Purple Martin	*Progne subis*	X	T		X
Northern Rough-winged Swallow	*Stelgidopteryx serripennis*	X	T		X
Barn Swallow	*Hirundo rustica*	X[16]			
Verdin	*Auriparus flaviceps*	XO	T		X
Cactus Wren	*Campylorhynchus brunneicapillus*	XO	X	X	X
Rock Wren	*Salpinctes obsoletus*				
Canyon Wren	*Catherpes mexicanus*				
Bewick's Wren	*Thryomanes bewickii*	X	T		

(continued)

Common Name	Scientific Name	Bendire 1892, 1895	Stephens 1882–1883	Howell 1916	Brandt 1951
Black-tailed Gnatcatcher	*Polioptila melanura*	X	X		X
Northern Mockingbird	*Mimus polyglottos*	XO	X		X
Bendire's Thrasher[17]	*Toxostoma bendirei*	XO	X	X	X
Curve-billed Thrasher	*T. curvirostre*	XO	X	X	X
Crissal Thrasher	*T. crissale*	XO	X		X
European Starling[18]	*Sturnus vulgaris*				
Phainopepla	*Phainopepla nitens*	XO	X		X
Lucy's Warbler	*Vermivora luciae*[19]***	XO	X	X	X
Yellow Warbler	*Dendroica petechia*[20]***	XO	T		X
Common Yellowthroat	*Geothlypis trichas*	XO	T[21]		X
Yellow-breasted Chat	*Icteria virens*	XO	T[22]		X
Canyon Towhee	*Pipilo fuscus*[23]***	XO			X
Abert's Towhee	*P. aberti*[24]***	XO	T		X
Rufous-winged Sparrow[25]	*Aimophila carpalis*[26]***	XO	X		
Black-throated Sparrow	*Amphispiza bilineata*	XO	X		X
Song Sparrow**[27]	*Melospiza melodia*	XO	T		
Summer Tanager	*Piranga rubra*	XO	X		X
Northern Cardinal	*Cardinalis cardinalis*	XO	T	X	X
Pyrrhuloxia	*C. sinuatus*	XO	T	X	X
Blue Grosbeak	*Passerina caerulea*	XO	T		X
Red-winged Blackbird	*Agelaius phoeniceus*	X	T		X
Western Meadowlark	*Sturnella neglecta*			X	X[28]
Great-tailed Grackle[29]	*Quiscalus mexicanus*				
Bronzed Cowbird[30]	*Molothrus aeneus*				X
Brown-headed Cowbird	*M. ater*	XO	T		X
Hooded Oriole	*Icterus cucullatus*	XO	T[31]	X	X
Bullock's Oriole	*I. bullockii*	XO			X
House Finch	*Carpodacus mexicanus*[32]***	XO	X		X
Lesser Goldfinch	*Carduelis psaltria*[33]***	T		X	X
House Sparrow[34]	*Passer domesticus*			X	X
Number of Species					
—Nesting Species[35]		70±	54±		54±
—Total Species[36]		77±	56±	?[37]	?[38]	60±[39]

Notes

1. Bendire's camp from 1871 to 1873 was situated along "Rillito Creek" at what would be the future site of Fort Lowell (Bendire 1872a).

2. Charles Bendire's information is from field notes in the archives of the US Natural History Museum, Washington, DC (1872a, 1872b) and publications (Bendire 1892, 1895); see appendix I.

3. Stephens's information, including quotes, was published by Brewster (1882, 1883); see appendix I.

4. A. Brazier Howell (1916) spent 7 December 1915 to 25 March 1916 along the Rillito River in the vicinity of Fort Lowell. Howell's winter work is included because he listed the permanent residents and was the only ornithologist who published his complete findings during the six decades between Bendire's and Brandt's work; see appendix I.

5. Herbert Brandt, an amateur Ohio ornithologist, began studies in southeastern Arizona in 1935 that lasted through eight seasons, scattered until 1948, but spent most of that time in other locations besides the GMF. His last work in the forest was in 1945; see appendix I.

6. After American Ornithologists' Union (1998, 2012).

7. Swarth (1914) and Phillips et al. (1964) discounted nesting of Black-crowned Night Heron along streams of southern Arizona, but newly fledged young have been found along the Salt/Gila River system south of Phoenix in a situation similar to that of the Rillito River (Rea 1983, Johnson and Simpson MS a); see appendix H.

8. From *Asturina nitida* to *Buteo nitidus.*

9. Zone-tailed Hawk no longer nests in the lowlands near Tucson but is restricted to the nearby mountains. A tale of a narrow escape from Apache Indians while taking an egg from a nest of this species along the Rillito River is told by Bendire (1892); also, a pair of Zone-tailed Hawks was seen by RRJ, circling over the Rillito River at Campbell Road in April or May 1953 (Johnson et al. 2000b); see appendix H.

10. Stephens watched as a Zone-tailed Hawk "attempted to catch some minnows in a shallow place, fluttering over the water and trying to snatch up the little fish with its feet" (Brewster 1883: 30), probably along the Rillito River.

11. From *Otus kennicottii* to *Megascops kennicottii.*

12. Ferruginous Pygmy-Owl was first found in the United States along the Rillito River by Bendire (Coues 1872a, Bendire 1892 [and see text]) and was relatively common along the Rillito River. Frank Stephens was the second person to take a specimen of this species in the United States, at Camp Lowell, on 3 June 1881 (Brewster 1883, Johnson et al. 2003). It was extirpated from the Tucson Basin by 2009; see appendix H.

13. *"E. pusillus"* was apparently the Willow Flycatcher, but it was not differentiated from *Empidonax traillii* by Bendire, Stephens, and Brewster and was considered a "[c]ommon bird about Tucson where it inhabited willow thickets near water" (Brewster 1882: 206). Southwestern Willow Flycatcher (*Empidonax traillii extimus*) is a federally endangered species (US Fish and Wildlife Service 2012) that was extirpated earlier; see appendix H.

14. The subspecies of Brown-crested Flycatcher nesting in Arizona, *Myiarchus tyrannulus magister,* was first discovered for the United States by Stephens (Brewster 1881) but not named until 1884 (Ridgway 1884; see appendix E). The type specimen was from Camp Lowell (American Ornithologists' Union 1957) and thus was apparently not differentiated by Bendire from the Ash-throated Flycatcher.

15. Tropical Kingbird was first recorded in Arizona (and the United States) by a specimen taken by H. H. Kimball near Fort Lowell on 12 May 1905 (Peters 1936, Phillips et al. 1964). Whether it was a solitary bird or a member of a breeding pair or colony is unknown; see appendix H.

16. Bendire (1872b: 95) wrote that the Barn Swallow was "a common summer visitor & breeds."

17. The type specimen for Bendire's Thrasher was taken near Camp Lowell by Bendire (Coues 1873).

18. European Starling was introduced into the United States from Europe; it was first reported for Arizona in 1946 (Phillips et al. 1964, Monson and Phillips 1981).

19. From *Vermivora luciae* to *Oreothlypis luciae.*

20. From *Dendroica petechia* to *Setophaga petechia.*

21. Common Yellowthroat was "abundant along streams" (Brewster 1882: 139).

22. Yellow-breasted Chat was "abundant in the vicinity of Tucson" (Brewster 1882: 139).

23. From *Pipilo fuscus* to *Melozone fusca.*

24. From *P. aberti* to *M. aberti.*

25. A species new to science, the type specimen for Rufous-winged Sparrow was taken near Fort Lowell by Bendire (Bendire 1872 b, Coues 1873); see appendix E.

26. From *Aimophila carpalis* to *Peucaea carpalis.*

27. Song Sparrow was "rather common about Tucson" (Brewster 1882: 196) but was extirpated as a breeding species in the Tucson Basin shortly after 1895 (Phillips et al. 1964); apparently it is now nesting in habitat created by Tucson sewage effluent; see appendixes C and H.

28. Western Meadowlarks were seen by Brandt in irrigated fields in the Rillito floodplain. This was the only species recorded from the Rillito River that was not recorded from the Great Mesquite Forest and adjacent ecosystems.

29. Great-tailed Grackle was first reported arriving in Arizona from Mexico in 1936 (Phillips et al. 1964, Monson and Phillips 1981).

30. Bronzed Cowbird was first reported arriving in Arizona from Mexico in 1909 (Visher 1909, Gilman 1914, Phillips et al. 1964).

31. Hooded Oriole was "found only in the valley, where it seemed to prefer cottonwoods" (Brewster 1882: 200).

32. From *Carpodacus mexicanus* to *Haemorhous mexicanus.*

33. From *Carduelis psaltria* to *Spinus psaltria.*

34. House Sparrow was introduced into North America from Europe; it was first reported in Tucson in 1902–1903 (Howard 1906, Phillips et al. 1964).

35. "Nesting species" excludes species not nesting on the study area or recorded as nesting only one year but includes species extirpated during study years.

(continued)

36. "Total species" includes the total number of species observed breeding, foraging, and (or) flying over the area during the breeding season from 1871 to 1948.

37. Because of the large number of species with no more specific locality than "Tucson," we provide no total numbers for Stephens's observations (see Brewster 1882, 1883).

38. Because Howell's work was done in winter, we provide no total numbers.

39. In comparing the Rillito River to the San Pedro River, Brandt (1951: 239) "found about 60 nesting species of birds common to both valleys," with six species for each not occurring in the other. Unfortunately, this is in narrative form, and those six exclusive species for each river are not named directly in his 725-page book.

Appendix C

Current status (1999–2012) of summer birds along the Santa Cruz River in the Tucson Basin, for comparison with the Great Mesquite Forest (GMF; see appendixes A and L). Species with no X formerly occurred in the Santa Cruz–Rillito River system but have not been recorded recently.

X?=Recorded during the summer but breeding status unclear.

Checklist abbreviations: a=abundant, c=common, u=uncommon, r=rare, v=very rare, i=irregular.

*Species probably not breeding in the immediate vicinity but flying over, foraging in, or passing through the area during the breeding season.

**Species recently established as summer birds; not reported historically as nesting in the Tucson Basin.

***Recent changes in classification or nomenclature that may or may not be accepted over time (American Ornithologists' Union 2012).

****Recorded (by US Fish and Wildlife Service personnel) in or near the Santa Cruz River from 29th Street to Grant Road between 2004 and 2011.

		Area		
		Sweetwater Wetlands Downriver to Marana[1]	Sweetwater Wetlands Checklists[2]	West Branch, Santa Cruz River[3]
Years of Work		Mid-May 1999–April 2000	2006–2012	Mid-May to August 2001
Common Name	Scientific Name[4]			
Mallard**[5]	*Anas platyrhynchos*	X	c	
Ruddy Duck**	*Oxyura jamaicensis*	X	c	
Gambel's Quail	*Callipepla gambelii*	X	c	X
Pied-billed Grebe**[6]	*Podylimbus podiceps*	X?	r/u	
Great Blue Heron[7]	*Ardea herodias*	X	u	
Green Heron	*Butorides virescens*	X	u	
*Black-crowned Night-Heron[8]	*Nycticorax nycticorax*	X?	u	
Black Vulture	*Coragyps atratus*			
*Turkey Vulture	*Cathartes aura*	X	u	X
Cooper's Hawk	*Accipiter cooperii*	X	u/r	X
Common Black-Hawk	*Buteogallus anthracinus*			

(continued)

Common Name	Scientific Name	Sweetwater Wetlands Downriver to Marana	Sweetwater Wetlands Checklists	West Branch, Santa Cruz River
Harris's Hawk	*Parabuteo unicinctus*	X	c	X
Gray Hawk	*Asturina nitida*[9]***			
Swainson's Hawk	*Buteo swainsoni*			
Zone-tailed Hawk	*B. albonotatus*			
Red-tailed Hawk	*B. jamaicensis*	X	u	X
Crested Caracara	*Caracara cheriway*			
American Kestrel	*Falco sparverius*	X	r	
*Peregrine Falcon	*F. peregrinus*	X	r	
*Prairie Falcon	*F. mexicanus*	X	v	X
Common Gallinule**	*Gallinula chloropus*[10]***	X	u/c	
American Coot**	*Fulica americana*	X	c	
Killdeer	*Charadrius vociferus*	X	c	X
Black-necked Stilt**	*Hemantopus mexicanus*	X	u	
Rock Dove**	*Columba livia*	X	u/c	X
White-winged Dove	*Zenaida asiatica*	X	c	X
Mourning Dove	*Z. macroura*	X	a/c	X
Inca Dove	*Columbina inca*	X	r	X
Common Ground-Dove****	*C. passerina*			
Yellow-billed Cuckoo	*Coccyzus americanus*			
Greater Roadrunner	*Geococcyx californianus*	X	u/r	X
Barn Owl	*Tyto alba*	X	v	
Western Screech-Owl[11]	*Otus kennicottii*[12]***			
Great Horned Owl	*Bubo virginianus*	X	u/r	
Ferruginous Pygmy-Owl[13]	*Glaucidium brasilianum*			
Elf Owl[14]	*Micrathene whitneyi*			
Burrowing Owl****	*Athene cunicularia*			
Lesser Nighthawk	*Chordeiles acutipennis*	X	c/u	
Common Poorwill	*Phalaenoptilus nuttallii*			
*White-throated Swift	*Aeronautes saxatalis*	X	r	
Black-chinned Hummingbird	*Archilochus alexandri*	X	c/u	X
Anna's Hummingbird**	*Calypte anna*	X	u	X
Costa's Hummingbird****	*C. costae*			
Gila Woodpecker	*Melanerpes uropygialis*	X	c	X
Ladder-backed Woodpecker	*Picoides scalaris*	X	u	X
Gilded Flicker	*Colaptes chrysoides*	X	r	X
Northern Beardless-Tyrannulet	*Camptostoma imberbe*			
Willow Flycatcher[15]	*Empidonax traillii*	X?	v	
Black Phoebe	*Sayornis nigricans*	X	u/c	
Say's Phoebe	*S. saya*	X	r	X
Vermilion Flycatcher	*Pyrocephalus rubinus*	X?	u/r	
Ash-throated Flycatcher	*Myiarchus cinerascens*	X	u/r	X
Brown-crested Flycatcher	*M. tyrannulus*	X	u/r	X
*Tropical Kingbird	*Tyrannus melancholicus*	X	v/u	
*Cassin's Kingbird	*T. vociferans*	X	v/u	X
Western Kingbird	*T. verticalis*	X	u	X
Rose-throated Becard	*Pachyramphus aglaiae*			
Loggerhead Shrike	*Lanius ludovicianus*	X	v	X
Bell's Vireo	*Vireo bellii*	X	r/u	
Chihuahuan Raven	*Corvus cryptoleucus*			
*Common Raven	*C. corax*	X	u/r	X
Purple Martin	*Progne subis*	X	u/r	

Common Name	Scientific Name	Sweetwater Wetlands Downriver to Marana	Sweetwater Wetlands Checklists	West Branch, Santa Cruz River
Northern Rough-winged Swallow	*Stelgidopteryx serripennis*	X	u/c	
Cliff Swallow**[16]	*Petrochelidon pyrrhonota*	X	c/u	
*Barn Swallow**[17]	*Hirundo rustica*	X	c	
Verdin	*Auriparus flaviceps*	X	c	X
Cactus Wren	*Campylorhynchus brunneicapillus*	X	u	X
Rock Wren	*Salpinctes obsoletus*			
Canyon Wren	*Catherpes mexicanus*			
Bewick's Wren	*Thryomanes bewickii*			
Black-tailed Gnatcatcher	*Polioptila melanura*	X	r	X
Northern Mockingbird	*Mimus polyglottos*	X	c/r	X
Bendire's Thrasher	*Toxostoma bendirei*			
Curve-billed Thrasher	*T. curvirostre*	X	u	X
Crissal Thrasher	*T. crissale*			
European Starling	*Sturnus vulgaris*	X	a/c	X
Phainopepla	*Phainopepla nitens*	X	i/r	X
Lucy's Warbler	*Vermivora luciae*[18]***	X	u/r	X
Yellow Warbler	*Dendroica petechia*[19]***	X	c	
Common Yellowthroat	*Geothlypis trichas*	X	c	
Yellow-breasted Chat	*Icteria virens*			X
Canyon Towhee	*Pipilo fuscus*[20]***			X
Abert's Towhee	*P. aberti*[21]***	X	c	X
Rufous-winged Sparrow****	*Aimophila carpalis*[22]***			X
Black-throated Sparrow	*Amphispiza bilineata*			
Song Sparrow[23]	*Melospiza melodia*	X	a/c	
Summer Tanager	*Piranga rubra*			
Northern Cardinal	*Cardinalis cardinalis*	X	r	X
Pyrrhuloxia	*C. sinuatus*	X	u	X
Blue Grosbeak	*Passerina caerulea*	X	v	
Red-winged Blackbird	*Agelaius phoeniceus*	X	a/c	
Great-tailed Grackle	*Quiscalus mexicanus*	X	a/c	X
Bronzed Cowbird	*Molothrus aeneus*	X	r	X
Brown-headed Cowbird	*M. ater*	X	u	X
Hooded Oriole	*Icterus cucullatus*	X?	r	
Bullock's Oriole	*I. bullockii*	X?	r	
House Finch	*Carpodacus mexicanus*[24]***	X	a/c	X
Lesser Goldfinch	*Carduelis psaltria*[25]***	X	c	X
House Sparrow	*Passer domesticus*	X	a/c	X
Number of Species				
—Breeding Species		66±	66±	40±
—Total Species		75±	75±	44±

Notes

1. Ponds formed by Tucson sewage effluent that then flows downstream in the Santa Cruz River past Marana (SWCA 2000, Stevenson 2007, Stejskal and Rosenberg 2011). Riparian vegetation along much of this stretch of the river is primarily tamarisk (*Tamarix* spp.), nonnative shrubs and trees that have largely replaced native riparian vegetation along much of the Santa Cruz River, as well as many other rivers throughout the Southwest. Two species occur in this region: the large shrub, saltcedar (*Tamarix ramosissima* or a hybrid complex), and the larger tree, Athel tamarisk (*Tamarix aphylla*).

(continued)

2. Combination of information from the Tucson Audubon Society online checklist (2012), the City of Tucson Water Department online checklist (https://sites.google.com/site/sweetwaterwetlands/checklist, updated 3 October 2006), and the City of Tucson Water Department hardcopy checklist (updated 14 April 2006).

3. Abandoned Santa Cruz River channel, roughly paralleling Mission Road from Ajo Way to Silverbell Road (Rosen 2001).

4. After American Ornithologists' Union (1998, 2012).

5. There is evidence that, in addition to *A. p. platyrhynchos*, the female-plumaged Mexican Duck (*A. p. diazi*) occurs in the Tucson area (Stevenson 2007, Stejskal and Rosenberg 2011); Mexican Duck has been reported from the Santa Cruz River basin between 29th Street and Grant Road.

6. Vorhies saw Pied-billed Grebes at Indian Dam in the GMF frequently in the 1930s, including three calling on open water, 15 April 1934 (Vorhies et al. 1935). The species may be nesting here, because nesting begins in December and January in the Arizona lowlands (Corman and Wise-Gervais 2005) and nests are reported for the Tucson region (Stevenson 2007, Stejskal and Rosenberg 2011).

7. A Great Blue Heron rookery was recorded during the SWCA (2000) study at the Sweetwater Wetlands.

8. Black-crowned Night-Heron is apparently currently breeding at the Tucson Zoo; young were seen in the summer of 2012 (RRJ). Swarth (1914) and Phillips et al. (1964) discounted this species' nesting along streams of southern Arizona, but newly fledged young have been found along the Salt/Gila River system south of Phoenix (Rea 1983, Johnson and Simpson MSa) in habitat similar to that of the former Santa Cruz River and of the current Sweetwater Wetlands.

9. From *Asturina nitida* to *Buteo nitidus.*

10. From *Gallinula chloropus* to *Gallinula galeata.*

11. The lack of records for Western Screech Owl may be due to the lack of nighttime work.

12. From *Otus kennicottii* to *Megascops kennicottii.*

13. Ferruginous Pygmy-Owl has been extirpated in the Tucson Basin; see appendixes B and H.

14. The lack of records for Elf Owl may be due to the lack of nighttime work.

15. Identification is difficult for Southwestern Willow Flycatcher (*Empidonax traillii extimus*), a federally endangered species, because of its similarity to several other *Empidonax* species. It was extirpated earlier (see appendix H).

16. Cliff Swallow is now "nesting under bridges, culverts, and on buildings" (Stevenson 2007: 281; Stejskal and Rosenberg 2011: 291); the species did not formerly nest in the Tucson region (Phillips et al. 1964).

17. The status of Barn Swallow is uncertain (Phillips et al. 1964). It is currently reported nesting in the Tucson area (Corman and Wise-Gervais 2005, Stejskal and Rosenberg 2011) and may have done so historically; see appendix B.

18. From *Vermivora luciae* to *Oreothlypis luciae.*

19. From *Dendroica petechia* to *Setophaga petechia.*

20. From *Pipilo fuscus* to *Melozone fusca.*

21. From *P. aberti* to *M. aberti.*

22. From *Aimophila carpalis* to *Peucaea carpalis.*

23. Song Sparrow, formerly a common riparian nesting species in the Tucson region (Bendire 1872b), was extirpated from the region shortly after 1895 (Phillips et al. 1964); it is now breeding in suitable riparian habitat formed by Tucson sewage effluent (Corman and Wise-Gervais 2005).

24. From *Carpodacus mexicanus* to *Haemorhous mexicanus.*

25. From *Carduelis psaltria* to *Spinus psaltria.*

Appendix D

Birds from Mexico that currently reach or formerly reached their northern breeding limits in southern Arizona. All species have been reported from the Santa Cruz River drainage unless indicated otherwise.

Abbreviations: R=riparian (including xeroriparian of desert washes); U=upland; [NO]=not reported from Santa Cruz River drainage.

*Species reported for the Great Mesquite Forest or nearby Santa Cruz River.

***Recent changes in classification or nomenclature that may or may not be accepted over time (American Ornithologists' Union 2012).

Common Name	Scientific Name[2]	Preferred Breeding Habitat[1]	
		Montane Woodland	Lowland and Canyon
*Gray Hawk[3]	Asturina nitida[4]***		R
Aplomado Falcon5 [NO]	Falco femoralis		U
Northern (Masked) Bobwhite[6]	Colinus virginianus ridgwayi		U
Thick-billed Parrot[7] [NO]	Rhynchopsitta pachyrhyncha	X	
Whiskered Screech-Owl[8]	Otus tricopsis[9]***	X	
*Ferruginous Pygmy-Owl[10]	Glaucidium brasilianum		R
Buff-collared Nightjar[8]	Caprimulgus ridgwayi		R&U
Broad-billed Hummingbird[11]	Cynanthus latirostris		R
White-eared Hummingbird[11]	Hylocharis leucotis	X	
Berylline Hummingbird	Amazilia beryllina	X	
Violet-crowned Hummingbird[8]	A. violiceps		R
Blue-throated Hummingbird[3]	Lampornis clemenciae	X	
Magnificent Hummingbird[11]	Eugenes fulgens	X	
Lucifer Hummingbird[11]	Colothorax lucifer		U
Elegant Trogon[8]	Trogon elegans	X	
Green Kingfisher[3]	Chloroceryle americana		R
Arizona Woodpecker[8]	Picoides stricklandi[12]***	X	
*Northern Beardless-Tyrannulet[11]	Camptostoma imberbe		R
Buff-breasted Flycatcher[8]	Empidonax fulvifrons	X	
Dusky-capped Flycatcher[8]	Myiarchus tuberculifer	X	
Sulphur-bellied Flycatcher	Myiodynastes luteiventris	X	
*Tropical Kingbird[3]	Tyrannus melancholicus		R
Thick-billed Kingbird[8]	T. crassirostris		R
*Rose-throated Becard[3]	Pachyramphus aglaiae		R
Mexican Chickadee[8]	Parus sclateri	X	
Black-capped Gnatcatcher	Polioptila nigriceps		U

(continued)

Common Name	Scientific Name	Preferred Breeding Habitat	
		Montane Woodland	Lowland and Canyon
*Rufous-winged Sparrow	*Aimophila carpalis*[13]***		U
Botteri's Sparrow[14]	*A. botterii*[15]***		U
Five-striped Sparrow	*A. quinquestriata*[16]***		U
Yellow-eyed Junco[8]	*Junco phaeonotus*	X	
Flame-colored Tanager	*Piranga bidentata*	X	
Varied Bunting[11]	*Passerina versicolor*		R&U
Streak-backed Oriole	*Icterus pustulatus*		R

Notes

1. After Marshall (1957), Phillips et al. (1964), Monson and Phillips (1981), *Birds of North America* accounts (1992–2002), American Ornithologists' Union (1998, 2012), Corman and Wise-Gervais (2005), and Stejskal and Rosenburg (2011).

2. After American Ornithologists' Union (1998); taxonomic order after American Ornithologists' Union (2012).

3. Species also occurs in southwestern Texas (Oberholser 1974, American Ornithologists' Union 1998).

4. From *Asturina nitida* to *Buteo nitidus*.

5. Aplomado Falcon was extirpated in the United States by the early 1900s; few if any substantiated recent records exist (Phillips et al. 1964, Monson and Phillips 1981). The species was recently reintroduced in southern Texas (Peregrine Fund online, http://www.peregrinefund.org/conserve_category.asp?category=aplomado%20falcon%20restoration [accessed 29 March 2011]).

6. Northern (Masked) Bobwhite was extirpated in the United States by the early 1900s; it was reintroduced into southeastern Arizona (Monson and Phillips 1981, Brown et al. 2012) with little success (see appendix H).

7. Thick-billed Parrot was extirpated by the early 1900s; it originally occurred north from Mexico in flight years (Monson and Phillips 1981). Recent introductions had little success (Snyder et al. 1999); it was extirpated again by the early 1990s (Corman and Wise-Gervais 2005).

8. Species also occurs in extreme southwestern New Mexico adjacent to southeastern Arizona (Hubbard 1978, American Ornithologists' Union 1998).

9. From *Otus tricopsis* to *Megascops trichopsis*.

10. Ferruginous Pygmy-Owl also occurs in southern Texas; it formerly occurred north to the New and Salt Rivers but now occurs only in southwestern Arizona (Millsap and Johnson 1988; Proudfoot and Johnson 2000; Johnson et al. 2000a, 2003). The Arizona population was listed as endangered in 1997 as *Glaucidium brasilianum cactorum* but delisted in 2007 (RRJ [member of species' recovery team]).

11. Species also occurs in southern New Mexico and southwestern Texas (Oberholser 1974, Hubbard 1978, American Ornithologists' Union 1998).

12. From *Picoides stricklandi* to *Picoides arizonae*.

13. From *Aimophila carpalis* to *Peucaea carpalis*.

14. Botteri's Sparrow also occurs in southern New Mexico and coastal Texas (American Ornithologists' Union 1998).

15. From *Aimophila botterii* to *Peucaea botterii*.

16. From *Aimophila quinquestriata* to *Amphispiza quinquestriata*.

Appendix E

Species and subspecies of birds new to science and discovered in the vicinity of Tucson. Included are several taxa from the Great Mesquite Forest (GMF) or Fort Lowell (formerly Camp Lowell) and the nearby Rillito River.

Abbreviations: sp. nov.=new species; ssp. nov.=new subspecies.

*Preferential riparian nesting species in the Tucson area.

**Obligate riparian nesting species in the Tucson area.[1]

***Recent changes in classification or nomenclature that may or may not be accepted over time (American Ornithologists' Union 2012).

Common Name	Scientific Name[2]	Collection Locality[3]	Published Account
**Northern Beardless-Tyrannulet	*Camptostoma imberbe ridgwayi* ssp. nov.	Tucson	Brewster (1882)
*Brown-crested Flycatcher	*Myiarchus tyrannulus magister* ssp. nov.	Camp Lowell (now Fort Lowell)	Ridgway (1884)
**Bell's Vireo	*Vireo bellii arizonae* ssp. nov.	Tucson	Ridgway (1903)
Bendire's Thrasher	*Toxostoma bendirei* sp. nov.	Tucson	Coues (1873)
Curve-billed Thrasher	*T. curvirostre palmeri* ssp. nov.	Tucson	Coues (1872b)
**Abert's Towhee	*Pipilo aberti vorhiesi* ssp. nov.[4]***	GMF	Phillips (1962)
Rufous-winged Sparrow	*Aimophila carpalis* sp. nov.[5]***	Tucson	Coues (1873)
Black-throated Sparrow	*Amphispiza bilineata deserticola* ssp. nov.	Tucson	Ridgway (1898)
*Northern Cardinal	*Cardinalis cardinalis superbus* ssp. nov.	Fort Lowell	Ridgway (1885a)
*Pyrrhuloxia	*C. sinuatus fulvescens* ssp. nov.	Fort Lowell	Van Rossem (1934)
**Blue Grosbeak	*Passerina caerulea interfusa* ssp. nov.	Fort Lowell	Dwight and Griscom (1927)
*Bronzed Cowbird	*Molothrus aeneus milleri* ssp. nov.	Fort Lowell	Van Rossem (1934)
*Hooded Oriole	*Icterus cucullatus nelsoni* ssp. nov.	Tucson	Ridgway (1885b)

(continued)

 Appendixes

Notes

1. Johnson et al. (1977, 1987).
2. After American Ornithologists' Union (1998, 2012).
3. See Brandt (1951) and American Ornithologists' Union (1957) for additional information. Some type specimens are annotated only as "Tucson" and were possibly from the vicinity of the GMF or Fort Lowell, but no specific locality information is available.
4. From *Pipilo aberti vorhiesi* to *Melozone aberti vorhiesi.*
5. From *Aimophila carpalis to Peucaea carpalis.*

Appendix F

Amphibians and reptiles (herpetofauna) of the Great Mesquite Forest (GMF), the Santa Cruz River adjacent to the bosque, and Sonoran desertscrub immediately adjacent to the bosque on the upland side.

A=Arnold (1940)

*CAN=Candidate species for federally threatened or endangered status.

***Recent changes in classification or nomenclature that may or may not be accepted over time (Society for the Study of Amphibians and Reptiles 2012).

****Recorded by US Fish and Wildlife personnel in or near the Santa Cruz River from 29th Street to Grant Road between 2004 and 2011.

Common Name	Scientific Name[1]	GMF (Arnold 1940)[2]	West Branch (Rosen 2001, 2003)[3]	Other Sources
Couch's Spadefoot [Toad]	*Scaphiopus couchii*	A	X	****
Mexican (Chihuahuan Desert) Spadefoot [Toad]	*Spea [Scaphiopus] multiplicata stagnalis***		X	
Sonoran Desert (Colorado River) Toad	*Bufo alvarius Incilius alvaria****	A	X	****
Southwestern Woodhouse's Toad	*B. woodhousii australis Anaxyrus w. australis****	A		
Red-spotted Toad	*B. punctatus A. punctatus****		X	****
Great Plains Toad	*B. cognatus A. cognatus****		X	****
Lowland Leopard Frog[4]	*Lithobates [Rana] yavapaiensis****	A		Extirpated[5]
American Bullfrog[6]	*L. [Rana] catesbeianus****			(Introduced)****
Western Narrow-mouthed Toad	*Gastrophryne olivacea*		X	
Sonoran Mud Turtle	*Kinosternon s. sonoriense*			Extirpated[7]
Desert Box Turtle	*Terrapene ornata luteola*	A		
Morafka's Desert Tortoise *CAN	*Gopherus [agassizii] morafkai****	A		
Tucson Banded Gecko[8]	*Coleonyx variegatus bogerti*	A	X	
Common Lesser Earless Lizard	*Holbrookia maculata* ssp.		X[9]	
Eastern Zebra-tailed Lizard	*Callisaurus draconoides ventralis*	A	X	
Long-nosed Leopard Lizard	*Gambelia wislizenii*	A		Extirpated[10]
Desert Spiny Lizard	*Sceloporus magister*	A	X	****
Sonoran Spiny Lizard	*S. c. clarkii*	A	X	
Southwestern Fence Lizard	*S. cowlesi****		X[11]	Extirpated[12]

(continued)

Common Name	Scientific Name	GMF (Arnold 1940)	West Branch (Rosen 2001, 2003)	Other Sources
Western Side-blotched Lizard	*Uta stansburiana elegans*	A		
Schott's Tree Lizard	*Urosaurus ornatus schottii*	A	X	****
Regal Horned Lizard	*Phrynosoma solare*	A	X	****
	*Anota solare****			
Giant Spotted Whiptail	*Cnemidophorus burti stictogrammus*		X	
	*Aspidoscelis stictogramma****			
Sonoran Spotted Whiptail[13]	*C.* cf. *sonorae*[14]		X	****
	*A. sonorae****			
Sonoran Tiger Whiptail	*C. tigris punctilinealus*		X	
	*A. tigris punctilinealus****			
Reticulate Gila Monster	*Heloderma s. suspectum*	A		
Southwestern Threadsnake (Blind Snake)	*Leptotyphlops h. humilis*	A		
	*Rena h. humilis****			
Regal Ring-necked Snake	*Diadophis punctatus regalis*	A		
Saddled (Pima) Leaf-nosed Snake	*Phyllorhynchus browni*	A		
Spotted Leaf-nosed Snake	*P. decurtatus*	A		
Red Racer (Coachwhip)	*Masticophis flagellum piceus*	A		
	*Coluber flagellum piceus****			
Sonoran Whipsnake	*M. bilineatus*	A		
	*C. bilineatus****			
Western Patch-nosed Snake	*Salvadora h. hexalepis*	A		
Arizona Glossy Snake	*Arizona elegans noctivaga*	A		
Sonoran Gophersnake	*Pituophis catenifer affinis*	A	X	****
Desert Kingsnake	*Lampropeltis splendida*	A	X	
Western Long-nosed Snake	*Rhinocheilus lecontei*	A		
(Northern) Mexican [Brown] Gartersnake *CAN	*Thamnophis eques megalops*			Extirpated[15]
Marcy's Checkered Gartersnake	*T. m. marcianus, T.* cf. *marcianus*	A	X[16]	Extirpated[17]
Smith's Black-headed Snake[18]	*Tantilla hobartsmithi*		X	
Variable Sandsnake	*Chilomeniscus cinctus*	A		
	*C. stramineus****			
Sonoran Nightsnake	*Hypsiglena torquata* ssp.	A		
	*Hypsiglena c. chlorophaea****			
Western Diamond-backed Rattlesnake	*Crotalus atrox*	A		
Northern Mojave Rattlesnake	*C. s. scutulatus*	A		

Total Species: 44

Notes

1. Nomenclature after Society for the Study of Amphibians and Reptiles (2012) and Jones and Lovich (2009); systematic order after Stebbins (1985, 2003); order is periodically changing owing to DNA analyses.

2. Reptiles identified by L. M. Klauber, an internationally recognized herpetologist. Among Arnold's specimens were lizards identified as Six-lined Racerunner (*Cnemidophorus sexlineatus* [*Aspidoscelis s. sexlineata*]) and Common Checkered Whiptail (*C. tesselatus* [*A. tesselata*]). Classification of this group has changed drastically, and the correct identification of those specimens is unknown; see appendix H.

3. Abandoned Santa Cruz River channel (see chapters 4–7) that roughly parallels Mission Road from Ajo Way to Silverbell Road (Rosen 2001, 2003).

4. Lowland Leopard Frog is listed as a special status species by the Arizona Game and Fish Department.

5. Lowland Leopard Frog became extinct by 1975 at the latest (Rosen 2001, 2003); see appendix H.

6. American Bullfrog is an introduced species from the eastern United States (Stebbins 1985).

7. Sonora Mud Turtle was not definitely reported from the GMF but widely inhabited perennial water along the Santa Cruz; it has been extirpated (Rosen 2001); see appendix H.

8. *Coleonyx variegatus bogerti* was named by L. M. Klauber (1945) from a specimen collected by Arold at the GMF on 17 July 1939 (holotype SDSNH 3286).

9. Common Lesser Earless Lizard was found on a "Tucson bajada adjoining West Branch floodplain" (Rosen 2003: 40).

10. Long-nosed Leopard Lizard was extirpated in the Tucson area by the 1950s or 1960s (Rosen 2008); see appendix H.

11. Southwestern Fence Lizard was found in "downtown Tucson, 1995" (Rosen 2003: 40).

12. Southwestern Fence Lizard has been extirpated in the Tucson area (Rosen 2008); see appendix H.

13. Sonoran Spotted Whiptail is a triploid unisexual species; females reproduce parthenogenetically (Jones and Lovich 2009).

14. Rosen (2003) discusses these parthenogenetically reproducing, unisexual species and refers to the Tiger Whiptail because of ongoing research into this complicated group of lizards.

15. Northern Mexican Gartersnake is known from the Rillito River and possibly also along the Santa Cruz River; it became extinct in Tucson Basin by the 1960s (Lowe 1985); see appendix H.

16. Marcy's Checkered Gartersnake is apparently a tentative identification by Rosen (2003).

17. Marcy's Checkered Gartersnake was reported from the GMF and other locations on the Santa Cruz floodplain; it was extirpated by the 1970s (Lowe 1985); see appendix H.

18. Smith's Black-headed Snake occurs in riparian situations and would have been expected in the GMF but would be difficult to detect. Secretive and mostly nocturnal, it hides under rocks and in crevices during the day (Stebbins 1985).

Appendix G

Mammals of the Great Mesquite Forest (GMF) and the adjacent Santa Cruz River.

A=Arnold (1940).

*Records in Cockrum (1960) from specific GMF localities (for example, "Santa Cruz River, 10 mi. S. of Tucson," "Indian Dam," and so forth).

**Records in Cockrum (1960) from 9 to 13 miles south of Tucson, distances without specific locality but approximating those of the GMF.

Common Name[1]	Scientific Name[2]	Great Mesquite Forest[3] (historic)	West Branch[4] (recent)	Other Sources (GMF)
Cave Myotis	*Myotis v. velifer*	A		Cockrum**
Long-legged Myotis[5]	*M. volans interior*	A		
California Myotis	*M. c. californicus*	A		Cockrum*
Big Brown Bat	*Eptesicus fuscus pallidus*	A		Cockrum*
Pallid Bat	*Antrozous p. pallidus*	A		Cockrum**
Brazilian Free-tailed Bat	*Tadarida brasiliensis mexicana*	A		
Big Free-tailed Bat[6]	*Nyctinomops macrotis*			Cockrum**
Desert Cottontail	*Sylvilagus auduboni minor*	A	X	
Black-tailed Jackrabbit	*Lepus californicus eremicus*	A	X	
Antelope Jackrabbit	*L. a. alleni*	A		
Harris's Antelope Squirrel	*Ammospermophilus h. harrisii*	A		Cockrum**
Rock Squirrel	*Spermophilus variegatus grammurus*	A		Cockrum*
Round-tailed Ground Squirrel	*S. tereticaudus*	A	X	Cockrum**
Botta's Pocket Gopher	*Thomomys bottae modicus*	A		Cockrum*
Silky Pocket Mouse	*Perognathus f. flavus*	A		Cockrum*
Bailey's Pocket Mouse	*Chaetodipus b. baileyi*	A		Cockrum**
Sonoran Desert Pocket Mouse**	*C. penicillatus pricei*	A	X	Cockrum**
Merriam's Kangaroo Rat	*Dipodomys m. merriami*	A		Cockrum**
Ord's Kangaroo Rat	*D. o. ordii*	A		Cockrum**
Fulvous Harvest Mouse	*Reithrodontomys f. fulvescens*	A		Cockrum*
Western Harvest Mouse	*R. m. megalotis*	A		Cockrum**
White-footed Mouse	*Peromyscus leucopus arizonae*	A		Cockrum*
Merriam's Mouse[7]	*P. m. merriami*	A		Cockrum*
Southern Grasshopper Mouse	*Onychomys t. torridus*	A		

Common Name	Scientific Name	Great Mesquite Forest (historic)	West Branch (recent)	Other Sources (GMF)
Hispid Cotton Rat *or* Arizona Cotton Rat[8]	*Sigmodon hispidus cienegae* *Sigmodon arizonae cienegae*	A	X	Cockrum**
Western White-throated Woodrat	*Neotoma a. albigula*	A		
Coyote	*Canis latrans mearnsi*	A	X	
Gray Wolf[9]	*C. lupus baileyi*			Davis (1982)
Kit Fox	*Vulpes macrotis arsipus*	A		
Gray Fox	*Urocyon cinereoargenteus scottii*	A		Cockrum**
Ringtail	*Bassariscus astutus arizonensis*	A		
Northern Raccoon	*Procyon lotor mexicanus*	A		
American Badger	*Taxidea taxus berlandieri*	A		
Western Spotted Skunk	*Spilogale gracilis arizonae*	A		
Hooded Skunk	*Mephitis macroura milleri*			Cockrum*
Striped Skunk	*M. mephitis estor*	A		
White-backed Hog-nosed Skunk	*Conepatus leuconotus venaticus*	A		
Bobcat	*Lynx rufus baileyi*	A		
Collared Peccary (Javelina)	*Pecari tajacu*			Marshall[10]

Total Species: 39

Notes

1. Taxonomic order after Hoffmeister (1986); names after Baker et al. (2003).

2. After Hoffmeister (1986).

3. Records from Arnold (1940) with specimens identified by Seth Benson, nationally noted mammalogist and curator of mammals, Museum of Vertebrate Zoology, University of California, Berkeley.

4. Abandoned Santa Cruz River channel, roughly paralleling Mission Road from Ajo Way to Silverbell Road (Rosen 2001).

5. Long-legged Myotis is not listed by Hoffmeister (1986); this is apparently either a migrant or a misidentification, since breeding localities in Arizona are generally 5,000 feet in elevation and higher (Cockrum 1960).

6. Big Free-tailed Bat is not listed by Hoffmeister (1986).

7. Merriam's Mouse is a mouse largely of mesquite bosques (Burt and Grossenheider 1964), occurring only in southern Arizona and Sonora, Mexico (Hall and Kelson 1959).

8. Identification is uncertain between Hispid Cotton Rat and Arizona Cotton Rat. This species was recorded in or near the Santa Cruz River between 29th Street and Grant Road between 2004 and 2011 by US Fish and Wildlife Service personnel.

9. Gray Wolf is considered a hypothetical species and has been extirpated (see appendix H). In 1850, Judge Benjamin Hayes of California wrote in his journal of the San Xavier bosque that wolves "were howling all around us, and one of very large size, was seen" (Davis 1982: 51).

10. Javelina often upset towhee traps in the GMF (J. T. Marshall, pers. comm. to RRJ); see Marshall citations in the reference list.

Appendix H

Special status of species for the Great Mesquite Forest (GMF) and the middle Santa Cruz River (Tucson Basin), including extirpations and hypothetical species, species of special concern, accidental species, and excluded species.[1]

*US Fish and Wildlife Service designations: *END=endangered; *CAN=candidate species for threatened or endangered status.
**Special status species, as listed by the Arizona Game and Fish Department; also see "Additional Special Species" section of this appendix.[2]

***Recent changes in classification or nomenclature that may or may not be accepted over time (American Ornithologists' Union 2012, Society for the Study of Amphibians and Reptiles 2012).

Hypothetical Species

We find no direct records of the following at the Great Mesquite Forest, but their former occurrence is highly probable.[3]

Common Name	Scientific Name	Evidence for Inclusion and References
Sonoran Mud Turtle	*Kinosternon s. sonoriense*	Present in Santa Cruz–Rillito drainage until at least 1990s;[4] also archaeological evidence[5]
Great Plains Skink	*Eumeces obsoletus* *Plestiodon obsoletus***	"Early reports place the Great Plains Skink (*Eumeces obsoletus*) in Tucson Basin's riparian assemblage"[6] (whether this included GMF is unclear)
Regal Ring-necked Snake	*Diadophis punctatus regalis*	"Former occurrence of the Regal Ringneck Snake (*Diadophis punctatus regalis*) at San Xavier"[7]
(Northern) Mexican [Brown] Gartersnake *CAN	*Thamnophis eques megalops*	Formerly in perennial water of Santa Cruz–Rillito drainage, where it fed on aquatic prey;[8] extirpated in Tucson Basin by 1985,[9] possibly as early as 1960s[10]
Brown Vinesnake	*Oxybelis aeneus*	"Long ago recorded from the 'outskirts of Tucson,'[11] a locality likely to have involved the Santa Cruz forest";[12] US occurrence only in Tucson-Nogales region[13]
**Black-bellied Whistling-Duck[14]	*Dendrocygna autumnalis*	Six killed on Santa Cruz, two miles south of Tucson, 5 May 1899;[15] also other Tucson records[16]
Mexican Duck (Mallard)	*Anas platyrhynchos diazi*	Female colored duck, considered a subspecies of Mallard;[17] has recently bred at sites around Tucson[18]

Common Name	Scientific Name	Evidence for Inclusion and References
Ruddy Duck	*Oxyura jamaicensis*	Now breeding on Santa Cruz in Sweetwater Wetlands;[19] "observed near Tucson in April, 1896,"[20] perhaps nesting; one of the most widely spread ducks nesting in Arizona during early 1900s;[21] also archaeological evidence[22]
**Northern (Masked) Bobwhite *END	*Colinus virginianus ridgwayi*	Discovered by Herbert Brown,[23] named by Brewster;[24] "[g]razed out of existence by early 1900s,"[25] reintroduction largely unsuccessful; see "Local Extirpations" section of this appendix
Pied-billed Grebe[26]	*Podylimbus podiceps*	Recorded recently at Sweetwater Wetlands;[27] Vorhies saw Pied-billed Grebes frequently at Indian Dam in GMF in 1930s and three calling on open water, 15 April 1934[28]
Least Bittern	*Ixobrychus exilis*	Early specimen taken at Tucson by Herbert Brown;[29] secretive bird of cattail marshes, nests along Salt River near Phoenix and other rivers of Colorado-Gila River system[30]
Black-crowned Night Heron	*Nycticorax nycticorax*	Often recorded (see appendix A); Swarth and Phillips et al. discounted this species' nesting along streams of southern Arizona,[31] but newly fledged young found along Salt-Gila River system south of Phoenix[32]
Common Gallinule	*Gallinula chloropus* *Gallinula galeata****	Historical summer records from Tucson region,[33] not specifically at GMF; now "regular at Sweetwater Wetlands,"[34] this species lives in cattail marshes and colonizes rapidly[35]
**Cactus Ferruginous Pygmy-Owl	*Glaucidium brasilianum cactorum*	135-year record of this diurnal owl finds only one specimen for Santa Cruz River (by Herbert Brown, 23 November 1884)[36] but numerous specimens from Rillito River drainage;[37] because of similarity of habitat and bird populations between Santa Cruz and Rillito,[38] this owl likely was extirpated from Santa Cruz River during earlier 1800s
Belted Kingfisher	*Megaceryl alcyon*	May have nested in Santa Cruz–Rillito drainage in 1870s when Bendire was here;[39] while at Camp Lowell, Bendire made several trips along Santa Cruz River[40]
**Tropical Kingbird[41]	*Tyrannus melancholicus*	First recorded in Arizona (and United States) by specimen taken by H. H. Kimball near Fort Lowell, 12 May 1905;[42] perhaps also seen by Brandt[43]
Barn Swallow	*Hirundo rustica*	Found nesting in San Xavier Mission School, adjacent to GMF, 8 June 1895, by Herbert Brown[44]
White-tailed Deer	*Odocoileus virginianus couesi*	Recorded in "the valley of the Santa Cruz" in large numbers[45]

(continued)

Notes

1. Nearly all species of riparian and wetland birds, as well as other riparian and wetland vertebrates, have been extirpated from the sparse remains of the GMF. This list includes only species that have been extirpated from both the GMF and most or all of the Santa Cruz–Rillito River system in the Tucson Basin. Some species have recently reestablished with water developments, such as Tucson sewage effluent from the Sweetwater Wetlands (see appendix C).

2. Arizona Game and Fish Department online lists of special status species, http://www.azgfd.com/w_c/edits/hdms _species_lists.shtml (updated 2 July 2012).

3. The determination of probable occurrence is based on distribution of species as delineated in pertinent literature and known records (especially specimen records) for the area.

4. Froebel (1859), Rosen (2001, 2003).

5. Sonoran Mud Turtle remains were recovered from a Classic Hohokam site (AD 1100–1450) at the I-19 San Xavier bridge (Ravesloot and Whittlesey 1987).

6. Rosen (2003: 40); see also Stebbins (1985).

7. Rosen (2003: 40).

8. Stebbins (1985).

9. Rosen (2001, 2003).

10. Lowe (1985).

11. Vorhies (1926).

12. Rosen (2003: 41).

13. Stebbins (1985).

14. Black-bellied Whistling-Duck, Mexican Duck, and Masked Bobwhite have all been recorded in the Santa Cruz valley and are hunted, so they may have been "shot out" the same way Wild Turkey was extirpated in the area with the arrival of early European settlers.

15. Brown (1906).

16. Phillips et al. (1964).

17. American Ornithologists' Union (1998, 2012). Long considered a species, the female plumaged Mexican Duck (*Anas diazi*) hybridizes with the ("green-head" male) Mallard (*Anas platyrhynchos*). The two populations are theorized to have been separated during past ice ages, then making secondary contact during warming interglacial periods (such as currently) and hybridizing (Hubbard 1977); see also endnote 14.

18. Corman and Wise-Gervais (2005).

19. Stevenson (2007), Stejskal and Rosenberg (2011); see appendix C.

20. Swarth (1914: 14).

21. Phillips et al. (1964).

22. Ruddy Duck remains were recovered from a Classic Hohokam site (AD 1100–1450) at the I-19 San Xavier bridge (Ravesloot and Whittlesey 1987).

23. Huels et al. (2013).

24. Brewster (1885).

25. Phillips et al. (1964: 28). This grassland species was discovered by Herbert Brown, who also published most of the early information on the species (Bendire 1892, Brown 1904, Phillips et al. 1964, Brown 1989). Specimens are from the upper Santa Cruz valley (Brown et al. 2012), but the species' early status in the middle Santa Cruz valley is unclear; see also endnote 14.

26. Because Pied-billed Grebe begins nesting in December and January in the Arizona lowlands (Corman and Wise-Gervais 2005), they may have been nesting here.

27. Stevenson (2007), Stejskal and Rosenberg (2011).

28. Vorhies et al. (1935).

29. Phillips et al. (1964).

30. Phillips et al. (1964), Johnson and Simpson (1971), Rea (1983), Corman and Wise-Gervais (2005), Johnson and Simpson (MS a).

31. Swarth (1914), Phillips et al. (1964).

32. Rea (1983), Johnson and Simpson (MS a).

33. Rhodes (1892), Swarth (1914).

34. Stevenson (2007: 256), Stejskal and Rosenberg (2011).

35. Phillips et al. (1964).

36. Johnson et al. (2003).

37. Ferruginous Pygmy-Owl was discovered for the United States by Bendire in mesquites near Fort Lowell on 24 January 1872 (Coues 1872a); Bendire (1872b, 1892), Johnson et al. (2003); see "Local Extirpations" section of this appendix.

38. See appendixes A and B.

39. Bendire (1895: 35) wrote"[I]n southern Arizona . . . I have found kingfishers breeding." Rea (1983) presents compelling evidence for breeding of the Belted Kingfisher along the Gila River in central Arizona prior to Anglo-American settlement despite reluctance of Swarth (1914) and Phillips et al. (1964) to accept the species as nesting in Arizona.

40. Bendire (1872a, 1872b, 1892, 1895).

41. Tropical Kingbird was first reported as a breeding species for the United States along the Santa Cruz between Tucson and the GMF (Phillips 1940, Phillips et al. 1964).

42. Peters (1936), Phillips et al. (1964). Whether Kimball's specimen was a solitary bird or a member of a breeding pair or colony is unknown.

43. Brandt (1951) questionably thought he saw a Tropical Kingbird in the GMF.

44. Huels et al. (2013).

45. Davis (1982: 82) quoting Kennerly's observations during a boundary survey in the mid-1800s and Cockrum (1960: 254) from Tucson specifically.

Local Extirpations

Common Name	Scientific Name	Description	Cause and Date of Extirpation	References
California Floater	*Anodonta californiensis*	3 inch edible clam (Unionidae)	Cessation of perennial flow by 1940s[1]	Bequaert and Miller (1973)
**Gila Chub *END	*Gila intermedia*	6–8 inch minnow (Cyprinidae)	Cessation of perennial flow by 1940s	Minckley (1973), USFWS (2012)
**Longfin Dace[2]	*Agosia chrysogaster*	3 1/2 inch minnow (Cyprinidae)	Cessation of perennial flow by 1940s	Minckley (1973), Wikipedia (2012)
Sonora Sucker	*Catostomus insignis*	Can reach 31.5 inches and weigh 4.4 pounds	Cessation of perennial flow by 1940s	Minckley (1973), Wikipedia (2012)
Santa Cruz (Monkey Springs) Pupfish[3]	*Cyprinodon arcuatus* (extirpated)	2 inch fish (Cyprinodontidae)	Cessation of perennial flow by 1940s	Minckley (1973), Minckley et al. (2002), USFWS (2011c)
Mosquitofish (nonnative)[4]	*Gambusia a. affinis*	2–3 inch fish (Cyprinodontidae)	Unknown	Minckley (1973), Minckley et al. (1977)
**Gila Topminnow *END	*Poeciliopsis o. occidentalis*	1–2 inch fish (Poeciliidae)	Cessation of perennial flow by 1940s	Minckley (1973), USFWS (2011c)
Desert Sucker	*Catostomus clarki*	4–16 inch fish (Catostomidae)	Cessation of perennial flow by 1940s	Minckley (1999)
**Lowland Leopard Frog	*Lithobates*** [Rana] yavapaiensis*		Lack of perennial water; extinct by 1975 at latest	Rosen (2001)
Sonoran Mud Turtle	*Kinosternon s. sonoriense*		Lack of perennial water; date of extirpation unknown	Present in Santa Cruz–Rillito drainage until at least 1990s;[5] also archaeological evidence[6]
Long-nosed Leopard Lizard	*Gambelia wislizenii*		Urbanization; apparently extinct by 1950s or 1960s	Rosen (2008)
Southwestern Fence Lizard	*Sceloporus undulatus cowlesi S. cowlesi****		Urbanization?[7]	Rosen (2008)
Great Plains Skink	*Eumeces obsoletus Plestiodon obsoletus****		Riparian habitat loss	See "Hypothetical Species" section in this appendix (Rosen 2003)
Regal Ring-necked Snake	*Diadophis punctatus regalis*		Riparian habitat loss	See "Hypothetical Species" section in this appendix (Rosen 2003)[8]
(Northern) Mexican [Brown] Gartersnake *CAN	*Thamnophis eques megalops*		Riparian habitat loss; extirpated in Tucson Basin by 1985	Lowe (1985), Rosen (2001); see "Hypothetical Species" section in this appendix

Common name	Scientific name	Cause/status	Reference
Marcy's Checkered Gartersnake	*T. m. marcianus*	Aquatic habitat loss; extirpated in Tucson Basin by 1970s	Lowe (1985)
Brown Vinesnake	*Oxybelis aeneus*	Urbanization?; largely tropical; no longer this far north	Stebbins (1985); see "Hypothetical Species" section in this appendix (Rosen 2003)
**Masked Bobwhite *END	*Colinus virginianus ridgwayi*	"Grazed out of existence by early 1900s,"[9] reintroduction largely unsuccessful	Phillips et al. (1964), Brown (1989), Brown et al. (2012)
Wild Turkey	*Meleagris gallopavo*	"Shot out" and last recorded in southern Arizona in 1907; later reintroduced[10]	Swarth (1914), Phillips et al. (1964: 30), Hamlin (1966)
Black Vulture	*Coragyps atratus*	Urbanization?[11]	Phillips et al. (1964)
**Common Black-Hawk[12]	*Buteogallus anthracinus*	Riparian habitat loss; extinct in Tucson Basin	Corman and Wise-Gervais (2005), Stevenson (2007)
**Gray Hawk	*Asturina nitida* *Buteo nitidus****	Riparian habitat loss; no longer nests near Tucson	Bibles et al. (2002), Corman and Wise-Gervais (2005)
**Zone-tailed Hawk[13]	*B. albonotatus*	Riparian habitat loss; no longer nests in Tucson lowlands	Johnson et al. (2000b), Stevenson (2007)
**Crested Caracara[14]	*Caracara cheriway*	Unknown; no longer nests near Tucson	Phillips et al. (1964)
Common Ground-Dove	*Columbina passerina*	Urbanization	Corman and Wise-Gervais (2005)[15]
**Yellow-billed Cuckoo *CAN	*Coccyzus americanus*	Riparian habitat loss; rare in Tucson Basin	Corman and Wise-Gervais (2005), Stevenson (2007)
**Cactus Ferruginous Pygmy-Owl[16]	*Glaucidium brasilianum cactorum*	Riparian habitat loss; no longer nests near Tucson; see "Hypothetical Species" section of this appendix	Proudfoot and Johnson (2000), Johnson et al. (2003)
**Southwestern Willow Flycatcher *END	*Empidonax traillii extimus*	Riparian habitat loss; "absent from the Santa Cruz River" by 1960s	Phillips et al. (1964: 89)
**Tropical Kingbird[17]	*Tyrannus melancholicus*	Riparian habitat loss, especially cottonwoods	See "Hypothetical Species" section of this appendix
**Rose-throated Becard[18]	*Pachyramphus aglaiae*	Riparian habitat loss; absent after 1959	Marshall (field notes), RRJ

(continued)

Common Name	Scientific Name	Description	Cause and Date of Extirpation	References
Song Sparrow	*Melospiza melodia*		Riparian/wetland habitat loss; extirpated shortly after 1895[19]	Phillips et al. (1964)
Muskrat	*Ondatra zibethicus*		Prehistoric;[20] unknown if present historically	Ravesloot and Whittlesey (1987)
Gray Wolf[21]	*Canis lupus*		Exterminated by government trappers et al.[22]	Davis (1982)

Notes

1. Long-time Tucson resident Allan R. Phillips stated that the Santa Cruz River ceased to flow in the mid- to late 1940s (Phillips et al. 1964).
2. Longfin Dace (*Agosia chrysogaster*) is "a species of concern" (US Fish and Wildlife Service 2012).
3. Monkey Springs Pupfish has been completely extirpated (Minckley et al. 2002).
4. A native of Atlantic and Gulf of Mexico drainages, this small fish was first found in Arizona in 1926 and has since been reported from throughout the Santa Cruz River drainage (Minckley 1973, Minckley et al. 1977). A voracious predator, introduced largely for mosquito control, it also feeds on other small animals, such as young fish, and has been largely successful in outcompeting fishes such as Gila Topminnow (*Poeciliopsis o. occidentalis;* Minckley et al. 1977).
5. Froebel (1859), Rosen (2001).
6. Sonoran mud turtle remains were recovered from a Classic Hohokam site (AD 1100–1450) at the I-19 San Xavier bridge (Ravesloot and Whittlesey 1987).
7. Southwestern fence lizard has been extirpated in Tucson area (Rosen 2008).
8. The phrasing "[f]ormer occurrence . . . at San Xavier" (Rosen 2003: 40) and the fact that it is not currently found in the West Branch of the Santa Cruz River suggest at least local extirpation.
9. Phillips et al. (1964: 28); Masked Bobwhite is a grassland species discovered by Herbert Brown who also published most of the early information on the species (Bendire 1892, Brown 1904, Phillips et al. 1964). Specimens are from the upper Santa Cruz valley (Brown et al. 2012), but the species' early status in the middle Santa Cruz valley is unclear; see also "Hypothetical Species" section in this appendix.
10. Wild Turkey was recorded calling at the GMF (1856 letter by Ewell *in* Hamlin 1966); also, a description of Wild Turkey occurring "in the valley of the Santa Cruz" (Swarth 1914: 23) was apparently based on records at Tubac in 1849 (Browne 1951), the 1850s (Poston 1951), 1864 (Harris 1960), and 1872 (Bendire 1872b).
11. Black Vulture occurs in surrounding areas but disappeared from the Tucson area with increasing urbanization; it is a common species in towns of Mexico but not Arizona (Phillips et al. 1964).
12. Common Black-Hawk is a species of concern (US Fish and Wildlife Service 2012).
13. A tale of a narrow escape from Apache Indians while taking an egg from a nest of this species along Rillito Creek is told by Bendire (1892); also, a pair of Zone-tailed Hawks was seen by RRJ, circling over Rillito Creek at Campbell Road in April or May 1953 (Johnson et al. 2000b).
14. Crested Caracara was reported by Bendire as "not at all uncommon about [his] camp" (1872b: 87) on the Rillito River. The last nest reported in the vicinity of Tucson was by Herbert Brown in 1889 (Bendire 1892, Phillips et al. 1964), and the last summer record was in 1917 by Dawson (1921).
15. During the Arizona Breeding Bird Atlas project, from 1993 to 2000, Common Ground-Doves were not detected near Tucson and Phoenix despite their historic occurrence in these areas (Corman and Wise-Gervais 2005).
16. Ferruginous Pygmy-Owl was first found in the United States along Rillito Creek by Bendire (Coues 1872a, Bendire 1892) and was relatively common along Rillito Creek (Johnson et al. 2003 [and see text]). The Arizona population was listed as federally endangered in 1997 but delisted in 2007; it was extirpated in the Tucson Basin by 2009 (RRJ [member of the species' recovery team]).

17. Tropical Kingbird was first reported as a breeding species for the United States along the Santa Cruz between Tucson and the GMF (Phillips 1940, Phillips et al. 1964). Brandt thought he saw one in the GMF but was uncertain.

18. A Rose-throated Becard nest was found in 1958, and a calling male was recorded in 1959 (Phillips et al. 1964; see chapter 7); see appendix A.

19. Song Sparrow, formerly a common riparian nesting species in the Tucson region (Bendire 1872b), was extirpated from the region shortly after 1895 (Phillips et al. 1964). Records by Arnold and Marshall (see appendix A) were probably migrating individuals; recently, breeding populations reestablished in the Sweetwater Wetlands; see appendix C.

20. The lack of recent muskrat records is puzzling, based on descriptions of the habitat of the Santa Cruz River at the time of arrival of Euro-Americans and on muskrat records from other sites in the Gila River drainage, including the San Pedro and other lower-elevation rivers (Cockrum 1960, Hoffmeister 1986). Muskrat remains recovered from a Classic Hohokam site (AD 1100–1450) at the I-19 San Xavier bridge (Ravesloot and Whittlesey 1987) suggest that the muskrat was considered a food item and thus was possibly "eaten out of existence" in this relatively small stream.

21. In 1850, Judge Benjamin Hayes of California wrote in his journal of the San Xavier bosque that wolves "were howling all around us, and one of very large size, was seen" (Davis 1982: 51); possibly these were coyotes instead, but because one was seen, they were probably wolves.

22. Brown (1983).

Additional Special Species

*CAN=Candidate for endangered or threatened species designation by the US Fish and Wildlife Service
sss=Special status species, as designated by the Arizona Game and Fish Department

Common Name	Scientific Name[1]	Status	Appendix
Lowland Leopard Frog[2]	*Lithobates [Rana] yavapaiensis***	sss	Appendix F
Western Narrow-mouthed Toad	*Gastrophryne olivacea*	sss	Appendix F
Morafka's Desert Tortoise	*Gopherus [agassizii] morafkai***	*CAN	Appendix F
Swainson's Hawk	*Buteo swainsoni*	sss	Appendix A
Western Burrowing Owl[3]	*Athene cunicularia hypugaea*	sss	Appendix A
Northern Beardless-Tyrannulet	*Camptostoma imberbe*	sss	Appendix A
Cave Myotis	*Myotis velifer*	sss	Appendix G
Big Free-tailed Bat[4]	*Nyctinomops macrotis*	sss	Appendix G
Merriam's Mouse[5]	*Peromyscus m. merriami*	sss	Appendix G

Notes

1. The "Additional Special Species" category includes special species that are not listed as hypothetical or extirpated.

2. Lowland Leopard Frog is listed as a special status species by the Arizona Game and Fish Department.

3. Western Burrowing Owl (*Athene cunicularia hypugea*) is "a species of concern" (US Fish and Wildlife Service 2012). Although the species has not been recorded along the Santa Cruz–Rillito River system since 1917 (Dawson 1921), a colony currently occurs near San Xavier Mission (B. Fontana, pers. comm.).

4. Big Free-tailed Bat is not listed by Hoffmeister (1986).

5. Merriam's Mouse is found largely in mesquite bosques (Burt and Grossenheider 1964), occurring only in southern Arizona and Sonora, Mexico (Hall and Kelson 1959).

Accidental Species

Common Name *Scientific Name[1]*	Record Information	Location	Where Published	Usual Breeding Grounds
Long-tailed Duck *Clangula hyemalis*	Specimen by Allan R. Phillips, 2 January 1950	Santa Cruz River sewer plant	Phillips et al. (1964)	Alaska, Canada, and offshore islands
Hooded Merganser *Lophodytes cucullatus*	Specimen by Herbert Brown, 1 June 1890	Calabasas (10 miles north of Nogales on Santa Cruz River)	Phillips et al. (1964)	Eastern and extreme northwestern United States, parts of Canada
Pacific Loon *Gavia pacifica*	Specimen by Herbert Brown, 27 August 1900	San Xavier Mission	Phillips et al. (1964)	Alaska and Canada
Anhinga *Anhinga anhinga*	Specimen by Herbert Brown, 12 September 1893	Silver Lake, Santa Cruz River	Brown (1906)	Southeastern United States and Mexico
Scarlet Ibis *Eudocimus ruber*	Observation by Herbert Brown, 17 September 1890	Rillito Creek, near Fort Lowell	Brown (1899)	South America
Black Rail *Laterallus jamaicensis*	Observation by Frank Stephens, 23 April 1881	Probably Santa Cruz or Rillito River[2]	Brewster (1882)	Lower Colorado River and locally throughout United States

Common Name *Scientific Name*	Record Information	Location	Where Published	Usual Breeding Grounds
Purple Gallinule *Porphyrio martinica*	Specimen by Herbert Brown, 20 October 1887	Santa Cruz River, Tucson	Brown (1888)	Locally throughout eastern United States, south into South America
Flammulated Owl *Otus flammeolus*	Observation by Richard Crossin (see appendix I)	Santa Cruz River, Tucson	Joe T. Marshall (pers. comm.)	Mountains of western North America into Mexico
Whiskered Screech-Owl *Megascops trichopsis*	Observation by Richard Crossin (see appendix I)	Santa Cruz River, Tucson	Joe T. Marshall (pers. comm.)	Mountains of extreme southwestern United States into Mexico
Spotted Owl *Strix occidentalis*	Observation by Richard Crossin (see appendix I)	Santa Cruz River, Tucson	Joe T. Marshall (pers. comm.)	Mountains of western North America into Mexico
Eastern Wood-Pewee *Contopus virens*	Specimen by Allan R. Phillips	Indian Dam, GMF[3]	Johnson et al. (1997a)	Eastern United States, Canada
Olive Warbler *Peucedramus taeniatus*	Observation by Earl and Virginia Morton, 5 May 1947	Indian Dam, GMF, Santa Cruz River	Phillips et al. (1964)	Mountains of southern Arizona and New Mexico into Mexico
Northern Parula *Parula americana* *Setophaga Americana****	Specimen by Allan R. Phillips, 26 March 1938	Santa Cruz River, near San Xavier	Monson (1942)	Eastern and central United States, casually in Southwest

Notes

1. Most of these "Accidental Species" are migratory species that do not normally occur in the lowlands of this region and for which there are usually no more than a couple of records for the area. Possibly more individuals of some of these species use rivers in the Tucson Basin as corridors for migration and other movements but are not detected because they are in such small numbers and (or) move at night.

2. Because Stephens spent much of his time in the Tucson area collecting along Rillito Creek and the Santa Cruz River, it is assumed that that is where he recorded the Black Rail.

3. Only one specimen of the Eastern Wood-Pewee is listed in Phillips et al. (1964); it was taken by Phillips at Tucson, but the more exact location is given by Johnson et al. (1997a) based on conversation with Phillips.

Excluded Species

Common Name	Scientific Name	Reporter and Reason for Exclusion
Six-lined Racerunner	*Cnemidophorus s. sexlineatus* *Aspidoscelis s. sexlineata****	Arnold (1940; identity of Arnold's specimen unknown; see appendix F); species does not occur in Arizona;[1] hybridizes with other species, sometimes resulting in unisexual (female) hybrid complexes[2]
Common Checkered Whiptail	*C. tesselatus A. tesselata****	Arnold (1940; identity of Arnold's specimen unknown; see Appendix F); species does not occur in Arizona; diploid parthenogenic species resulting from hybridization between other closely related species

(continued)

Common Name	Scientific Name	Reporter and Reason for Exclusion
Common Nighthawk	*Chordeiles minor*	Arnold (1940; see appendix A); species does not nest in Arizona lowlands, where it is replaced by Lesser Nighthawk (*C. acutipennis*; Phillips et al. 1964)
Bank Swallow	*Riparia riparia*	Swarth (1905; see appendix A); species does not nest in Arizona,[3] probably mistaken identity of Northern Rough-winged Swallow (*Stelgidopteryx serripennis*)
Blue-gray Gnatcatcher	*Polioptila caerulea*	Arnold (1940; see appendix A); species does not nest in Arizona lowlands, where it is replaced by Black-tailed Gnatcatcher (*P. melanura*; Phillips et al. 1964)

Notes

1. Jones and Lovich (2009).

2. Jones and Lovich (2009) discuss numerous female species of *Aspidoscelis* (*Cnemidophorus*) that reproduce parthenogenetically. They vary from diploid to triploid species, and most or all have been derived from hybridization between living and (or) extinct species. Studies in molecular systematics are currently under way to further determine ancestors of these female species. The complexities in this group of lizards were not understood until the 1960s, when some groups were found to be unisexual (females).

3. Phillips et al. (1964), Corman and Wise-Gervais (2005).

Appendix I

Ornithologists who conducted studies in the Santa Cruz–Rillito River system, especially in the Great Mesquite Forest (GMF). The list of scientific and educational institutions with which these ornithologists were associated illustrates the importance of work in the GMF to development of the science of ornithology.

*Co-authors of *The Birds of Arizona*.[1]

Ornithologist	Institutions	Area of Study	Dates	Results
Primary Investigators				
Charles E. Bendire	US Army Medical Corps; US National Museum	Rillito Creek (now Rillito River) with side trips, including camping, in the GMF	October 1871 to March 1873	Extensive notes (Bendire 1872a, 1872b); large collection of birds, nests, and eggs; publications; see appendix B
Frank Stephens	Collected for William Brewster, Harvard Museum of Comparative Zoology; San Diego Natural History Museum[2]	SE Arizona, especially Tucson area, spent time at Fort Lowell[3] and in mesquites along Santa Cruz River[4]	Spring and summer 1881	Extensive collection of museum specimens; see appendix B and Brewster citations in reference list
Herbert Brown	Arizona State Museum, University of Arizona; "First resident Arizona ornithologist"; editor and owner of Tucson newspaper	Tucson, especially Santa Cruz and Rillito Rivers; took only Ferruginous Pygmy-Owl found along Santa Cruz River, south of Tucson[5]	1873 until death in 1913; began avian studies in 1880s after talking to Nelson[6]	Bird records from Santa Cruz and Rillito Rivers and contributed first avian specimens to Arizona State Museum; University of Arizona publications
Harry S. Swarth	Field Columbian Museum, Chicago; Museum of Vertebrate Zoology, University of California, Berkeley; California Academy of Sciences	First ornithologist to publish studies from GMF (see appendix A); also worked in Santa Rita Mountains	In GMF 17–23 May 1902 and 1–14 June 1903[7]	Publications; published first annotated checklist of Arizona birds;[8] see appendix A
Frank Willard	Teacher at Tombstone, Arizona	Early resident amateur ornithologist and independent collector, especially of eggs,[9] in SE Arizona	At least 1898–1923 (see Anderson 1972)	Accompanied A. C. Bent in 1922 expedition (see below); publications on birds of SE Arizona; see appendix A
A. Brazier Howell	US Biological Survey; Johns Hopkins Medical School[10]	Rillito Creek in the vicinity of Fort Lowell (see appendix B); Howell visited GMF in spring 1916 and 1918	7 December 1915 to 25 March 1916	Field notes of visit to GMF in archives of US Natural History Museum; publication; see appendix B
William L. Dawson	Museum of Comparative Oology, Santa Barbara, California[11]	Conducted most intensive short-term study of GMF (see appendix A)	7–26 May and 2–11 June 1917	Museum specimens of birds, nests, and eggs; publication; see appendix A

Arthur Cleveland Bent	US National Museum, Smithsonian Institution; Harvard Museum of Comparative Zoology[12]	Spent several days in GMF while on a field trip in SE Arizona with Frank Willard[13] (see above)	Spring and early summer 1922	Findings of Bent/Willard expedition were not published separately; GMF information in several species accounts; see appendix A and chapter 12
Lee W. Arnold	University of Arizona; Arizona Game and Fish 1941–1942; USFWS 1947–1952[14]	Only person to list all classes of terrestrial vertebrates for GMF	1938–1940	Unpublished master's thesis; see appendix A
Herbert Brandt	Amateur ornithologist and author who founded the Bird Research Foundation, Cleveland, Ohio	Made several trips to GMF during eight breeding seasons, some with professional ornithologists	1935–1948	Two chapters plus additional information on GMF in 725-page book, *Arizona and Its Bird Life*, detailing Brandt's work in SE Arizona; see appendixes A and B
*Joe T. Marshall	University of California, Berkeley; ornithology professor, University of Arizona; US National Museum	Leading expert on Brown Towhees and Screech-Owls; conducted studies in GMF and Sabino Creek, a tributary of Rillito River	1957–1963	Widely published, including two major papers on Abert's and Canyon Towhees and one on Screech-Owls; see appendix A[15]

Secondary Investigators

W. E. D. Scott	Princeton University Museum	Pinal, Pima, and Gila Counties	1881–March 1886	Discussed the Santa Cruz River and, specifically, GMF; publications
O. W. Howard	Unknown (amateur ornithologist)	SE Arizona	1899–1906[16]	Assisted Swarth in GMF, 17–23 May 1902; publications
Charles T. Vorhies	University of Utah; University of Arizona	SE Arizona, accompanied Brandt (see above) in fieldwork	1915–1949[17]	Published records from GMF; see appendix H and appendix A notes
*Allan R. Phillips	University of Arizona; Denver Natural History Museum; Delaware Natural History Museum; numerous Mexican universities	Arizona's premier ornithologist, worked throughout United States and Mexico	Moved to Tucson in 1931 and from there to Mexico in 1957[18]	Obtained several important records from GMF; lead author of *The Birds of Arizona*, containing records from GMF[19]

(continued)

Ornithologist	Institutions	Area of Study	Dates	Results
Anders H. Anderson, Anne Anderson	Amateur ornithologists (Anders an electrician, Anne a schoolteacher)[20]	SE Arizona	1933–1972	Published records from Santa Cruz and Rillito Rivers; publications; also see Anderson (1972)[21]
*Gale Monson	US Soil Conservation Service; US Fish and Wildlife Service	Southern Arizona	1940–1950s	Published records;[22] involved in White-winged Dove studies in GMF[23]
Harry C. Oberholser	US Biological Survey; US Fish and Wildlife Service; Cleveland Museum of Natural History, Ohio	Accompanied Brandt (see above) in fieldwork in GMF and Rillito River	1945	Unknown whether Oberholser published any of his findings in addition to assisting Brandt
William G. George	University of Arizona; Southern Illinois University	Responsible for reclassification of Olive Warbler (*Peucedramus taeniatus*); records from GMF	1958–1961	Master's thesis and PhD dissertation at University of Arizona under Joe T. Marshall; publications
Patrick G. Gould	University of Arizona; Moore Laboratory of Zoology, Occidental College; US Fish and Wildlife Service	Studied Cardinals and Pyrrhuloxias in GMF	1958–1960	Master's thesis at University of Arizona under Joe T. Marshall; publication[24]
R. Roy Johnson	University of Arizona; University of Kansas; University of Texas at El Paso; Prescott College	Assisted Marshall with Brown Towhee studies; records from GMF	1958–1960	Master's thesis at University of Arizona under Joe T. Marshall; publications[25]
Richard S. Crossin	Neotropical Ornithological Foundation, independent collector	Conducted fieldwork along Santa Cruz River;[26] collected birds in Mexico	1950s–1960s	Large collection of tropical birds; records from Santa Cruz and GMF; publications

Notes

1. A review of *The Birds of Arizona* in *The Auk* by one of the nation's leading ornithologists states, "[T]he Birds of Arizona is a landmark among state ornithologies, and the authoritativeness of its distributional information will seldom, if ever, be exceeded" (Parkes 1966: 487).

2. Brewster hired Stephens to travel from California to collect birds in Arizona. Brewster was also largely responsible for the establishment of the Nuttall Ornithological Club, which later became the American Ornithologists' Union. Later, Stephens was largely responsible for establishment of the San Diego Natural History Museum and served as its first director.

3. Brewster (1881, 1882, 1883).

4. Assisted Swarth in first study in GMF 1–14 June 1903; also see Gray Hawk account in Bendire (1892).

5. Johnson et al. (2003).

6. Edward W. Nelson (1913a, 1913b), chief of the US Biological Survey, spent several months in Tucson in 1883 and interested Herbert Brown in birds.

7. Swarth, a professional ornithologist, worked 17–23 May 1902 with O. W. Howard, an amateur Arizona ornithologist, and 1–14 June 1903 with Frank Stephens (also see note 4, above).

8. Swarth (1914).

9. Bent (1930), Palmer et al. (1954).

10. Howell was a California biologist and independent researcher in 1916 and later joined the US Biological Survey and Johns Hopkins Medical School (chrono-biographical sketch online, http://people.wku.edu/charles.smith/chronob/HOWE1886.htm).

11. Dawson was largely responsible for the establishment of the Museum of Comparative Oology, served as its first director, and wrote *The Birds of California* (Dawson 1923), one of the most complete books on California birds.

12. Taber (1955).

13. Perhaps the best-known American ornithologist after John J. Audubon, Bent took time from writing *Life Histories of North American Birds* for the 1922 trip to southern Arizona. Bent was accompanied by Frank Willard, a local amateur ornithologist (see appendix A and the reference list for the present volume).

14. In addition to studying all vertebrates in the GMF, Arnold worked on a White-winged Dove study for the Arizona Game and Fish Department (Arnold 1941, 1943). He later served in the US Navy between working for Arizona Game and Fish and the US Fish and Wildlife Service; he died in a Los Angeles Veterans Health Administration hospital in 1956 (Kalmbach 1958).

15. Marshall used information from studies in the GMF in these publications (Marshall 1960, 1964, 1967).

16. Howard published eight papers on birds in southeastern Arizona in *The Condor* between 1899 and 1906 (Anderson 1972).

17. Phillips (1950), Palmer et al. (1954).

18. Dickerman (1997).

19. GMF was one of Phillips's favorite study areas (pers. comm. to RRJ, spring 1954); he obtained several important records in the GMF and described the new subspecies *Pipilo* (now *Melozone*) *aberti vorhiesi* from there.

20. Anderson and Anderson (1973).

21. Anderson published at least twenty avian papers in journals and a series of seven papers on the Cactus Wren with his wife, Anne Anderson, in *The Condor*; these articles were later published as a book (Anderson and Anderson 1973). Additionally, Anderson (1972) published *A Bibliography of Arizona Ornithology*.

22. Records from GMF in Monson (1942).

23. Arnold (1941, 1943).

24. Gould (1960, 1961).

25. Marshall and Johnson (1968), Johnson and Carothers (1982), Johnson and Haight (1996).

26. Crossin (1965).

Appendix J

Comparison between regularly occurring summer birds of the Great Mesquite Forest (GMF), Rillito River, and Blue Point Cottonwoods (BPC) near the confluence of the Salt and Verde Rivers east of Phoenix, Arizona.

*Species probably not breeding in the immediate vicinity but flying over, foraging in, or passing through the area during the breeding season.

*** Recent changes in classification that may or may not be accepted over time (American Ornithologists' Union 2012).

E=Formerly nested but locally extirpated during study period or thereafter; see appendix H.

1X=Species recorded only one summer, with date.

X?=Tucson record possibly from this area but actual location unclear.

		Areas of Studies		
		Great Mesquite Forest[1]	Rillito River[2]	Blue Point Cottonwoods[3]
Years of Work		1902–1963	1871–1948	1930s–1990s
Common Name	Scientific Name[4]			
Gambel's Quail	*Callipepla gambelii*	X	X	X
Great Blue Heron	*Ardea herodias*	*X	*X	*X
Green Heron	*Butorides virescens*	X	X	X
Black-crowned Night-Heron	*Nycticorax nycticorax*	*X		
Black Vulture	*Coragyps atratus*	*X		
Turkey Vulture	*Cathartes aura*	*X	*X	*X
Osprey	*Pandion haliaetus*			*E
Bald Eagle	*Haliaeetus leucocephalus*			X
Cooper's Hawk	*Accipiter cooperii*	X	X	X
Common Black-Hawk	*Buteogallus anthracinus*	E	E	E
Harris's Hawk	*Parabuteo unicinctus*	X	X?	X
Gray Hawk	*Asturina nitida*[5]***	E	E	
Swainson's Hawk	*Buteo swainsoni*	E	X?	
Zone-tailed Hawk	*B. albonotatus*	*E	E	
Red-tailed Hawk	*B. jamaicensis*	X	X	X
Crested Caracara	*Caracara cheriway*	*E	E	
American Kestrel	*Falco sparverius*	X	X	X
Peregrine Falcon	*F. peregrinus*	*X		
Prairie Falcon	*F. mexicanus*	*X	*X?	
Common Gallinule	*Gallinula chloropus*[6]***			X
American Coot	*Fulica americana*			X
Killdeer	*Charadrius vociferus*	X	X	X

Common Name	Scientific Name	Great Mesquite Forest	Rillito River	Blue Point Cottonwoods
White-winged Dove	*Zenaida asiatica*	X	X	X
Mourning Dove	*Z. macroura*	X	X	X
Inca Dove	*Columbina inca*	X	X	
Common Ground-Dove	*C. passerina*	X	X	
Yellow-billed Cuckoo	*Coccyzus americanus*	E	E	X
Greater Roadrunner	*Geococcyx californianus*	X	X	X
Barn Owl	*Tyto alba*	X	X	X
Western Screech-Owl	*Otus kennicottii*[7]***	X	X	X
Great Horned Owl	*Bubo virginianus*	X	X	X
Ferruginous Pygmy-Owl[8]	*Glaucidium brasilianum*		E	E
Elf Owl	*Micrathene whitneyi*	X	X	X
Burrowing Owl	*Athene cunicularia*	E		
Lesser Nighthawk	*Chordeiles acutipennis*	X	X	X
Common Poorwill	*Phalaenoptilus nuttallii*	X	X	X
White-throated Swift	*Aeronautes saxatalis*	*X	*X?	*X
Black-chinned Hummingbird	*Archilochus alexandri*	X	X	X
Costa's Hummingbird	*Calypte costae*			X
Gila Woodpecker	*Melanerpes uropygialis*	X	X	X
Ladder-backed Woodpecker	*Picoides scalaris*	X	X	X
Gilded Flicker	*Colaptes chrysoides*	X	X	X
Northern Beardless-Tyrannulet	*Camptostoma imberbe*	X	X	
Willow Flycatcher	*Empidonax traillii*	E	E	
Black Phoebe	*Sayornis nigricans*	X	X	X
Say's Phoebe	*S. saya*	X		
Vermilion Flycatcher	*Pyrocephalus rubinus*	X	X	X
Ash-throated Flycatcher	*Myiarchus cinerascens*	X	X	X
Brown-crested Flycatcher	*M. tyrannulus*	X	X	X
Tropical Kingbird[9]	*Tyrannus melancholicus*	X?	1X 1905	1X 1956[10]
Cassin's Kingbird	*T. vociferans*	X	X	
Western Kingbird	*T. verticalis*	X	X	X
Rose-throated Becard	*Pachyramphus aglaiae*	1X? 1958[11]		
Loggerhead Shrike	*Lanius ludovicianus*	X	X	
Bell's Vireo	*Vireo bellii*	X	X	X
Chihuahuan Raven	*Corvus cryptoleucus*	*X	X	
Common Raven	*C. corax*	*X	*X	X
Purple Martin	*Progne subis*	X	X	
Northern Rough-winged Swallow	*Stelgidopteryx serripennis*	X	X	X
Cliff Swallow	*Petrochelidon pyrrhonota*			X
Barn Swallow	*Hirundo rustica*		X[12]	
Verdin	*Auriparus flaviceps*	X	X	X
Bushtit	*Psaltriparus minimus*			1X 1973
Cactus Wren	*Campylorhynchus brunneicapillus*	X	X	X
Rock Wren	*Salpinctes obsoletus*	X		X
Canyon Wren	*Catherpes mexicanus*	X		
Bewick's Wren	*Thryomanes bewickii*	X	X	X
Black-tailed Gnatcatcher	*Polioptila melanura*	X	X	X
Northern Mockingbird	*Mimus polyglottos*	X	X	
Bendire's Thrasher	*Toxostoma bendirei*	X	X	

(continued)

Common Name	Scientific Name	Great Mesquite Forest	Rillito River	Blue Point Cottonwoods
Curve-billed Thrasher	*T. curvirostre*	X	X	X
Crissal Thrasher	*T. crissale*	X	X	X
European Starling[13]	*Sturnus vulgaris*	X		X
Phainopepla	*Phainopepla nitens*	X	X	X
Lucy's Warbler	*Vermivora luciae*[14]***	X	X	X
Yellow Warbler	*Dendroica petechia*[15]***	X	X	X
Common Yellowthroat	*Geothlypis trichas*	X	X	X
Yellow-breasted Chat	*Icteria virens*	X	X	X
Canyon Towhee	*Pipilo fuscus*[16]***	X	X	X
Abert's Towhee	*P. aberti*[17]***	X	X	X
Rufous-winged Sparrow	*Aimophila carpalis*[18]***	X	X	
Black-throated Sparrow	*Amphispiza bilineata*	X	X	X
Song Sparrow[19]	*Melospiza melodia*		E	X
Summer Tanager	*Piranga rubra*	X	X	X
Northern Cardinal	*Cardinalis cardinalis*	X	X	X
Pyrrhuloxia	*C. sinuatus*	X	X	
Blue Grosbeak	*Passerina caerulea*	X	X	X
Red-winged Blackbird	*Agelaius phoeniceus*	X	X	X
Western Meadowlark	*Sturnella neglecta*		X	
Great-tailed Grackle[20]	*Quiscalus mexicanus*	*X		*X
Bronzed Cowbird	*Molothrus aeneus*	X	X	X
Brown-headed Cowbird	*M. ater*	X	X	X
Hooded Oriole	*Icterus cucullatus*	X	X	E[21]
Bullock's Oriole	*I. bullockii*	X	X	X
House Finch	*Carpodacus mexicanus*[22]***	X	X	X
Lesser Goldfinch	*Carduelis psaltria*[23]***	X	X	X
House Sparrow	*Passer domesticus*	X	X	
Number of Species				
—Regular Nesting Species[24]		73±	70±	65±
—Total Species		85±	77±	70±

Notes

1. See also appendix A.

2. See also appendix B.

3. Blue Point Cottonwoods (BPC) study area consists of 250 acres along the Salt River immediately upstream from its confluence with the Verde River. This cottonwood-mesquite plot more nearly approximates the habitat and avifauna of the GMF than any other we have found besides the former Rillito Creek. However, it lacks the large mature mesquites and an adjacent cottonwood-willow gallery forest of the GMF. BPC was also used by Rea (1983) for comparisons with the avifauna of the Gila River Reservation. More details are known about BPC than either the GMF or Rillito River from trip reports by several ornithologists who conducted more than 200 field trips between 1933 and the early 1990s, including L. L. Hargrave, A. R. Phillips, R. R. Johnson, and J. M. Simpson (Johnson et al. 2000a, Johnson and Simpson MS b). More than 610 pairs of nesting birds/100 acres were censused annually at BPC during the 1970s and 1980s (Johnson et al. 2000a). Today, the average breeding bird population is approximately 400–500 pairs/100 acres in southwestern lowland riparian habitats (Johnson et al. 1977). This contrasts with upland breeding bird populations in surrounding Sonoran desertscrub of less than 100 to 150 breeding pairs/100 acres (Johnson et al. 1977).

4. After American Ornithologists' Union (1998, 2012).

5. From *Asturina nitida* to *Buteo nitidus.*

6. From *Gallinula chloropus* to *Gallinula galeata.*

7. From *Otus kennicottii* to *Megascops kennicottii.*

8. Ferruginous Pygmy-Owl was first found in the United States along the Rillito River by Bendire (Coues 1872a, Bendire 1892) and was relatively common along the Rillito River (Johnson et al. 2003 and see text). It was extirpated from BPC by the early 1970s and the Tucson Basin by 2009; see appendix H.

9. Tropical Kingbird was first recorded in Arizona (and the United States) from a specimen taken by H. H. Kimball on 12 May 1905 near Fort Lowell (Peters 1936, Phillips et al. 1964). Whether it was a solitary bird or a member of a breeding pair or colony is unknown.

10. A female Tropical Kingbird was taken from a pair building a nest in a cottonwood tree, 19 May 1956 (Simpson and Werner 1958).

11. Rose-throated Becard possibly nested only one year; a nest was found in 1958, and a calling male was recorded in 1959.

12. Whether the Barn Swallow was migrating or nesting here is not clear.

13. European Starling was introduced into the United States from Europe; it was first reported for Arizona in 1946 (Phillips et al. 1964, Monson and Phillips 1981).

14. From *Vermivora luciae* to *Oreothlypis luciae.*

15. From *Dendroica petechia* to *Setophaga petechia.*

16. From *Pipilo fuscus* to *Melozone fusca.*

17. From *P. aberti* to *M. aberti.*

18. From *Aimophila carpalis* to *Peucaea carpalis.*

19. Song Sparrow, formerly a common riparian nesting species in the Tucson region (Bendire 1872a, 1872b), was extirpated from the region shortly after 1895 (Phillips et al. 1964).

20. Great-tailed Grackle was first reported arriving in Arizona from Mexico in 1936 (Phillips et al. 1964, Monson and Phillips 1981).

21. Local extirpation only; Hooded Oriole continues to nest in the Phoenix region.

22. From *Carpodacus mexicanus* to *Haemorhous mexicanus.*

23. From *Carduelis psaltria* to *Spinus psaltria.*

24. The category "Regular Nesting Species" excludes species not nesting on study area or recorded as nesting in only one year but includes species extirpated during study years.

Appendix K

Common names and Latin equivalent names for plants (after Lehr 1978 and Turner et al. 1995).

Common Name	Latin Name	Common Name	Latin Name
agave	*Agave* spp.	saltcedar	*Tamarix* sp.
Arizona ash	*Fraxinus velutina*	screwbean mesquite	*Prosopis pubescens*
Arizona sycamore	*Platanus wrightii*	seep willow	*Baccharis salicifolia*
Arizona walnut	*Juglans major*	snakeweed	*Gutierrezia sarothrae*
arrowweed	*Pluchea sericea*	soapberry	*Sapindus saponaria*
Athel tamarisk	*Tamarix aphylla*	staghorn cholla	*Opuntia versicolor*
barrel cactus	*Ferocactus wislizenii*	three-awn	*Aristida* spp.
black willow	*Salix gooddingii*	tobosa	*Hilaria mutica*
blue paloverde	*Cercidium floridum*	tree tobacco	*Nicotiana glauca*
buffelgrass	*Pennisetum ciliare*	triangle leaf bursage	*Ambrosia deltoidea*
burrobrush	*Hymenoclea monogyra*	turpentine bush	*Ericameria laricifolius*
catclaw	*Acacia greggii*	velvet ash	*Fraxinus velutina*
cholla	*Opuntia* spp.	velvet mesquite	*Prosopis velutina*
common reed	*Phragmites australis*	whitethorn	*Acacia constricta*
cottonwood	*Populus fremontii*	wolfberry	*Lycium berlandieri*
coyote willow	*Salix exigua*	yucca	*Yucca* spp.
creosotebush	*Larrea tridentata*		
desert willow	*Chilopsis linearis*		
Engelmann prickly pear	*Opuntia engelmannii*		
foothill paloverde	*Cercidium microphyllum*		
giant cactus	*Carnegiea gigantea*		
Goodding willow	*Salix gooddingii*		
graythorn	*Ziziphus obtusifolia*		
jumping cholla	*Opuntia fulgida*		
juniper	*Juniperus* sp.		
mesquite	*Prosopis velutina*		
Mexican elder	*Sambucus mexicana*		
netleaf hackberry	*Celtis reticulata*		
oak	*Quercus* spp.		
ocotillo	*Fouquieria splendens*		
palmilla	*Yucca elata*		
paloverde	*Cercidium* spp.		
prickly pear	*Opuntia* spp.		
sacaton	*Sporobolus airoides*		
saguaro	*Carnegiea gigantea*		
saltbush	*Atriplex* spp.		

Appendix L

Summer bird occurrence in nonnative saltcedar, or tamarisk (*Tamarix* spp.), stands at various places in the southwestern United States compared to summer birds that have been recorded historically for the Santa Cruz River. Tamarisk is now an abundant riparian plant along much of the river. Many species listed here no longer occur along the middle Santa Cruz River (see appendixes A and H).

Abbreviations: N=nesting; R=roosting; X?=species occurring in saltcedar but activity not clarified by source(s).

**Species recently established as summer birds; not reported historically as nesting in the Tucson Basin.

***Recent changes in classification that may or may not be accepted over time (American Ornithologists' Union 2012).

Common Name	Scientific Name[1]	Breeding[2]	Foraging	Other	Source[3]
			Activity		
Ruddy Duck**	*Oxyura jamaicensis*	X			13
Gambel's Quail	*Callipepla gambelii*		X?		10,11
Pied-billed Grebe**	*Podylimbus podiceps*				
Great Blue Heron	*Ardea herodias*	N			5
Green Heron	*Butorides virescens*				
Black-crowned Night-Heron	*Nycticorax nycticorax*	N			1,12
Black Vulture	*Coragyps atratus*				
Turkey Vulture	*Cathartes aura*				
Cooper's Hawk	*Accipiter cooperii*				
Common Black-Hawk	*Buteogallus anthracinus*				
Harris's Hawk	*Parabuteo unicinctus*				
Gray Hawk	*Asturina nitida*[4]***				
Swainson's Hawk	*Buteo swainsoni*				
Zone-tailed Hawk	*B. albonotatus*				
Red-tailed Hawk	*B. jamaicensis*				
Crested Caracara	*Caracara cheriway*				
American Kestrel	*Falco sparverius*				
Peregrine Falcon	*F. peregrinus*				
Prairie Falcon	*F. mexicanus*				
Common Gallinule**	*Gallinula chloropus*[5]***	X			12,13
American Coot**	*Fulica americana*	X			13
Killdeer	*Charadrius vociferus*				
White-winged Dove	*Zenaida asiatica*	N		R	2,8,10
Mourning Dove	*Z. macroura*	N		R	2,11
Inca Dove	*Columbina inca*				

(continued)

Common Name	Scientific Name	Breeding	Foraging	Other	Source
Common Ground-Dove	*C. passerina*				
Yellow-billed Cuckoo	*Coccyzus americanus*		X		2,3,7
Greater Roadrunner	*Geococcyx californianus*	X			3,12
Barn Owl	*Tyto alba*			R	2
Western Screech-Owl	*Otus kennicottii*[6]***			R	2
Great Horned Owl	*Bubo virginianus*				
Ferruginous Pygmy-Owl	*Glaucidium brasilianum*				
Elf Owl	*Micrathene whitneyi*		X?		12
Burrowing Owl	*Athene cunicularia*				
Lesser Nighthawk	*Chordeiles acutipennis*		X		6,13
Common Poorwill	*Phalaenoptilus nuttallii*				
White-throated Swift	*Aeronautes saxatilis*				
Black-chinned Hummingbird	*Archilochus alexandri*	N			1,2
Anna's Hummingbird**	*Calypte anna*				
Gila Woodpecker	*Melanerpes uropygialis*				
Ladder-backed Woodpecker	*Picoides scalaris*	X			2,3
Gilded Flicker	*Colaptes chrysoides*				
Northern Beardless-Tyrannulet	*Camptostoma imberbe*				
Willow Flycatcher	*Empidonax traillii*	N	X		4,9
Black Phoebe	*Sayornis nigricans*		X		6,12
Say's Phoebe	*S. saya*				
Vermilion Flycatcher	*Pyrocephalus rubinus*	N			2
Ash-throated Flycatcher	*Myiarchus cinerascens*	X			2,3
Brown-crested Flycatcher	*M. tyrannulus*		X?		12
Tropical Kingbird	*Tyrannus melancholicus*		X?		9
Cassin's Kingbird	*T. vociferans*				
Western Kingbird	*T. verticalis*		X?		3,10
Rose-throated Becard	*Pachyramphus aglaiae*				
Loggerhead Shrike	*Lanius ludovicianus*		X?		3
Bell's Vireo	*Vireo bellii*	N			1,11
Chihuahuan Raven	*Corvus cryptoleucus*		X?		12
Common Raven	*C. corax*				
Purple Martin	*Progne subis*				
Northern Rough-winged Swallow	*Stelgidopteryx serripennis*		X		6,12,13
Cliff Swallow**[7]	*Petrochelidon pyrrhonota*		X?		12
Verdin	*Auriparus flaviceps*	N			2,10
Cactus Wren	*Campylorhynchus brunneicapillus*	X			3,12
Rock Wren	*Salpinctes obsoletus*				
Canyon Wren	*Catherpes mexicanus*				
Bewick's Wren	*Thryomanes bewickii*	X			2,11
Black-tailed Gnatcatcher	*Polioptila melanura*	N			10,12
Northern Mockingbird	*Mimus polyglottos*		X		3,11,7
Bendire's Thrasher	*Toxostoma bendirei*				
Curve-billed Thrasher	*T. curvirostre*		X?		7
Crissal Thrasher	*T. crissale*	X			3,9,12
European Starling	*Sturnus vulgaris*				
Phainopepla	*Phainopepla nitens*				
Lucy's Warbler	*Vermivora luciae*[8]***	N			1,2
Yellow Warbler	*Dendroica petechia*[9]***	N			1,2
Common Yellowthroat	*Geothlypis trichas*	X			11,12
Yellow-breasted Chat	*Icteria virens*	N			1,3
Canyon Towhee	*Pipilo fuscus*[10]***				
Abert's Towhee	*P. aberti*[11]***	N			2,11

Common Name	Scientific Name	Breeding	Foraging	Other	Source
Rufous-winged Sparrow	*Aimophila carpalis*[12]***				
Black-throated Sparrow	*Amphispiza bilineata*				
Song Sparrow	*Melospiza melodia*	X			2,6
Summer Tanager	*Piranga rubra*		X?		2,9,12
Northern Cardinal	*Cardinalis cardinalis*	X			9
Pyrrhuloxia	*C. sinuatus*	N			3,12
Blue Grosbeak	*Passerina caerulea*	N			1,2,3
Red-winged Blackbird	*Agelaius phoeniceus*	N		R	2,8
Great-tailed Grackle	*Quiscalus mexicanus*		X?		11
Bronzed Cowbird	*Molothrus aeneus*				
Brown-headed Cowbird	*M. ater*	X			1,2,3
Hooded Oriole	*Icterus cucullatus*				
Bullock's Oriole	*I. bullockii*	N			2,3
House Finch	*Carpodacus mexicanus*[13]***	N?			10,11
Lesser Goldfinch	*Carduelis psaltria*[14]***		X?		11
House Sparrow	*Passer domesticus*				

Total Species: 51±

Sources:
1. Brown et al. (1987).
2. Rosenberg et al. (1991).
3. Hunter et al. (1988).
4. Brown (1988).
5. Stevens et al. (1997).
6. Johnson and Simpson (MS b).
7. Livingston and Schemnitz (1996).
8. Rea (2007).
9. Hunter et al. (1987).
10. Anderson et al. (1977).
11. Engel-Wilson and Ohmart (1979).
12. Sogge et al. (2008).
13. Johnson and Simpson (MS a).

Notes

1. After American Ornithologists' Union (1998, 2012).

2. Species breeding in the area but may or may not actually nest in saltcedar; X indicates unknown use of saltcedar. Most species that nest in saltcedar also forage in it.

3. Not every source is listed for each species.

4. From *Asturina nitida* to *Buteo nitidus*.

5. From *Gallinula chloropus* to *Gallinula galeata*.

6. From *Otus kennicottii* to *Megascops kennicottii*.

7. Cliff Swallow did not formerly nest in the Tucson region (Phillips et al. 1964) but now is reported as "nesting under bridges, culverts, and on buildings" (Stevenson 2007: 281, Stejskal and Rosenberg 2011: 291).

8. From *Vermivora luciae* to *Oreothlypis luciae*.

9. From *Dendroica petechia* to *Setophaga petechia*.

10. From *Pipilo fuscus* to *Melozone fusca*.

11. From *P. aberti* to *M. aberti*.

12. From *Aimophila carpalis* to *Peucaea carpalis*.

13. From *Carpodacus mexicanus* to *Haemorhous mexicanus*.

14. From *Carduelis psaltria* to *Spinus psaltria*.

Appendix M

Comparison among the species of historically, regularly occurring summer birds of the Great Mesquite Forest (GMF) and the middle Santa Cruz River, as well as the Rillito River, and summer birds recorded along the present-day Santa Cruz River in the reach augmented by Tucson and Marana wastewater effluent (see also appendixes A, B, C, and L).

*Species probably not breeding in the immediate vicinity but flying over, foraging in, or passing through the area during the breeding season.

**Species not reported historically for the Santa Cruz–Rillito River system as breeding.

***Recent changes in classification that may or may not be accepted over time (American Ornithologists' Union 2012).

E=Formerly nested but locally extirpated during study period or thereafter (see appendix H).

1X=Species recorded only one summer, with date.

X?=Tucson record possibly from this area but not positive.

		Areas of Studies	
		Great Mesquite Forest	Rillito River[1]	Present-Day Santa Cruz River[2]
Years of Work		1902–1963	1871–1948	2000–2012
Common Name	Scientific Name[3]			
Mallard**[4]	*Anas platyrhynchos*			X
Ruddy Duck**	*Oxyura jamaicensis*			X
Gambel's Quail	*Callipepla gambelii*	X	X	X
Pied-billed Grebe**[5]	*Podylimbus podiceps*	X?		X
Great Blue Heron	*Ardea herodias*	*X	*X	*X
Green Heron	*Butorides virescens*	X	X	X
Black-crowned Night-Heron[6]	*Nycticorax nycticorax*	*X		X
Black Vulture	*Coragyps atratus*	*E		
Turkey Vulture	*Cathartes aura*	*X	*X	*X
Cooper's Hawk	*Accipiter cooperii*	X	X	X
Common Black-Hawk	*Buteogallus anthracinus*	E	E	
Harris's Hawk	*Parabuteo unicinctus*	X	X?	X
Gray Hawk	*Asturina nitida*[7]***	E	E	
Swainson's Hawk	*Buteo swainsoni*	X	X?	
Zone-tailed Hawk	*B. albonotatus*	*E	E	
Red-tailed Hawk	*B. jamaicensis*	X	X	X
Crested Caracara	*Caracara cheriway*	*E	E	
American Kestrel	*Falco sparverius*	X	X	X

Common Name	Scientific Name	Great Mesquite Forest	Rillito River	Present-Day Santa Cruz River
*Peregrine Falcon	*F. peregrinus*	X		X
*Prairie Falcon	*F. mexicanus*		*X?	X
Common Gallinule**	*Gallinula chloropus*[8]***			X
American Coot**	*Fulica americana*			X
Killdeer	*Charadrius vociferus*	X	X	X
Black-necked Stilt**	*Himantopus mexicanus*			X
Rock Dove**	*Columba livia*			X
White-winged Dove	*Zenaida asiatica*	X	X	X
Mourning Dove	*Z. macroura*	X	X	X
Inca Dove	*Columbina inca*	X	X	X
Common Ground-Dove	*C. passerina*	E	E	
Yellow-billed Cuckoo	*Coccyzus americanus*	E	E	
Greater Roadrunner	*Geococcyx californianus*	X	X	X
Barn Owl	*Tyto alba*	X	X	X
Western Screech-Owl[9]	*Otus kennicottii*[10]***	X	X	
Great Horned Owl	*Bubo virginianus*	X	X	X
Ferruginous Pygmy-Owl[11]	*Glaucidium brasilianum*		E	
Elf Owl[12]	*Micrathene whitneyi*	X	X	
Burrowing Owl[13]	*Athene cunicularia*	E		
Lesser Nighthawk	*Chordeiles acutipennis*	X	X	X
Common Poorwill	*Phalaenoptilus nuttallii*	X	X	
White-throated Swift	*Aeronautes saxatalis*	*X	*X?	*X
Black-chinned Hummingbird	*Archilochus alexandri*	X	X	X
Anna's Hummingbird**	*Calypte anna*			X
Gila Woodpecker	*Melanerpes uropygialis*	X	X	X
Ladder-backed Woodpecker	*Picoides scalaris*	X	X	X
Gilded Flicker	*Colaptes chrysoides*	X	X	X
Northern Beardless-Tyrannulet	*Camptostoma imberbe*	X	X	
Willow Flycatcher[14]	*Empidonax traillii*	E	E	X?
Black Phoebe	*Sayornis nigricans*	X	X	X
Say's Phoebe	*S. saya*	X		X
Vermilion Flycatcher	*Pyrocephalus rubinus*	X	X	X
Ash-throated Flycatcher	*Myiarchus cinerascens*	X	X	X
Brown-crested Flycatcher	*M. tyrannulus*	X	X	X
*Tropical Kingbird[15]	*Tyrannus melancholicus*	X?	1X 1905	X
*Cassin's Kingbird	*T. vociferans*	X	X	X
Western Kingbird	*T. verticalis*	X	X	X
Rose-throated Becard	*Pachyramphus aglaiae*	1X? 1958[16]		
Loggerhead Shrike	*Lanius ludovicianus*	X	X	X
Bell's Vireo	*Vireo bellii*	X	X	X
Chihuahuan Raven	*Corvus cryptoleucus*	*X	X	
Common Raven	*C. corax*	*X	*X	*X
Purple Martin	*Progne subis*	X	X	X
Northern Rough-winged Swallow	*Stelgidopteryx serripennis*	X	X	X
Cliff Swallow**	*Petrochelidon pyrrhonota*			X
Barn Swallow	*Hirundo rustica*		X[17]	X
Verdin	*Auriparus flaviceps*	X	X	X
Cactus Wren	*Campylorhynchus brunneicapillus*	X	X	X
Rock Wren	*Salpinctes obsoletus*	X		

(continued)

Common Name	Scientific Name	Great Mesquite Forest	Rillito River	Present-Day Santa Cruz River
Canyon Wren	*Catherpes mexicanus*	X		
Bewick's Wren	*Thryomanes bewickii*	X	X	
Black-tailed Gnatcatcher	*Polioptila melanura*	X	X	X
Northern Mockingbird	*Mimus polyglottos*	X	X	X
Bendire's Thrasher	*Toxostoma bendirei*	X	X	
Curve-billed Thrasher	*T. curvirostre*	X	X	X
Crissal Thrasher	*T. crissale*	X	X	
European Starling[18]	*Sturnus vulgaris*	X		X
Phainopepla	*Phainopepla nitens*	X	X	X
Lucy's Warbler	*Vermivora luciae*[19]***	X	X	X
Yellow Warbler	*Dendroica petechia*[20]***	X	X	X
Common Yellowthroat	*Geothlypis trichas*	X	X	X
Yellow-breasted Chat	*Icteria virens*	X	X	X
Canyon Towhee	*Pipilo fuscus*[21]***	X	X	X
Abert's Towhee	*P. aberti*[22]***	X	X	X
Rufous-winged Sparrow	*Aimophila carpalis*[23]***	X	X	X
Black-throated Sparrow	*Amphispiza bilineata*	X	X	
Song Sparrow[24]	*Melospiza melodia*		E	X
Summer Tanager	*Piranga rubra*	X	X	
Northern Cardinal	*Cardinalis cardinalis*	X	X	X
Pyrrhuloxia	*C. sinuatus*	X	X	X
Blue Grosbeak	*Passerina caerulea*	X	X	X
Red-winged Blackbird	*Agelaius phoeniceus*	X	X	X
Western Meadowlark	*Sturnella neglecta*		X	
Great-tailed Grackle[25]	*Quiscalus mexicanus*	*X		X
Bronzed Cowbird	*Molothrus aeneus*	X	X	X
Brown-headed Cowbird	*M. ater*	X	X	X
Hooded Oriole	*Icterus cucullatus*	X	X	X
Bullock's Oriole	*I. bullockii*	X	X	X
House Finch	*Carpodacus mexicanus*[26]***	X	X	X
Lesser Goldfinch	*Carduelis psaltria*[27]***	X	X	X
House Sparrow	*Passer domesticus*	X	X	X
Number of Species				
—Regular Nesting Species[28]		73±	70±	66±
—Total Species		85±	77±	75±[29]

Notes

1. Formerly Rillito Creek.

2. Includes the Santa Cruz River through Tucson, downriver to Marana; see appendix C.

3. After American Ornithologists' Union (1998, 2012).

4. There is evidence that, in addition to *A. p. platyrhynchos*, the female-plumaged Mexican Duck (*A. p. diazi*) occurs in the Tucson area (Stevenson 2007, Stejskal and Rosenberg 2011); Mexican Duck has been reported from the Santa Cruz River basin between 29th Street and Grant Road.

5. Vorhies saw Pied-billed Grebes at Indian Dam in the GMF frequently in the 1930s, including three calling on open water on 15 April 1934 (Vorhies et al. 1935). The species may be nesting here, because nesting begins in December and January in the Arizona lowlands (Corman and Wise-Gervais 2005) and nests are reported for the Tucson region (Stevenson 2007, Stejskal and Rosenberg 2012).

6. Swarth (1914) and Phillips et al. (1964) discounted this species' nesting along streams of southern Arizona, but newly fledged young have been found along the Salt/Gila River system south of Phoenix (Rea 1983, Johnson and Simpson MS a) in habitat similar to that of the former Santa Cruz and of the current Sweetwater Wetlands.

7. From *Asturina nitida* to *Buteo nitidus.*

8. From *Gallinula chloropus* to *Gallinula galeata.*

9. The lack of records for Western Screech-Owl may be due to a lack of nighttime work.

10. From *Otus kennicottii* to *Megascops kennicottii.*

11. Ferruginous Pygmy-Owl was first found in the United States along the Rillito River by Bendire (Coues 1872a, Bendire 1892) and was relatively common along the Rillito River (Johnson et al. 2003 and see text). It was extirpated from central Arizona by the early 1970s and the Tucson Basin by 2009; see appendixes H and J.

12. The lack of records for Elf Owl may be due to a lack of nighttime work.

13. Burrowing Owl is considered "a species of concern" (US Fish and Wildlife Service 2012). Although the species has not been recorded along the Santa Cruz–Rillito River system since 1917 (Dawson 1921), a colony currently occurs near San Xavier Mission (B. Fontana, pers. comm.); see appendix H.

14. Identification is difficult for Southwestern Willow Flycatcher (*Empidonax traillii extimus*), a federally endangered species, because of its similarity to several other *Empidonax* species. It was extirpated earlier; see appendix H.

15. Tropical Kingbird was first recorded in Arizona (and the United States) from a specimen taken by H. H. Kimball on 12 May 1905 near Fort Lowell (Peters 1936, Phillips et al. 1964). Whether it was a solitary bird or a member of a breeding pair or colony is unknown.

16. Rose-throated Becard possibly nested in only one year; a nest was found in 1958, and a calling male was recorded in 1959.

17. Whether the Barn Swallow was migrating or nesting here is not clear.

18. European Starling was introduced into the United States from Europe; it was first reported for Arizona in 1946 (Phillips et al. 1964, Monson and Phillips 1981).

19. From *Vermivora luciae* to *Oreothlypis luciae.*

20. From *Dendroica petechia* to *Setophaga petechia.*

21. From *Pipilo fuscus* to *Melozone fusca.*

22. From *P. aberti* to *M. aberti.*

23. From *Aimophila carpalis* to *Peucaea carpalis.*

24. Song Sparrow, formerly a common riparian nesting species in the Tucson region (Bendire 1872b), was extirpated from the region shortly after 1895 (Phillips et al. 1964); it is now breeding in suitable riparian habitat formed by Tucson sewage effluent (Corman and Wise-Gervais 2005).

25. Great-tailed Grackle was first reported arriving in Arizona from Mexico in 1936 (Phillips et al. 1964, Monson and Phillips 1981).

26. From *Carpodacus mexicanus* to *Haemorhous mexicanus.*

27. From *Carduelis psaltria* to *Spinus psaltria.*

28. The category "Regular Nesting Species" excludes species not nesting on the study area or recorded as nesting in only one year but includes species extirpated during study years.

29. The large number of species is at least partly an artifact of the large distance, approximately twenty-five miles of riparian habitat along the Santa Cruz River.

Notes

Chapter 1

1. Smith (1975).
2. Powell (1878).
3. Davis (1903).
4. Gregory (1917).
5. Scientific publications concerning arroyo downcutting began with Dodge's (1902) observations. William Morris Davis (1903) was the first widely recognized geomorphologist to report on the issue.
6. Cooke and Reeves (1976).
7. Examples include Burkham (1970), Cooke and Reeves (1976), Williams (1978), Webb (1985), Webb et al. (1991), Hereford (1993).
8. Much of this book derives from Betancourt (1990), Webb and Betancourt (1992), and Parker (1995b), who began their studies of the Santa Cruz River following the 1983 floods in southern Arizona.
9. Sheridan (1995: 56).
10. Betancourt (1990).
11. Thornber (1909), Cooke and Reeves (1976), Betancourt (1990), Wood et al. (1999), Mauz (2002).
12. Hendrickson and Minckley (1985), Webb et al. (2007a).
13. Huntington (1914), Bryan (1925), Antevs (1952), Hastings (1959), Hastings and Turner (1965), Cooke and Reeves (1976), Dobyns (1981), Hendrickson and Minckley (1985).
14. Graf (1983), Webb (1985), Webb and Hereford (2010).
15. For example, see Haynes (1968), Euler et al. (1979).
16. Walker et al. (2009).
17. Waters (1988).
18. Webb et al. (2002).
19. Betancourt (1990).
20. For example, see Burkham (1972), Cooke and Reeves (1976), Webb and Baker (1987), Webb et al. (1991).
21. Eidenbach and Wimberly (1980).
22. Bryan (1954).
23. Hendrickson and Minckley (1985).
24. Cooke and Reeves (1976).
25. Rogers et al. (1984).
26. Webb et al. (2007a).
27. The historical research in this book largely derives from Betancourt (1990), who used much more extensive direct quotations than we use in this book.
28. Rasmusson (1985).
29. Wells et al. (1988).
30. Rich (1911).

Chapter 2

1. Seymour (2012).
2. Pope et al. (1998: 542).
3. Data on watershed areas, elevations, and precipitation comes from the US Geological Survey, e.g., Pope et al. (1998), http://waterdata.usgs.gov/az/nwis/inventory/?site_no=09480000&agency_cd=USGS& (accessed 12 November 2010).
4. Webb et al. (2007b).
5. Pope et al. (1998: 398).
6. Applegate (1981).
7. Condes de la Torre (1970) estimated transmission losses in this reach to be about 10,500–17,900 acre-feet per mile.
8. The Rillito River, a multilingual contradiction, is the official name approved by the Board of Geographic Names, but this watercourse is also known as Rillito Creek, a multilingual redundancy.
9. Bailey (1979). See also http://www.pima.gov/wwm/about/div/trtmnt/ (accessed 3 October 2013).
10. http://waterdata.usgs.gov/az/nwis/uv?cb_00065=on&cb_00060=on&format=gif_default&period =60&site_no =09486500 (accessed 20 July 2009).
11. See appendix K for a conversion table of common to Latin names for plants as used in this book.
12. The occurrence and effects of many large floods on the Santa Cruz River have been described previously in the literature, e.g., Knapp (1937), Lewis (1963), Aldridge (1970), Aldridge and Eychaner (1984), Saarinen et al. (1984), Roeske et al. (1989), Webb and Betancourt (1992).

13. Pashley (1966), Nations and Stump (1981: 147), Dickinson (1991).

14. Force (1997).

15. Houser et al. (2004) describes the general physiographic characteristics and broad geologic history of the Tucson Basin.

16. Houser et al. (2004: 5).

17. Houser et al. (2004: 7).

18. Dickinson (1991: 90).

19. Anderson (1987).

20. Houser et al. (2004: 17).

21. Davidson (1973: E20–E30).

22. Pashley (1966).

23. Helmick (1986).

24. Halpenny and Halpenny (1988).

25. McFadden (1978).

26. Smith (1938).

27. Waters (1988).

28. Davidson (1970).

29. Haynes and Huckell (1986), Waters (1988), Waters and Haynes (2001), Waters and Ravesloot (2001), Mabry (2006a, 2006b).

30. Haynes and Huckell (1986).

31. Haynes (2007).

32. Waters (1992: 98).

33. Waters (1992: 97).

34. Waters and Haynes (2001).

35. Mabry (2006b).

36. Mabry (2006b).

37. Waters (1988).

38. Haynes and Huckell (1986), Waters (1988).

39. Mabry (2006b).

40. Mabry (2006b).

41. Haynes (1968), Euler et al. (1979), Haynes (2007).

42. Haynes and Huckell (2007).

43. Waters (1992: 99).

44. Waters (1988: 489).

45. Mabry (2006a).

46. Thiel and Mabry (2006).

47. Huckleberry (1995, 1999), Waters and Ravesloot (2001).

48. Haury (1976).

49. Waters and Ravesloot (2001).

50. Waters (1988), Mabry (2006b).

51. Betancourt (1990).

52. http://www.wrcc.dri.edu/cgi-bin/cliMAIN.pl?az 8820 (accessed 17 July 2009).

53. Beginning with Hastings and Turner (1965), several authors have examined the stability of various metrics of seasonal and annual climate in southern Arizona, e.g., Cooke and Reeves (1976), Webb and Betancourt (1992), and Turner et al. (2003).

54. Updated from Turner et al. (2003).

55. Webb and Betancourt (1992), Hereford et al. (2002), Turner et al. (2003), Woodhouse et al. (2005), Webb et al. (2005).

56. Woodhouse et al. (2005).

57. Webb et al. (2005).

58. Hereford et al. (2002).

59. Hirschboeck (1985), Webb and Betancourt (1992).

60. Webb et al. (2008).

61. Hansen et al. (1977), Maddox et al. (1980), Hansen and Shwarz (1981), Hirschboeck (1985, 1987), Smith (1986), Webb and Betancourt (1992).

62. Hansen et al. (1977).

63. Ely et al. (1993).

64. Hales (1974), Douglas (1983), Reyes and Cadet (1988), Douglas et al. (1993), Adams and Comrie (1997).

65. McDonald (1956).

66. Webb and Betancourt (1992).

67. Corbosiero et al. (2009), Ritchie et al. (2011).

68. Smith (1986), Corbosiero et al. (2009).

69. Roeske et al. (1978), Roeske et al. (1989).

70. Webb and Betancourt (1992).

71. Aldridge and Hales (1984).

72. Webb and Betancourt (1992).

73. Sellers and Hill (1974), Pyke (1972), Hansen and Shwarz (1981).

74. Hansen and Shwarz (1981).

75. Corbosiero et al. (2009), Ritchie et al. (2011).

76. Rosendal (1962), Cross (1988), http://www.aoml.noaa .gov/hrd/tcfaq/E10.html (accessed 17 November 2010).

77. Webb and Betancourt (1992).

78. Eidemiller (1978), Smith (1986), Cross (1988).

79. Corbosiero et al. (2009), Ritchie et al. (2011).

80. Smith (1986), Corbosiero et al. (2009), Ritchie et al. (2011).

81. Smith (1986).

82. Smith (1986), Corbosiero et al. (2009).

83. Roeske et al. (1989), Webb and Betancourt (1992).

84. Webb and Betancourt (1992).

85. Eidemiller (1978).

86. Tang and Reiter (1984).

87. Maddox et al. (1980), Hansen and Shwarz (1981).

88. Reitan (1960), Rasmusson (1967), Tang and Reiter (1984), Hales (1974), Pyke (1972), Hansen and Schwarz (1981).

89. Webb and Betancourt (1992).

90. Hales (1974), McCollum et al. (1995), Maddox et al. (1995), Webb et al. (2008).

91. Hirschboeck (1985).

92. Webb and Betancourt (1992).

93. Dzerdzeevskii (1969), Kalnicky (1974).

94. Horel and Wallace (1981), Philander (1983), Rasmusson (1984).

95. Philander (1985).

96. Horel and Wallace (1981), Philander (1983), Rasmusson (1984).

97. Bjerknes (1969).

98. Ramage (1983).

99. Quinn et al. (1987).

100. Gergis and Fowler (2005).

101. Gergis and Fowler (2009).
102. Douglas and Englehart (1984).
103. Ropelewski and Halpert (1986).
104. Cayan and Webb (1992), Ely et al. (1993).
105. Webb et al. (2007b).
106. Webb and Betancourt (1992).
107. Schwalen (1942).
108. Wilson and Garrett (1989).
109. Webb and Betancourt (1992).
110. Pope et al. (1998).
111. Webb and Betancourt (1992).
112. Pope et al. (1998).
113. Smith (1910).
114. Smith (1936).
115. Webb and Betancourt (1992).
116. House and Hirschboeck (1997).
117. Lewis (1963), Roeske et al. (1978), Roeske et al. (1989).
118. Saarinen et al. (1984).
119. Hyndman et al. (1991).
120. Roberts (1990).
121. Beverage and Culbertson (1964).
122. Parker (1995a).
123. Foothill paloverde trees occupy the uplands and minor ridges between arroyos and runnels in most of the Sonoran Desert. Blue paloverde trees line the ephemeral streams in this region, except in northeastern Baja California, where the species does not occur, and foothill paloverde trees line the channels (Turner et al. 1995).
124. Webb et al. (2007b).
125. Johnson et al. (1984).
126. Olson (1940), Dobyns (1981).
127. Johnson et al. (1984: 379).
128. Johnson et al. (1984: 377–378).
129. http://en.wikipedia.org/wiki/Ecosystem_services (accessed 30 November 2012).
130. Johnson (1971), Carothers et al. (1974), Rosenberg et al. (1982), Ohmart (1996).
131. Stromberg (1993a, 1993b).
132. Carothers et al. (1974), Hanson (2001: 14).
133. Hardy et al. (2004).
134. Anderson and Ohmart (1982).
135. Johnson et al. (1977).
136. McGrath and van Riper (2005).
137. Hardy et al. (2004).
138. Skagen et al. (1998).
139. Webb et al. (2007b).
140. Willis (1939).
141. Hastings and Turner (1965), Turner et al. (2003), Webb et al. (2007b).
142. References on changing riparian ecosystems in this region are numerous, and many are summarized in Stromberg and Tellman (2009).
143. Webb and Leake (2006), Webb et al. (2007b).
144. Hastings (1959), Hastings and Turner (1965), Hendrickson and Minckley (1985), Webb et al. (2007b).

145. Johnson and Carothers (1982), Webb and Leake (2006), Webb et al. (2007b).
146. Haney et al. (2008).
147. Hoffmeister (1986).
148. Swarth (1905: 24).
149. Cottam and Trefethen (1968).
150. Webb (1959: 77).
151. Neff (1940a, 1940b). Komatke Thicket ultimately was destroyed by groundwater pumping, and a grove of nonnative tamarisk occurs in this area today (Webb et al. 2007b).
152. Anderson (1972), Fischer (2001), see appendixes.
153. Phillips et al. (1964), Monson and Phillips (1981).
154. Webb and Leake (2006).
155. Brown and Lowe (1980), Brown et al. (1980), Brown (1982).
156. Corman and Wise-Gervais (2005); see appendix D.

Chapter 3

1. Hastings and Turner (1965), Cooke and Reeves (1976), Graf (1983), Turner et al. (2003).
2. Gerrard (1984).
3. Bailey (1935), Cooke and Reeves (1976), Graf (1983), Webb (1985).
4. For Arizona, Turner et al. (2003); for Utah, Gregory and Moore (1931), Gregory (1945).
5. Grove (1988).
6. Bryan (1925).
7. Schumm and Hadley (1957).
8. Webb (1985).
9. *State of California, Marin County versus E. Richeletti and others*, 1969, cited in Wolman (1977).
10. Several summaries of arroyo research have been published in the past, notably Hastings and Turner (1965) and the update by Turner et al. (2003), Cooke and Reeves (1976), Graf (1983), and Webb (1985).
11. Examples range from Dodge (1902) to Alford (1982); see Graf (1983) for more examples.
12. Several extensive general reviews of grazing impacts are available, notably, Fleischner (1994) and Ohmart (1996).
13. See the extensive discussion of livestock effects on desert grasslands in Turner et al. (2003).
14. Turner et al. (2003).
15. Lusby et al. (1971).
16. Gregory and Moore (1931).
17. Koch and Barnosky (2006).
18. Tuan (1966), Cooke and Reeves (1976), Graf (1983), Webb (1985).
19. Hastings and Turner (1965), Tuan (1966), Turner et al. (2003).
20. Denevan (1967).
21. Dobyns (1981).
22. Hastings (1959), Hastings and Turner (1965).

23. Melton (1965).

24. Schumm (1979), Schumm et al. (1984).

25. Branson et al. (1981).

26. Cooke and Reeves (1976).

27. Dobyns (1981).

28. Bryan (1927), Miller and Wendorf (1958), Leopold (1976), Tuan (1966).

29. Diamond (1986).

30. Tuan (1966: 573–574).

31. Calkins (1941: 77–78) in reviewing Bryan (1941).

32. Dutton (1882), Davis (1902).

33. Huntington (1914) was one of the first proponents of drought-driven watershed changes leading to arroyo downcutting.

34. Tuan (1966), Hall (1977), Knox (1983), Love (1983).

35. Cooke and Reeves (1976), Graf (1983), Webb (1985), Hereford (1993).

36. Bryan (1928).

37. Bryan (1925).

38. Huntington (1914).

39. Haynes (1968), Euler et al. (1979), Haynes (2007).

40. Hack (1939), Leopold and Miller (1954), Haynes (1968), Euler et al. (1979).

41. Antevs (1952).

42. Antevs (1955).

43. Dean (1988).

44. Hack (1939).

45. Haynes (1968).

46. Leopold (1976).

47. Webb et al. (2007b).

48. Thornthwaite et al. (1942).

49. Leopold (1951).

50. Leopold (1951), Leopold et al. (1966).

51. Leopold and Miller (1954).

52. Martin (1963).

53. Cooke and Reeves (1976). Knox (1978) noted that a significant discontinuity in the 1890s was missed because of the way Cooke and Reeves grouped data for statistical analyses. Turner et al. (2003) replicated the no-trend results of Cooke and Reeves (1976).

54. Webb (1985), Hereford and Webb (1992).

55. Bull (1964).

56. Kimball et al. (2010).

57. Cable (1975).

58. Thornthwaite et al. (1942), Schumm and Hadley (1957).

59. Schumm and Hadley (1957), Patton and Schumm (1975, 1981), Schumm (1979).

60. Elliott et al. (1999).

61. Cooke and Reeves (1976), Graf (1983), Webb and Baker (1987), Webb et al. (1991).

62. Davis (1903), Huntington (1914), Gregory (1917).

63. Thornthwaite et al. (1942), Tuan (1966), Love (1983), Webb (1985), Hereford (1984).

64. Webb and Baker (1987), Webb and Rathburn (1989).

65. Webb (1985), Ely et al. (1993).

66. Knox (1983), Hirschboeck (1987), Macklin et al. (2005).

67. Webb et al. (1988), Webb and Rathburn (1989), Ely et al. (1993), Webb et al. (2002), Harden et al. (2010).

68. Ely (1997).

69. Harden et al. (2010) applied this type of test and found clustering of radiocarbon dates for floods and channel change, albeit in broad time periods.

70. Pielke and Downton (2000).

71. Webb and Betancourt (1992).

72. Pielke and Downton (2000).

73. Ely (1996).

74. Webb et al. (2007b), Webb and Hereford (2010).

75. Hereford (1984, 1986, 1987, 1993, 2004), Hereford et al. (1996).

76. Hereford (1984), Hereford and Webb (1992), Hereford et al. (2002).

77. Webb et al. (2007b), Webb and Hereford (2010).

78. Hereford (1993), Hereford et al. (1996), Webb et al. (2007b), Webb and Hereford (2010).

79. Webb and Hereford (2010). The Fremont River in southern Utah is a notable exception owing to repeated extremely large floods, notably, one in 2005.

80. Hirschboeck (1985), Webb and Betancourt (1992), Webb et al. (2007b), Webb and Hereford (2010).

81. Aldridge and Eychaner (1984), Aldridge and Hales (1984), Roeske et al. (1989), House and Hirschboeck (1997).

82. Webb et al. (2007b).

83. Haynes (1968), Hack (1942), Lance (1963), Leopold (1976), Euler et al. (1979), Hereford (2002).

84. Waters (1985), Waters (1988), Waters and Haynes (2001), Waters and Ravesloot (2001).

85. Cooke and Reeves (1976).

86. Tuan (1966: 595).

87. Webb et al. (2007b), Webb and Hereford (2010).

Chapter 4

1. Fontana (1971).

2. Mooney (1928), http://jeff.scott.tripod.com/Tohono.html (accessed 21 November 2010).

3. Seymour (1989).

4. http://www.accessgenealogy.com/native/tribes/pima/sobaipuriindianhist.htm (accessed 17 February 2011).

5. Logan (2002: 50n23, and references within).

6. Jackson (1951).

7. Bolton (1919: 76).

8. Manje (1954: 92–93).

9. Bolton (1919: 205).

10. Manje (1954: 168).

11. Fontana (1996).

12. Bolton (1931: 26–27).

13. Bolton (1931).

14. McCarty (1976: 84).

15. McCarty (1976: 87).

16. The word *teraque*, the meaning of which is disputed, was used here. It can be interpreted as meaning "tamarisk," a nonnative species that did not arrive in the United States until the late nineteenth century; "scrub vegetation"; or perhaps a more specific native species such as "coyote willow." Betancourt (1990).

17. US Court of Private Land Claims (1881).

18. Cooke (1873: 154).

19. Cooke (1873: 161).

20. Couts (1961: 67).

21. Couts (1961: 70).

22. Durivage (1937: 209, 211). Argonaut William P. Huff in his 1850 diary likewise commented on the fertility of the Santa Cruz valley, as well as its abundance of water and grass; see Hosmer et al. (1991).

23. Clarke (1852).

24. Aldrich (1950: 52).

25. Powell (1931: 141, 143).

26. US Court of Private Land Claims (1882).

27. Bartlett (1854b).

28. Bartlett (1854b: 292–302).

29. Powell (1931: 145).

30. Even though the Gadsden Purchase was transacted in 1853, the US Senate did not ratify it until 1854.

31. We use the name Great Mesquite Forest, coined in 1921, in recognition of its unusually large size and its dominant species (Dawson [1921: 30]). Other observers referred to it as the Grand Mesquite Forest (Brandt [1951]), the Giant Mesquite Forest (Howell [1918a]), the Tucson mesquite forest (Phillips et al. [1964: xvi]), or the San Xavier Bosque (Bibles et al. [2002]).

32. Hayes (1850) *in* Davis (1982: 51).

33. Eccleston (1950) *in* Davis (1982: 52).

34. Evans (1945) *in* Davis (1982: 48).

35. Bartlett (1854b).

36. Parke (1855) *in* Davis (1982: 106).

37. Froebel (1859) *in* Davis (1982: 106).

38. Ewell (1856) *in* Hamlin (1966).

39. Phillips et al. (1964).

40. Emory (1857: 19).

41. McGuire (1979: 5).

42. Betancourt (1990).

43. Testimony of Juan Romero *in* Drake (1885).

44. Betancourt (1990) placed these springs in the southern half of Section 26, T14S, R13E.

45. Way (1960: 160).

46. Corle (1951: 203).

47. Hughes (1885).

48. *Arizona Mining Index*, 27 February 1886.

49. Froebel (1859: 503).

50. Drake (1885).

51. Thompson (2008).

52. Durrenberger and Ingram (1978), Engstrom (1996), Webb et al. (2007b).

53. McGuire (1979: 7).

54. Fergusson (1863: 14).

55. Nicolson (1974: 166–167, 171–172).

56. Browne (1951: 132–133).

57. Browne (1951: 144).

58. Bell (1869: 99–100).

59. Spring (1966: 47).

60. US Surgeon General's Office (1870: 462–463).

61. Bourke (1891: 53–63).

62. *Weekly Arizonan*, 22 May 1869.

63. *Arizona Daily Star*, 28 February 1891.

64. Drake (1885).

65. *Tucson Citizen*, 12 September 1887.

66. *Arizona Daily Star*, 28 February 1891.

67. Foreman (1872). Betancourt (1990: fig. 10B) placed this spring in the southeast corner of Section 20, T13S, R13E.

68. Hastings (1959).

69. *Arizona Weekly Citizen*, 17 November 1883.

70. Hinderlider (1913).

71. Foreman (1872).

72. This is shown on the USGS topographic map, 15′ Tucson quadrangle, edition of 1905; Betancourt (1990) places it in the northern half of Section 2, T15S, R13E.

73. Smith (1910) claims that in 1858 the Rillito River was impounded by a series of beaver dams. This claim is not corroborated by any other historical accounts.

74. Mearns (1907: 350).

75. Lucero (1928).

76. Fish and Gillespie (1987).

77. Clarke (1852: 85).

78. Foreman (1872).

79. Betancourt (1990) places the spring at Punta de Agua at the northern boundary of Section 2, T16S, R13E.

80. Foreman (1872).

81. *Arizona Daily Star*, 24 September 1882.

82. Betancourt (1990).

83. Waters (1988).

84. Betancourt (1990: fig. 10A, notes 46–68).

85. Betancourt (1990: fig. 10B, note 2).

86. US Court of Private Land Claims (1881); see note 12, this chapter.

87. Various reviews of the tamarisk introduction have been published; see Webb et al. (2007b).

88. Robinson (1965). Although the taxonomy remains questionable, saltcedar generally is considered to be *Tamarix ramosissima* or a hybrid of that species and various other *Tamarix*.

89. Webb et al. (2007b). Athel tamarisk is *Tamarix aphylla* (see appendix K).

90. Robinson (1965).

Chapter 5

1. http://www.pagnet.org/documents/Population/histpop20091.pdf (accessed 21 November 2010).

2. T12S, R12E was the only one not completed in the 1870s.

3. Betancourt (1990).

4. See appendix D.

5. Henshaw (1875a, 1875b).

6. Brandt (1951), American Ornithologists' Union (1957). See appendix E.

7. Henshaw (1875a, 1875b).

8. Hume (1978), Fischer (2001).

9. Bendire (1892, 1895).

10. Bendire (1872a, 1872b, 1892, 1895). See appendix B,

11. Coues (1873).

12. Coues (1872a), Bendire (1892).

13. Bendire (1872b, 1892).

14. Bendire (1872a, 1892).

15. Phillips et al. (1964).

16. Bendire (1892, 1895).

17. Bendire (1895: 128).

18. Bendire (1895: 63).

19. A. C. Bent, who often visited the Great Mesquite Forest in the 1930s (see chapter 7), did not mention the Ladder-backed Woodpecker during his visit in 1922 but stated that it was "fairly common about Tombstone and near Fairbanks on the San Pedro River" (1939: 85).

20. Bendire (1872b). See Anderson (1972).

21. Bendire (1892). The Spotted Owl is *Strix occidentalis lucida*.

22. Phillips et al. (1964), Gutierrez et al. (1995).

23. Bendire (1892), Bent (1937), Johnson et al. (2000b).

24. Bendire (1892: 253). Stephens's observations are reported in Brewster (1881); see appendix A.

25. Fischer (2001).

26. Brown field notes, University of Arizona bird collection. Brown started this collection.

27. Brown (1906).

28. Phillips et al. (1964), Monson and Phillips (1981), Frederick and Siegel-Causey (2000).

29. Brown (1906).

30. Phillips et al. (1964: 11), James and Thompson (2001: 20).

31. Phillips et al. (1964), Monson and Phillips (1981), West and Hess (2002).

32. Brown (1899).

33. Phillips et al. (1964), Monson and Phillips (1981), Fischer (2001).

34. Phillips et al. (1964).

35. Brown (1884, 1885, 1904).

36. US Fish and Wildlife Service (2011a). Masked Bobwhite have been reintroduced in the Santa Cruz River drainage, on the Buenos Aires National Wildlife Refuge, with little success.

37. Bequaert and Miller (1973: 221).

38. Bequaert and Miller (1973).

39. Bequaert and Miller (1973).

40. Hovingh (2004).

41. Dobyns (1981).

42. *Arizona Citizen*, 18 July 1874.

43. *Arizona Citizen*, 22 August 1874. In July 1875, McKay had to sue Warner to recover the cost of building the millrace.

44. Bahre (1985).

45. *Arizona Weekly Star*, 23 June 1877.

46. *Arizona Weekly Star*, 9 August 1877.

47. *Arizona Weekly Star*, 18 July 1878.

48. Because monsoonal storms have a limited extent, determining the wettest summer months is very uncertain. For example, even though Tucson experienced extremely large floods in July 2006, this month is not particularly wet in the main Tucson Airport record for the city. The wettest July at the airport was in 1991, a year without significant floods and relatively low rainfall at the University of Arizona.

49. Artesian wells, also known as flowing wells, generally represent a pressurized water table beneath a confining layer. The pressure comes from a higher water level upslope, and when the confining layer is breached by a well or a natural cause, water can flow up and out of the surface. Prior to extensive groundwater development in the Tucson Basin, artesian wells were reported at several sites.

50. *Tucson Magazine*, December 1948.

51. *Arizona Weekly Star*, 17 April, 24 March, 1 May, and 8 May 1879; 24 March and 11 August 1881.

52. *Arizona Citizen*, 2 May 1879.

53. *Arizona Daily Star*, 9 August 1880.

54. Barter (1881).

55. *Tombstone Daily Nugget*, 27 July 1881.

56. *Arizona Daily Star*, 24 September 1882.

57. Warner (1884).

58. Warner (1884).

59. *Arizona Citizen*, 18 November 1883.

60. Warner (1884).

61. Betancourt (1990).

62. The court case of *Dalton et al. v. Carrillo et al.* reveals a complex scenario of water and land use in the bottomlands west of Tucson. The plaintiffs were several landowners north of the road, including W. A. Dalton, Emilio Carrillo, Joaquin Telles, E. N. Fish, Lauterio Acedo, Ramón Pacheco, Cerilio León, and Francisco Munguia. Their attorney was C. C. Stephens. The primary defendants were Leopoldo Carrillo, Sam Hughes, and W. C. Davis, all of whom owned agricultural lands south of the road.

63. Betancourt (1990).

64. Drake (1885).

65. *Arizona Daily Star*, 11 June 1885.

66. *Arizona Mining Index*, 13 February 1886.

67. *Arizona Mining Index*, 20 February 1886.

68. *Arizona Daily Star*, 12 and 26 May 1886; *Arizona Citizen*, 2 June 1886.

69. *Arizona Mining Index*, 19 June 1886.

70. *Arizona Mining Index*, 27 February 1886.

71. *Arizona Mining Index*, 7 August 1886.

72. *Arizona Daily Star*, 14 August 1886.

73. *Arizona Daily Star*, 17 August 1886.

74. DuBois and Smith (1980).

75. Goodfellow (1888).

76. Dubois and Smith (1980).

77. Bennett (1977).

78. *Tucson Weekly Citizen*, 7 May 1887.

79. Olberg and Schanck (1913), Castetter and Bell (1942).

80. Tevis (1954).

81. *Arizona Daily Star*, 12 July 1887.

82. *Arizona Daily Star*, 13 July 1887.

83. *Arizona Daily Star*, 11 September 1887.

84. *Tucson Citizen*, 12 September 1887.

85. Hastings (1959).

86. *Arizona Daily Star*, 3 September 1887.

87. *Arizona Daily Star*, 3 November 1887.

88. Castetter and Bell (1942).

89. *Arizona Daily Star*, 12 February 1884.

90. Gelt et al. (1999).

91. Kupel (2003: 42). Schwalen and Shaw (1957: 90–92) also review the water-development history of Tucson.

92. Kupel (2003: 45).

Chapter 6

1. *Arizona Citizen*, 30 July and 1 August 1890.

2. *Tucson Citizen*, 12 September 1887.

3. *Arizona Daily Star*, 1 August 1890.

4. Cameron (1890), quoted in Betancourt (1990: 130).

5. *Arizona Daily Star*, 6 August 1890.

6. *Arizona Citizen*, 1 August 1890.

7. *Arizona Daily Star*, 8 August 1890.

8. *Arizona Daily Star*, 9 August 1890.

9. *Arizona Daily Star*, 8 August 1890.

10. *Arizona Daily Star*, 5 and 6 August 1890.

11. *Arizona Daily Star*, 7 August 1890.

12. *Arizona Daily Star*, 13 August 1890.

13. *Arizona Daily Star*, 17 August 1890.

14. *Arizona Daily Star*, 20 August 1890.

15. *Arizona Daily Star*, 26 August 1890.

16. *Arizona Daily Star*, 28 and 29 August 1890.

17. *Arizona Daily Star*, 9 August 1890.

18. *Arizona Daily Star*, 5 October 1890.

19. *Arizona Daily Star*, 25 October 1899.

20. Spalding (1909: 9).

21. *Arizona Daily Star*, 10 June 1934.

22. León (n.d.).

23. Kitt Family Papers and Business Records, Arizona Historical Society, Tucson, http://cip.azlibrary.gov/Collection.aspx?CollID=616.

24. *Arizona Daily Star*, 2 July 1891.

25. *Arizona Daily Star*, 19 August 1891.

26. *Arizona Daily Star*, 29 August 1891.

27. *Arizona Daily Star*, 16 July 1891.

28. Durrenberger and Ingram (1978).

29. *Arizona Daily Star*, 22 October 1891.

30. *Arizona Daily Star*, 24 October 1891.

31. *Arizona Daily Star*, 26 August 1891.

32. *Arizona Daily Star*, 5 February 1891. This farm was in Section 6, T17S, R14E.

33. *Arizona Daily Star*, 14 April 1891. This farm was in Section 34, T13S, R13E.

34. *Arizona Daily Star*, 6 October 1891.

35. *Arizona Daily Star*, 16 January 1892.

36. *Arizona Daily Star*, 21 July 1892.

37. *Arizona Daily Star*, 20 January 1893.

38. *Arizona Daily Star*, 20 December 1892.

39. *Arizona Daily Star*, 15 February 1893.

40. *Arizona Daily Star*, 17 February 1893.

41. *Arizona Daily Star*, 11 June 1893.

42. *Arizona Daily Star*, 17 January 1895.

43. *Arizona Daily Star*, 13 June 1895.

44. *Arizona Daily Star*, 17 November 1895.

45. *Arizona Daily Star*, 9 August 1895.

46. *Arizona Daily Star*, 8 March 1895.

47. Allison (n.d.).

48. This smelter was in the NW 1/4 of the SE 1/4 of Section 2, T14S, R13E.

49. Allison (n.d.).

50. *Arizona Daily Star*, 1 May 1902.

51. *Arizona Daily Star*, 10 July 1902.

52. *Arizona Daily Star*, 30 July 1902.

53. Pearthree and Baker (1987: 20).

54. Smith (1910).

55. Smith (1910).

56. Graf (1984), using the dates of aerial photography, gives 1937 as the year arroyo downcutting ceased.

57. Sykes (1967).

58. Sykes (1967).

59. Bendire (1872, 1892, 1895), Brewster (1882, 1883).

60. Mearns (1907: 108).

61. Largely as a result of work by Swarth (1905).

62. Phillips et al. (1964), University of Arizona avian collection.

63. Brown (1888, 1906), Phillips et al. (1964).

64. See appendixes A and I.

65. Bent (1930), Palmer et al. (1954).

66. See appendix A.

67. Brown (1906).

68. Corman and Wise-Gervais (2005).

69. Johnson and Barlow (1971) found a pair nesting in a hollow in a lone cottonwood tree along an irrigation ditch near Peoria, Arizona. See Monson and Phillips (1981).

70. Bryan (1928) describes this phenomenon in general for the Santa Cruz River but does not specifically name the Great Mesquite Forest.

71. Swarth (1904).

72. Fischer (2001).

73. Bowers (2010), Webb and Turner (2010).

74. Spalding (1909).

75. Swarth (1905: 22).

76. Anderson (1972).

77. Anderson (1972).

78. Swarth (1905).

79. Swarth (1905: 25).

80. Willard (1912: 57–58).

81. Willard (1912: 58).

82. See appendix H.

83. *Arizona Daily Star*, 19 June 1903.

84. *Arizona Daily Star*, 12 March 1905. This bridge was located on the section line between Sections 7 and 18, T13S, R13E.

85. *Arizona Daily Star*, 21 March 1905.

86. Schwalen (1942).

87. *Arizona Daily Star*, 24 June 1905.

88. Smith (1910: 177), Olberg and Schank (1913: 8).

89. Olberg and Schanck (1913: 9).

90. Olberg and Schanck (1913: 9).

91. James (1917).

92. Schwalen and Shaw (1957: 94).

93. *Arizona Daily Star*, 1 January 1922.

94. Hinderlider (1913: 200–201, 244).

95. Jones (1973).

96. Fuller (1913: 8–9, 28–29).

97. Olberg and Schanck (1913: 10).

98. Olberg and Schanck (1913: 10).

99. Olberg and Schanck (1913: 11).

100. More precisely, the dike was built on a north-south line between sections 14 and 13 of T16S, R13E, north of Pima Mine Road.

101. Olberg and Schanck (1913: 12).

102. *Arizona Daily Star*, 23 December 1914; *Tucson Citizen*, 24 December 1914.

103. *Arizona Daily Star*, 1 February 1915.

104. *Arizona Daily Star*, 3 February 1915.

105. Fuller (1913).

106. Hays (1984).

107. This dam was located in the SW 1/4 of the SE 1/4 of Section 26, T15S, R13E.

108. C. A. Engle to E. W. Kramer Jr., 29 March 1937, Records of the Supervising Engineer, 1912–1942, Records of the Indian Irrigation Service District Four, Arizona, US National Archives (http://www.archives.gov/research/guide -fed-records/groups/075.html#75.21.7).

109. Spalding (1909: 8).

110. Spalding (1909: plate 4).

111. Spalding (1909: 10).

112. D. T. McDougal to W. L. Tower, 4 February 1915. D. T. McDougal to J. H. Harris, 6 February 1915, reports similar changes. Both in University of Arizona Libraries, Special Collections, Desert Botanical Laboratory of the Carnegie Institution Records, 1903–1985 (hereafter cited as "Desert Botanical Laboratory records").

113. Huntington (1914: plate 1A).

114. D. T. McDougal to Ellsworth Huntington, 20 March 1915, Desert Botanical Laboratory records.

115. Godfrey Sykes to D. T. McDougal, 16 and 21 August 1916; D. T. McDougal to Godfrey Sykes, 23 August 1916; all in Desert Botanical Laboratory records.

116. Godfrey Sykes to D. T. McDougal, 24 August 1916, Desert Botanical Laboratory records.

Chapter 7

1. The peak discharge for the September 1926 flood on the San Pedro River is controversial because it was so large; see Hirschboeck (2009).

2. *Arizona Daily Star*, 24 and 25 September 1929.

3. Schwalen and Shaw (1957).

4. Youngs (1931: 46).

5. Gelt et al. (1999).

6. *Arizona Citizen*, 3 April 1925.

7. Phillips et al. (1964: xvi).

8. Gardner (1962:1).

9. United States Senate (1931: 8347).

10. Baker (1935: 3).

11. Baker (1935: 3).

12. Betancourt (1990).

13. Aldridge and Eychaner (1984).

14. Knapp (1937).

15. Kupel (2006).

16. Webb and Leake (2006).

17. Betancourt (1990) places this road in Section 25, T16S, R17E.

18. Knapp (1937).

19. Swarth (1905) and Dawson (1921), both from California, were the only two professional ornithologists to conduct studies and publish their findings from the Great Mesquite Forest prior to Bent's visit here in 1922. Others who published information from the forest and nearby Santa Cruz River, such as Willard (see appendix A) and Herbert Brown, were amateur ornithologists.

20. Vorhies saw Green Heron nests at Indian Dam, near the former source of the Spring Branch and the Great Mesquite Forest, for "three years in a row" in the mid-1930s (Vorhies et al. 1935: 244). Vorhies also saw Pied-billed Grebes at Indian Dam frequently in the 1930s, including three calling on open water on 15 April 1934 (Vorhies et al. 1935). They may have been nesting here, because the species begins nesting by December and January in Arizona lowlands (Corman and Wise-Gervais 2005).

21. Cannon (1911: 81) established the edge of the Great Mesquite Forest as "9 miles south of Tucson"; Swarth (1905) and Brandt (1951: 71) described it as beginning ten miles from Tucson; Willard (1912: 56) said that it was "eleven miles south of the city," with which Brandt (1951: 77) agreed; and Dawson (1921: 30) placed it "some twelve miles south of Tucson," with which Howell (1918a: 1) agreed.

22. Spalding (1909: 9).

23. Willard (1912: 57).

24. Brandt (1951: 71).

25. Johnson et al. (1997a).

26. Webb and Leake (2006).

27. Johnson et al. (2003).

28. Herbert Brandt (1951), an amateur ornithologist, spent eight seasons in southeastern Arizona between 1935 and 1948 collecting information for his book.

29. Although Brandt discusses specific species, he did not provide lists for either of the two rivers, nor did he name the species that he considered unique to each. For specific species that Brandt discussed, see appendix B.

30. Howell (1916).

31. Howell (1916: 210).

32. Howell (1918b).

33. Dawson (1921: 34).

34. Dawson (1921: 30).

35. See appendix I.

36. See Bent (1919–1968) and appendix A.

37. Bent (1937b: 259).

38. Howell (1918b: 2).

39. Bent (1953: 130).

40. Phillips et al. (1964: xvi).

41. These ornithologists included L. L. Hargrave, H. C. Oberholser, A. R. Phillips, and C. T. Vorhies.

42. Brandt (1951: 71).

43. Arnold (1940); see also appendixes A, F, and G.

44. Johnson et al. (1997).

45. Dickerman (1997).

46. Phillips (1939, 1946), Phillips and Monson (1964), Phillips et al. (1964).

47. A. R. Phillips, pers. comm. to R. R. Johnson, spring 1954.

48. Castetter and Bell (1942).

49. Examples of publications that specifically report observations in the Great Mesquite Forest include Bendire (1895), Swarth (1905), Dawson (1921), and Bent (1937a, 1937b). Other publications on the region that specifically mention the Santa Cruz River near San Xavier include those by Vorhies et al. (1935) and Monson (1942).

50. Bendire (1872a, 1872b), Arnold (1940).

51. Swarth (1905: 22) was the first professional ornithologist to study the avifauna of the Great Mesquite Forest.

52. Bent (1937b: 259; 1953: 130) was one of America's leading ornithologists at that time.

53. Brandt (1951: 71).

54. Willard (1912: 57).

55. Dawson (1921: 30).

56. Dobyns (1981).

57. Brown et al. (1980).

58. Brandt (1951: 71–72).

59. Bent (1937a).

60. A nest of this species was discovered in a large cottonwood tree in 1959 in the remains of the Great Mesquite Forest (RRJ field notes); see appendixes A and H.

61. All three owls—Flammulated Owl (*Otus flammeolus*), Whiskered Screech-Owl (*Megascops trichopsis*), and Spotted Owl (*Strix occidentalis*)—were recorded by Richard S. Crossin (J. T. Marshall, pers. comm. to R. R. Johnson); see appendix I.

62. The old nest was observed by W. George and R. Johnson in the fall of 1958.

63. The Rose-throated Becard was observed by W. George, Marshall, Phillips, and others in 1959; Joe T. Marshall Field Notes, 1932–2001, Division of Birds, Smithsonian Institution, NCDC532.

64. RRJ personal observation, 1959.

65. Phillips (1940).

66. Phillips et al. (1964: xvi).

67. Dawson (1921).

68. Bent (1953), Monson (1979), Johnson et al. (1997b).

69. Dawson (1921: 31).

70. http://www.iucnredlist.org/apps/redlist/search (accessed 6 March 2011).

71. Brown (1993: 3).

72. Howell (1918b).

73. Bent (1953: 130).

74. Bent (1937b: 259).

75. Johnson et al. (1977, 1987).

76. Arnold (1941, 1943), Cottam and Trefethen (1968).

77. Brown (1989).

78. Arnold (1943: 63).

79. Bent (1937b: 259).

80. Arnold (1941).

81. Neff (1940a, 1940b).

82. Cottam and Trefethen (1968). They referred to the Komatke Thicket by its other name, the New York Thicket.

83. Johnson et al. (1997b).

84. For the Albert's Towhee (*Pipilo aberti vorhiesi*) subspecies, see Phillips (1962); also see appendix E.

85. Minckley (1973); also see appendix H.

86. The mesquite mouse is *Peromyscus merriami*; see Hall and Kelson (1959).

87. Arnold (1940), Cockrum (1960), Hoffmeister (1986).

88. Appendix H lists species known to have been extirpated as well as some speculation on what else might have lived in the Great Mesquite Forest prior to its destruction.

89. Willis (1939).

90. Lowe (1985); see also appendix H.

91. Swarth (1905), Willard (1912), Dawson (1921), Brandt (1951).

92. Dawson (1921).

93. Bent (1937b: 259).

94. Arnold (1940).

95. Arnold (1940).

96. Brandt (1951), Johnson and Carothers (1982).

97. Brandt (1951).

98. Observations and publications by Joe T. Marshall, students, and associates.

99. RRJ personal observations.

100. Dawson (1921: 30).

101. Willard (1912).

102. Bent (1953: 130).

103. Brandt (1951: 76).

104. Arnold (1940: 7).

105. Brandt (1951: 76).

106. Lumholtz (1912: 14–15).

107. A. R. Phillips, pers. comm. to R. R. Johnson, 1954.

108. Marshall (1960, 1964); Marshall and Johnson (1968).

109. Phillips et al. (1964: xvi).

110. Kupel (2003: 196).

111. Johnson and Carothers (1982), Brown and Minckley (1982), Johnson and Haight (1983).

Chapter 8

1. Sources for data used to compile figure 8.1 are as follows. For 1900–1920 data for Pima County, see http://www.census.gov/population/cencounts/az190090.txt (accessed 9 November 2010). For 1870–1890, 1935, 1945, 1955, 1965, 1975, 1985, 1995, and 2005 records generally and for Tucson 1900–1950 and 2008, and Pima County 2007 data, see http://www.pagnet.org/documents/Population/histpop20091.pdf (accessed 9 November 2010). For 1930–1980 data for Pima County and for 1960–1980 data for Tucson, see Bureau of the Census 1982. For 1990, 2000, and 2007 Tucson and 2008 Pima County data, we used the current US Census web page: http://www.census.gov/main/www/cen2000.html (accessed 9 November 2010). Livestock data are from http://www.ag-census.usda.gov/Publications/2007/Full_Report/Volume_1,_Chapter_2_County_Level/Arizona/index.asp (accessed 8 November 2010).

2. http://www.cdc.gov/malaria/about/history/ (accessed 7 March 2011).

3. http://www.ci.tucson.az.us/water/docs/wp-ch02.pdf#page=7 (accessed 27 July 2009).

4. http://www.pagnet.org/documents/Population/histpop20091.pdf (accessed 15 July 2009).

5. Federal Emergency Management Agency (2002).

6. Knapp (1937: 11).

7. Galloway et al. (1999).

8. Evans and Pool (2000).

9. Anderson (1989).

10. Evans and Pool (2000), Carruth et al. (2007).

11. Baum et al. (2008).

12. Hoffmann et al. (1998).

13. Carpenter (1999).

14. Bull and Scott (1974).

15. Schrader et al. (1917: 29).

16. Bull and Scott (1974).

17. http://www.ag.ndsu.edu/pubs/ageng/irrigate/ae1057w.htm#Centrifugal (accessed 7 March 2011).

18. Kupel (2003: 92).

19. Gelt et al. (1999).

20. Konieczki and Heilman (2004).

21. Schwalen and Shaw (1961).

22. Kupel (2003: 135).

23. Kupel (2003: 197).

24. Bryan (1922).

25. Examples of G. E. P. Smith's work include Smith (1910, 1940).

26. Schwalen and Shaw (1957).

27. Schwalen and Shaw (1957).

28. Durrenberger and Ingram (1978: 29).

29. Durrenberger and Ingram (1978: 25).

30. Lewis (1963).

31. Roeske et al. (1978).

32. Aldridge and Eychaner (1984).

33. Betancourt (1990). Parker (1995b) provides some measurements of channel downcutting at bridges in the Tucson area.

34. Phillips et al. (1964).

35. Arnold (1940: 7).

36. The bosque (and its demise) was documented by Marshall (1960, 1964), Gould (1961), and Marshall and Johnson (1968).

37. R. Roy Johnson field notes.

38. Johnson and Carothers (1982).

39. Marshall (1960, 1964), Marshall and Johnson (1968). Additional information on birds was obtained by us from R. S. Crossin, W. G. George, P. J. Gould, A. R. Phillips, and other associates who had worked in the Great Mesquite Forest.

40. Marshall and Johnson (1968).

41. Johnson and Carothers (1982), Johnson et al. (1997b).

42. Marshall (1967).

43. Gould (1960, 1961).

44. Crossin (1965).

45. George (1958) 1961). The bird observations by the late 1950s observers appear in appendix A.

46. Phillips et al. (1964), Phillips (1968).

47. Phillips et al. (1964), Monson and Phillips (1981).

48. Phillips et al. (1964), Monson and Phillips (1981).

49. Phillips et al. (1964), Lowther et al. (1999).

50. Phillips et al. (1964), Phillips (1968).

51. Goldman records, US Fish and Wildlife Service files cited in Phillips et al. (1964) and Phillips (1968).

52. Swarth (1929).

53. American Ornithologists' Union (1931).

54. Phillips et al. (1964).

55. Phillips et al. (1964).

56. The Gray Hawk has been moved back and forth by raptor experts between the genera *Buteo* and *Asturina* several times (Bibles et al. 2002).

57. Bendire (1892), Swarth (1905), Marshall 1958 field notes (Joe T. Marshall Field Notes, 1932–2001, Division of Birds, Smithsonian Institution, NCDC532; hereafter cited as "Marshall field notes," with year).

58. Willard (1912).

59. Marshall 1959 field notes.

60. Phillips et al. (1964).

61. Bendire (1892: 316).

62. Monson and Phillips (1981).

63. The Rose-throated Becard was observed by W. George, Marshall, Phillips, and others in 1959 (Marshall 1959 field notes).

64. R. Roy Johnson, personal observation, 1959.

65. Mauz (2002) compared late twentieth-century distributions with those discussed by Thornber (1909).

66. Phillips et al. (1964: xvi).

67. Brandt (1951: 76).

68. Marshall and Johnson (personal observations).

69. Johnson and Carothers (1982).

70. Hastings (1959).

71. Webb (1996), Turner et al. (2003), and Webb et al. (2007a) are the most prominent examples of the combined use of repeat photography and historical information to evaluate landscape change.

Chapter 9

1. Soil cement is a mixture of Portland cement and aggregate developed by the Pima County Department of Transportation in the early 1980s; http://www.fhwa.dot .gov/engineering/hydraulics/pubs/09112/page07.cfm. The amount of stabilizer—"cementitious portion"—is variable and determined through laboratory testing.

2. Saarinen et al. (1984).

3. Baker (1984), Hirschboeck (1985), Ponce et al. (1985), Reich (1984), Saarinen et al. (1984), US Geological Survey (1985), Zeller (1984), Webb and Betancourt (1992).

4. Turner et al. (2003), Hereford et al. (2006).

5. Federal Emergency Management Agency (2002).

6. See review in Hirschboeck (1985).

7. For a history of this process, see Thomas (1985).

8. The skew coefficient has the greatest control on the tail of the log-Pearson type III distribution, but this moment—known as the third moment—is too poorly sampled in most annual flood series to be evaluated for stationarity. Bulletin 17B employs a "generalized skew coefficient" to stabilize this moment (US Water Resources Council 1981).

9. Roeske (1978).

10. US Water Resources Council (1981).

11. Knapp (1937).

12. Federal Emergency Management Agency (1982).

13. Eychaner (1984).

14. Aldridge and Eychaner (1984).

15. Federal Emergency Management Agency (1982).

16. After the 1983 flood, the setbacks along the Santa Cruz were increased to five hundred feet from the primary channel bank of the 100-year floodway.

17. Smith (1986: 177).

18. Aldridge and Eychaner (1984).

19. Aldridge and Eychaner (1984).

20. Saarinen et al. (1984: 62).

21. Aldridge and Eychaner (1984).

22. Aldridge and Eychaner (1984: 13).

23. City of Tucson (1981).

24. Smith (1986: 196–203).

25. Saarinen et al. (1984), Roeske et al. (1989).

26. Most discharge measurements for the 1983 floods were originally reported by Roeske et al. (1989).

27. Slezak-Pearthree and Baker (1988).

28. Hansen et al. (2011).

29. Saarinen et al. (1984), Roeske et al. (1989).

30. Parker (1995b).

31. Kresan (1988).

32. Baker (1984), Slezak-Pearthree and Baker (1988).

33. Baker (1984).

34. Kresan (1988: 470).

35. Halpenny and Halpenny (1988).

36. Ponce et al (1985); this model was criticized by Hjalmarson (1987).

37. Kresan (1988: 484).

38. Guttman et al. (1993).

39. House and Hirschboeck (1997).

40. House and Hirschboeck (1997: 13).

41. McHugh (1995).

42. Aldridge and Eychaner (1984), Aldridge and Hales (1984), Roeske et al. (1989), House and Hirschboeck (1997).

43. Webb and Betancourt (1992).

44. See Burkham (1981) for extensive discussion of the effects of channel change on flow conveyance for the Gila River.

45. Parker (1995b).

46. Aldridge and Eychaner (1984).

47. Reich (1984), Zeller (1984).

48. Hirschboeck (1985), Betancourt and Turner (1988), Webb and Betancourt (1992).

49. Webb and Betancourt (1992).

50. Hirschboeck (1985).

51. Webb and Betancourt (1992).

52. Hansen et al. (2011).

53. Webb et al. (2007b: fig. 30.1).

54. Webb et al. (2007b).

55. Konieczki and Heilman (2004).

56. Carruth et al. (2007).

57. Stevens (1988: 18).

58. Reisner (1986: 267–268).

59. Sheridan (1995: 341).

60. Webb et al. (2005).

61. http://www.cap-az.com/AboutUs/History.aspx (accessed 15 March 2010).

62. Gelt et al. (1999).

63. http://www.cap-az.com/static/index.cfm?content ID=20 (accessed 9 February 2009).

64. Kupel (2003: 192) provides an extensive review of the Central Arizona Project and its checkered history with Tucson water users.

65. Gelt et al. (1999: 11–12).

66. Gelt et al. (1999: 28).

67. http://cms3.tucsonaz.gov/water/consider_reclaim (accessed 28 October 2013).

Chapter 10

1. Griffiths et al. (2009).

2. Magirl et al. (2007), Webb et al. (2008).

3. Griffiths et al. (2009).

4. Webb et al. (2008).

5. Hansen et al. (2011) report some local problems with soil cement having been undercut or failing but note that the soil cement has performed very well.

6. Webb and Betancourt (1992).

7. Glennon (2002: 49).

8. Guber (1988).

9. Webb and Hereford (2010).

10. Webb et al. (2007b).

11. Phillips and Ingersoll (1998), Phillips et al. (1998).

12. Fabre and Cayla (2009) discuss the various water sources used in restoration attempts.

13. Webb et al. (2007b).

14. Gelt et al. (1999: 14).

15. Gelt et al. (1999: 15).

16. http://www.pima.gov/wwm/div/trtmnt.htm (accessed 28 July 2009).

17. Turns out it was wishful thinking that riparian vegetation alone could fix water-quality problems: in the 2010s, the wastewater treatment plants are undergoing a multimillion dollar retrofit to improve discharged water quality, mostly in decreased nitrogen compounds as well as odor and suspended solids (http://www.pima.gov/wwm/pubs/pdf/Bill percent20Inserts/February2011.pdf [accessed 10 February 2013]).

18. Fabre and Cayla (2009).

19. Bruce Prior, personal communication to R. Roy Johnson, 19 August 2011.

20. Mauz (2002), Gormally (2011).

21. The State of Arizona regulates groundwater under a 1980 law that created the Tucson Active Management Area (AMA). The goal of the AMA is to regulate groundwater withdrawals so that they equal recharge into the basin. Effluent infiltrating into the Santa Cruz River counts as recharge under the AMA.

22. Gaylean (1996), cited in memorandum from Evan Canfield to Bill Zimmerman, Pima County Regional Flood Control District, 11 May 2012.

23. The clogging layer is called Schmutzdecke (Treese et al. 2009).

24. E. C. Canfield, unpublished data, 22 June 2012.

25. Eden et al. (2008).

26. The river restoration projects in the Tucson Basin are described by Eden et al. (2008).

27. Bailey (1979).

28. Gelt et al. (1999: 23).

Chapter 11

1. Betancourt (1990).

2. Betancourt (1990), Webb and Betancourt (1992), Turner et al. (2003).

3. Ropelewski and Halpert (1986).

4. Walker (1924), Walker and Bliss (1932).

5. McEwen (1925), Namias (1960), Bjerknes (1969), Pyke (1972), Douglas (1976), Douglas and Englehart (1984), Ropelewski and Halpert (1986), Andrade and Sellers (1988).

6. Mooley and Parthasarathy (1984), Ojo (1987), Folland et al. (1986).

7. Betancourt (1990).

8. Hereford and Webb (1992), Webb and Betancourt (1992), Hereford et al. (2002), Turner et al. (2003).

9. Horel and Wallace (1981), Elliott and Angell (1988).

10. Andrade and Sellers (1988), Douglas and Engelhart (1984), Betancourt (1990), Webb and Betancourt (1992).

11. Nicholls (1988).

12. Rasmusson (1984, 1985), Yarnal and Diaz (1986).

13. Reyes and Cadet (1988).

14. Gergis and Fowler (2009).

15. Smith (1986).

16. Webb and Betancourt (1992).

17. Andrade and Sellers (1988), Ropelewski and Halpert (1986), Douglas and Englehart (1984).

18. Betancourt (1990), Webb and Betancourt (1992).

19. These years are 1862, 1868, 1891, 1905, 1914–1915, 1926, 1940–1941, 1965–1966, 1972, 1977, 1983, and 1993; Betancourt (1990), Webb (1985), Webb and Betancourt (1992).

20. Betancourt (1990).

21. Douglas and Englehart (1984).

22. Betancourt (1990).

23. Betancourt (1990).

24. Douglas (1983).

25. Ropelewski and Halpert (1986). Their analysis did not include any stations from Arizona.

26. Leopold (1951).

27. Leopold (1951), Leopold et al. (1966).

28. Webb (1985), Webb and Baker (1987), Hereford and Webb (1992), Hereford et al. (2002).

29. Cooke and Reeves (1976).

30. Bull (1964), Cooke and Reeves (1976).

31. Leopold et al. (1966).

32. Hereford and Webb (2002). Late nineteenth-century precipitation was not examined, because too few places had records.

33. Cooke and Reeves (1976).

34. Bull (1964), Cooke and Reeves (1976).

35. Betancourt (1990).

36. Quinn et al. (1987).

37. Quinn et al. (1987).

38. Betancourt (1990).

39. Gergis and Fowler (2009).

40. Quinn et al. (1987).

41. US Signal Service (1890).

42. Betancourt (1990).

43. Green and Sellers (1964).

44. Wright (1989).

45. Emery and Hamilton (1985).

46. Webb and Betancourt (1992).

47. Miller et al. (1994).

48. Webb and Betancourt (1992).

49. Mantua et al. (1997), Bond and Harrison (2000). The Pacific Decadal Oscillation (PDO) is defined as the leading principal component of monthly sea-surface temperature variability in the North Pacific Ocean (north of 20°N) and informally is considered to be an indicator of decadal-scale changes in the frequency of ENSO conditions.

50. Nigam et al. (1999).

51. Minobe (1997, 1999).

52. Webb and Betancourt (1992).

53. Intercept ditches are referred to as infiltration galleries by Cooke and Reeves (1976).

54. Betancourt (1990).

55. Brandt (1951), Johnson and Carothers (1982), Monson (1998).

56. Johnson and Carothers (1982), Minckley and Brown (1982).

57. Sonoran Institute (2008).

58. Minckley and Brown (1982: 229).

59. These practices had originally been brought to America, for the most part, from northern Europe, a relatively mesic region with which the eastern United States was more closely allied, a concept discussed in greater detail by social anthropologist Henry Dobyns (1981).

60. For riparian ecosystems, see Webb et al. (2007b). For avifauna, see Brandt (1951), Phillips and Monson (1964), and Johnson and Carothers (1982).

61. Johnson et al. (1997b).

62. Phillips (1968).

63. Webb et al. (2007b).

64. Webb et al. (2007b).

65. More information is available about the early biology of the Santa Cruz River than the nearby San Pedro River and even the larger riverine ecosystems of the Gila, Salt, and Verde rivers of central Arizona (Johnson and Simpson MSa, MSb).

66. Kendeigh (1944).

67. Carothers et al. (1974).

68. Browne (1951, reprint of 1869 book), Emory (1857).

69. Johnson et al. (1977, 1987); see appendixes A and B.

70. General early ornithological information about the middle Santa Cruz valley published in the late 1800s and early 1900s suggests that an additional four or five species may have also nested here. This includes species such as the Black-bellied Whistling-Duck (formerly the Black-bellied Tree-Duck; Brown 1906); see appendix H.

71. See especially Bendire (1895) and Brewster (1882, 1883).

72. Brandt (1951). He did not have access to several of the records studied by us, and unfortunately, he did not list either the sixty species common to both rivers or the six that he found unique to each river system.

73. Swarth (1905). Vorhies et al. (1935) saw Pied-billed Grebes at Indian Dam (in the Great Mesquite Forest) frequently in the 1930s, including three calling on open water on 15 April 1934. They may have been nesting here, since the species begins nesting in December and January in Arizona lowlands (Corman and Wise-Gervais 2005). They also saw Green Heron nests at Indian Dam for "three years in a row" in the mid-1930s (Vorhies et al. 1935: 244).

74. Corman and Wise-Gervais (2005), Stevenson (2007).

75. See appendix C; water from the Tucson sewer system has resulted in the establishment of ponds, cattail marshes, and riparian habitat, including cottonwoods and willows, saltcedars (introduced), and other woody riparian plants.

76. Phillips et al. (1964), Monson and Phillips (1981).

77. Rea (2007).

78. Scott (1886).

79. Swarth (1914: 21).

80. Browne (1951), Harris (1960).

81. Harris (1960: 77).

82. Observations on both domestic turkey and Wild Turkey (R. Roy Johnson).

83. See Phillips et al. (1964) for a discussion of Wild Turkey records in Arizona.

84. Johnson et al. (2003).

85. Terres (1980), Proudfoot and Johnson (2000).

86. Coues (1872b), Bendire (1872b, 1892).

87. Johnson et al. (2003).

88. An ecological sink occurs when a species reproduces, etc., and is apparently in a suitable ecological situation, but over time the population continues to diminish and, as in the case of the Ferruginous Pygmy-Owl in Sonoran desertscrub in suburban Tucson, is finally extirpated.

89. Breninger (1898).

90. Millsap and Johnson (1988), Johnson et al. (2003).

91. Confirmed by R. Roy Johnson, Federal Endangered Ferruginous Pygmy-Owl Recovery Team member.

92. Bendire (1892). The Spotted Owl is *Strix occidentalis lucida*.

93. Bendire (1892), Bent (1937), Johnson et al. (2000b).

94. Phillips et al. (1964), Monson and Phillips (1981).

95. Johnson et al. (2000b), Corman and Wise-Gervais (2005), Stevenson (2007).

96. Bibles et al. (2002).

97. Corman and Wise-Gervais (2005).

98. Bibles et al. (2002) discuss the loss of habitat and extirpation of the Gray Hawk in the middle Santa Cruz Valley; Phillips et al. (1964) state that the Common Black-Hawk needs permanent streams.

99. Dawson (1921), Phillips et al. (1964).

100. No successful nesting was recorded for the Rose-throated Becard.

101. Phillips et al. (1964), Marshall (field notes), Johnson (personal observations). This bird was again observed on 18 July 1959.

102. Johnson et al. (1984).

103. Stebbins (1985), Burt and Grossenheider (1964), Hoffmeister (1986).

104. Stebbins (1985).

105. Haney et al. (2008), Perla and Stevens (2008).

106. Information is most complete for birds with other vertebrate information derived from field observations, historic notes, and museum specimens.

107. Bequaert and Miller (1973: 221).

108. According to Rosen (2001: 31), the lowland leopard frog was "gone from Tucson proper by 1975 at the latest," the Sonoran mud turtle was "present here until the 1990s," and the northern Mexican gartersnake was "extinct in Tucson Basin by 1985 or much earlier."

109. Hoffmeister (1986).

110. Johnson and Carothers (1982).

111. Kerlinger (1993).

112. http://www.azgfd.gov/wc/surveyresults.shtml.

113. Ohmart et al. (1988), Rosenberg et al. (1991), Stromberg (1993b).

114. Gavin and Sowls (1975), Anderson et al. (1977), Stamp (1978).

115. Carothers et al. (1974), Anderson et al. (1977), Stamp (1978).

116. Johnson (1971), Carothers et al. (1974).

117. MacArthur and MacArthur (1961), MacArthur et al. (1966).

118. Rosenberg et al. (1991).

119. Krieger (2001).

120. Di Tomaso (1998).

121. Horton and Campbell (1974), Turner (1974).

122. Horton and Campbell (1974).

123. Turner (1974).

124. The original Greek words from which the term *phreatophyte*, which originated with Meinzer (1923), derives literally mean "pump plant," making *phreatophyte* an apt term because of the characteristic of uptake of groundwater (like a pump) by plant roots, followed by transpiration of this water into the air, thereby "losing" it from the stream system.

125. Horton and Campbell (1974).

126. Shafroth et al. (2005), Webb et al. (2007b).

127. Wilkinson (1966).

128. Lesica and DeLuca (2004).

129. Brown et al. (1987), Rosenberg et al. (1991), Webb et al. (2007b).

130. Brown (1989).

131. Johnson (1971), Carothers et al. (1974), Rosenberg et al. (1991).

132. Turner (1974: H19).

133. Stevens (1987: 99).

134. Turner and Karpiscak (1980), Brown et al. (1987), Webb et al. (2007b).

135. Brown et al. (1987), Johnson (1991).

136. Johnson (1979: 45).

137. The tamarisk bark beetle (*Diorhabda carinulata*) has been described in numerous publications, and its ecology and efficacy are described in DeLoach et al. (2003) and Lewis et al. (2003).

138. http://www.nps.gov/grca/parknews/tamarisk-beetle.htm.

139. Paxton et al. (2011: 255).

140. Treese et al. (2009).

References

Adams, D. K., and A. C. Comrie. 1997. The North American monsoon. Bulletin of the American Meteorological Society 78: 2197–2213.

Aldrich, L. D. 1950. A Journal of the Overland Route to California and the Gold Mines. Los Angeles, CA: Dawson's Book Shop.

Aldridge, B. N. 1970. Floods of November 1965 to January 1966 in the Gila River Basin, Arizona and New Mexico, and Adjacent Basins in Arizona. US Geological Survey Water-Supply Paper 1850-C.

Aldridge, B. N., and J. H. Eychaner. 1984. Floods of October 1977 in Southern Arizona and March 1978 in Central Arizona. US Geological Survey Water-Supply Paper 2223.

Aldridge, B. N., and T. A. Hales. 1984. Floods of November 1978 to March 1979 in Arizona and West-Central New Mexico. US Geological Survey Water-Supply Paper 2241.

Alford, J. 1982. San Vicente arroyo. Annals of the Association of American Geographers 72: 398–403.

Allison, W. No date. Arizona, the last frontier. Tucson: Arizona Historical Society Library, unpublished manuscript.

Amadon, D., and A. R. Phillips. 1939. Notes on the Mexican Goshawk. Auk 56: 183–184.

American Ornithologists' Union (AOU). 1931. Check-List of North American Birds. 4th edition. Lancaster, PA: American Ornithologists' Union.

American Ornithologists' Union (AOU). 1957. Check-List of North American Birds. 5th edition. Baltimore, MD: American Ornithologists' Union.

American Ornithologists' Union (AOU). 1998. Check-List of North American Birds. 7th edition. Washington, DC: American Ornithologists' Union.

American Ornithologists' Union (AOU). 2012. AOU Checklist of North and Middle American Birds. Online edition, http://checklist.aou.org (accessed 1 September 2012).

Anderson, A. H. 1972. A Bibliography of Arizona Ornithology: Annotated. Tucson: University of Arizona Press.

Anderson, A. H., and A. Anderson. 1947. Bird notes from southeastern Arizona. Condor 49: 89–90.

Anderson, A. H., and A. Anderson. 1948. Notes on two nests of the Beardless Flycatcher near Tucson, Arizona. Condor 50: 163–164.

Anderson, A. H., and A. Anderson. 1973. The Cactus Wren. Tucson: University of Arizona Press.

Anderson, B. W., and R. D. Ohmart. 1982. Revegetation for Wildlife Enhancement along the Lower Colorado River. Final report. Boulder City, NV: US Bureau of Reclamation, Lower Colorado Region.

Anderson, B. W., A. Higgins, and R. D. Ohmart. 1977. Avian use of saltcedar communities for the lower Colorado River valley. Pages 128–136 in R. R. Johnson and D. A. Jones, technical coordinators, Importance, Preservation, and Management of Riparian Habitats. US Forest Service General Technical Report RM-43. Fort Collins, CO: Rocky Mountain Forest and Range Experiment Station.

Anderson, B. W., R. D. Ohmart, and J. Rice. 1983. Avian and vegetation community structure and their seasonal relationships in the lower Colorado River valley. Condor 85: 392–405.

Anderson, S. R. 1987. Cenozoic Stratigraphy and Geologic History of the Tucson Basin, Pima County, Arizona. US Geological Survey Water-Resources Investigations Report 87-4190.

Anderson, S. R. 1989. Potential for aquifer compaction, land subsidence, and earth fissures in Avra Valley, Pima and Pinal Counties, Arizona. US Geological Survey Hydrologic Investigations Atlas 718. 3 sheets, scale 1:250,000.

Andrade, E. R., and W. D. Sellers. 1988. El Niño and its effect on precipitation in Arizona. Journal of Climatology 8: 403–410.

Antevs, E. 1952. Arroyo-cutting and filling. Journal of Geology 60: 375–385.

Antevs, E. 1955. Geologic-climatic dating in the West. American Antiquity 20: 317–335.

Applegate, L. H. 1981. Hydraulic effects of vegetation changes along the Santa Cruz River near Tumacácori, Arizona. Tucson: University of Arizona, unpublished MS thesis.

Arnold, L. W. 1940. An ecological study of the vertebrate animals of the mesquite forest. Tucson: University of Arizona, unpublished MS thesis.

Arnold, L. 1941. The mesquite forest and the White-wing. Arizona Wildlife and Sportsman 3 (11): 5–6.

Arnold, L. W. 1943. The White-winged Dove in Arizona. Pittman-Robertson Project Arizona 9-R. Phoenix: Arizona Game and Fish Commission, Federal Aid Division.

Bahre, C. J. 1985. Wildfire in southeastern Arizona between 1859 and 1890. Desert Plants 7: 190–194.

Bailey, E. D. 1979. City of Tucson waste water effluent delivery facility. Water Resources Bulletin, AWRA 15: 801–811.

Bailey, R. W. 1935. Epicycles of erosion in the valleys of the Colorado Plateau Province. Journal of Geology 43: 337–355.

Baker, R. C. 1935. Reports of outstanding WPA projects. Phoenix: State of Arizona Library and Archives, unpublished manuscripts.

Baker, R. J., L. C. Bradley, R. D. Bradley, J. W. Dragoo, M. D. Engstrom, R. S. Hoffmann, C. A. Jones, F. Reid, D. W. Rice, and C. Jones. 2003. Revised Checklist of North American Mammals North of Mexico, 2003. Occasional Paper 229. Lubbock: Museum of Texas Tech University.

Baker, V. R. 1984. Questions raised by the Tucson flood of 1983. Journal of the Arizona-Nevada Academy of Science 14: 211–219.

Baker, V. R. 1987. Paleoflood hydrology and extraordinary flood events. Journal of Hydrology 96: 79–99.

Balling, R. C., and S. G. Wells. 1990. Historical rainfall patterns and arroyo activity within the Zuni River drainage basin, New Mexico. Annals of the American Association of Geographers 80: 603–617.

Barnes, W. C. 1935. Arizona Place Names. Tucson: University of Arizona Press. Reissued 1988.

Barter, G. W. 1881. Directory of City of Tucson for the Year 1881. San Francisco: G. W. Barter.

Bartlett, J. R. 1854a. The Progress of Ethnology: An Account of Recent Archaeological, Philological and Geographical Researches in Various Parts of the Globe, Tending to Elucidate the Physical History of Man. New York: American Ethnological Society (William Van Holden, printer).

Bartlett, J. R. 1854b. Personal Narrative of Explorations and Incidents in Texas, New Mexico, California, and Chihuahua. New York: D. Appleton and Company.

Baum, R. L., D. L. Galloway, and E. L. Harp. 2008. Landslide and Land Subsidence Hazards to Pipelines. US Geological Survey Open-File Report 2008-1164.

Bell, W. A. 1869. New Tracks in North America. London: Chapman and Hall.

Bendire, C. E. 1872a. [Nests and/or eggs from 1872] Camp on Rillitto [*sic*] Creek present site of Camp Lowell, Arizona Terr. Field Books I: 9–83 (SIA Acc 12-279), Smithsonian Institution Archives Capital Gallery, Washington, DC (http://siarchives.si.edu/collections /siris_arc_360520).

Bendire, C. E. 1872b. List of birds shot or observed & seen in the vicinity of Tucson and Rillitto [*sic*] Creek Arizona in the years 1871 and 1872. Field Books I: 84–112 (SIA Acc 12-279), Smithsonian Institution Archives Capital Gallery, Washington, DC (http://siarchives.si .edu/collections/siris_arc_360520).

Bendire, C. E. 1888. Notes on the habits, nests and eggs of the genus *Glaucidium boie*. Auk 5: 366–372.

Bendire, C. E. 1892. Life Histories of North American Birds with Special Reference to Their Breeding Habits and Eggs. US National Museum Special Bulletin 1.

Bendire, C. E. 1895. Life Histories of North American Birds, from the Parrots to the Grackles, with Special Reference to Their Breeding Habits and Eggs. US National Museum Special Bulletin 3.

Bennett, E. F. 1977. An afternoon of terror: The Sonoran earthquake of May 1877. Arizona and the West (Summer): 107–120.

Bent, A. C. 1919–1968. Life Histories of North American Birds. US National Museum Bulletins. Vols. 1–23 covering all species regularly occurring in the United States with information from the San Xavier Mesquite Forest scattered among several accounts, all written by Bent and collaborators except Bulletin 237, parts 1–3 (1968), edited by O. L. Austin Jr.

Bent, A. C. 1930. Obituaries: Francis Cottle Willard. Auk 47: 455–456.

Bent, A. C. 1932. Life Histories of North American Gallinaceous Birds. US National Museum Bulletin 162.

Bent, A. C. 1937a. Life Histories of North American Birds of Prey: Order Falconiformes. US National Museum Bulletin 167 (part 1): 12–28.

Bent, A. C. 1937b. *Urubitinga anthracinus anthracinus* (Lichtenstein): Mexican Goshawk. Pages 259–264 *in* Life Histories of North American Birds of Prey, part 1. US National Museum Bulletin 167.

Bent, A. C. 1939. Life Histories of North American Woodpeckers. US National Museum Bulletin 174.

Bent, A. C. 1948. Life Histories of North American Nuthatches, Wrens, Thrashers and Their Allies. US National Museum Bulletin 195.

Bent, A. C. 1953. *Vermivora luciae* (Cooper): Lucy's Warbler. Pages 129–134 *in* Life Histories of North American Wood Warblers. US National Museum Bulletin 203.

Bequaert, J. C., and W. B. Miller. 1973. The Mollusks of the Arid Southwest, with an Arizona Check List. Tucson: University of Arizona Press.

Betancourt, J. L. 1990. Tucson's Santa Cruz River and the arroyo legacy. Tucson: University of Arizona, PhD dissertation.

Betancourt, J. L., and R. M. Turner. 1988. Historic arroyocutting and subsequent channel changes at the Congress

Street crossing, Santa Cruz River. Pages 1353–1371 *in* E. E. Whitehead, C. F. Hutchinson, B. N. Timmermann, and R. G. Varady, editors, Arid Lands Today and Tomorrow. Boulder, CO: Westview Press.

Beverage, J. P., and J. K. Culbertson. 1964. Hyperconcentrations of suspended sediment. American Society of Civil Engineers, Journal of the Hydraulics Division 90: 117–126.

Bibles, B. D., R. L. Glinski, and R. R. Johnson. 2002. Gray Hawk (*Asturina nitida*). *In* A. Poole and F. Gill, editors, The Birds of North America, no. 652. Philadelphia, PA: Birds of North America, Inc.

Birds of North America series. 1992–2002. Individual accounts for 716 species of birds occurring regularly in North America. Philadelphia, PA: Academy of Natural Sciences.

Bjerknes, J. 1969. Atmospheric teleconnections from the equatorial Pacific. Monthly Weather Review 97: 163–172.

Bolton, H. E. 1919. Kino's Historical Memoir of Pimería Alta. Cleveland, OH: Arthur H. Clark Company.

Bolton, H. E. 1931. Anza's California Expeditions: The Diary of Pedro Font. Berkeley: University of California Press.

Bond, N., and D. E. Harrison. 2000. The Pacific Decadal Oscillation, air-sea interaction and central North Pacific winter atmospheric regimes. Geophysical Research Letters 27: 731–734.

Boughton, W. C., and K. G. Renard. 1984. Flood frequency characteristics of some Arizona watersheds. Water Resources Bulletin, AWRA 20: 761–769.

Bourke, J. G. 1891. On the Border with Crook. New York: Charles Scribner's Sons.

Bowers, J. E. 2010. A debt to the future: Achievements of the Desert Laboratory, Tumamoc Hill, Tucson, Arizona. Desert Plants 26: 25–39.

Brandt, H. 1951. Arizona and Its Bird Life. Cleveland, OH: Bird Research Foundation.

Branson, F. A., G. F. Gifford, K. G. Renard, and R. F. Hadley. 1981. Rangeland Hydrology. Range Management Series 1. Denver, CO: Society for Range Management.

Breninger, G. F. 1898. The Ferruginous Pygmy-Owl. Osprey 2:128.

Breninger, G. F. 1901. A list of birds observed on the Pima Indian Reservation, Arizona. Condor 3: 44–46.

Breninger, G. F. 1905. The English Sparrow at Tucson, Arizona. Auk 22: 417.

Brewster, W. 1881. Additions to the avi-fauna of the United States. Bulletin of the Nuttall Ornithological Club 6: 252.

Brewster, W. 1882. On a collection of birds lately made by Mr. F. Stephens in Arizona. Bulletin of the Nuttall Ornithological Club 7: 65–86, 135–147, 193–212.

Brewster, W. 1883. On a collection of birds lately made by Mr. F. Stephens in Arizona. Bulletin of the Nuttall Ornithological Club 8: 21–36.

Brewster, W. 1885. Additional notes on some birds collected in Arizona and the adjoining province of Sonora, Mexico, by Mr. F. Stephens in 1884; with a description of a new species of *Ortyx*. Auk 2: 196–200.

Brown, Bryan T. 1988. Breeding ecology of a Willow Flycatcher population in Grand Canyon, Arizona. Western Birds 19: 25–33.

Brown, B. T. 1993. Bell's Vireo (*Vireo bellii*). *In* A. Poole, P. Stettenheim, and F. Gill, editors, The Birds of North America, no. 35. Philadelphia: Academy of Natural Sciences; Washington, DC: American Ornithologists' Union.

Brown, B. T., S. W. Carothers, and R. R. Johnson. 1987. Grand Canyon Birds: Historical Notes, Natural History, and Ecology. Tucson: University of Arizona Press.

Brown, D. E., editor. 1982. Biotic communities of the American Southwest—United States and Mexico. Special issue of Desert Plants, 4: 1–342.

Brown, D. E., editor. 1983. The Wolf in the Southwest: The Making of an Endangered Species. Tucson: University of Arizona Press.

Brown, D. E. 1989. Arizona Game Birds. Tucson: University of Arizona Press; Phoenix: Arizona Game and Fish Department.

Brown, D. E., and C. H. Lowe. 1980. Biotic Communities of the Southwest. US Forest Service General Technical Report RM-78. 1-page map. Fort Collins, CO: Rocky Mountain Forest and Range Experiment Station.

Brown, D. E., and W. L. Minckley. 1982. Wetlands. Pages 223–287 *in* D. E. Brown, editor, Biotic communities of the American Southwest—United States and Mexico. Special issue of Desert Plants, 4: 1–342.

Brown, D. E., C. H. Lowe, and C. P. Pase. 1980. A Digitized Systematic Classification for Ecosystems with an Illustrated Summary of the Natural Vegetation of North America. US Forest Service General Technical Report RM-73. Fort Collins, CO: Rocky Mountain Forest and Range Experiment Station.

Brown, D. E., K. B. Clark, R.A. Babb, and G. Harris. 2012. An analysis of Masked Bobwhite collection locales and habitat characteristics. Proceedings of the National Quail Symposium 7.

Brown, H. 1884. *Ortyx virginianus* in Arizona. Forest and Stream 22 (6): 104.

Brown, H. 1885. Arizona quail notes. Forest and Stream 25 (23): 445.

Brown, H. 1888. *Ionornis martinica* in Arizona. Auk 5: 109.

Brown, H. 1899. The Scarlet Ibis (*Guara rubra*) in Arizona. Auk 16: 270.

Brown, H. 1904. Masked Bobwhite (*Colinus ridgwayi*). Auk 21: 209–213.

Brown, H. 1906. The Water Turkey and tree ducks near Tucson, Arizona. Auk 23: 217–218.

Browne, J. R. 1951. A Tour Through Arizona 1864; or, Adventures in the Apache Country. Tucson: Arizona

Silhouettes. Originally published in 1869 as A Tour Through Arizona and Sonora: Adventures in the Apache Country, by Harper and Brothers, New York.

Brush, T. P. 2000. Couch's Kingbird (*Tyrannus couchii*). *In* A. Poole and F. Gill, editors, The Birds of North America, no. 437. Philadelphia, PA: Birds of North America, Inc.

Bryan, K. 1922. Routes to Desert Watering Places in the Papago Country. US Geological Survey Bulletin 490.

Bryan, K. 1925. Date of channel trenching (arroyo cutting) in the arid Southwest. Science 62: 338–344.

Bryan, K. 1927. Channel erosion of the Rio Salado, Socorro County, New Mexico. US Geological Survey Bulletin 790: 17–19.

Bryan, K. 1928. Historic evidence on changes in the channel of the Rio Puerco, a tributary of the Rio Grande in New Mexico. Journal of Geology 36: 265–282.

Bryan, K. 1941. Precolumbian agriculture in the Southwest as conditioned by periods of alluviation. Annals of the Association of American Geographers 31: 219–242.

Bryan, K. 1954. The Geology of Chaco Canyon, New Mexico, in Relation to the Life and Remains of the Prehistoric Peoples of Pueblo Bonito. Smithsonian Miscellaneous Collections no. 5. Washington, DC: Smithsonian Institution.

Bryson, R. A., and W. P. Lowry. 1955. The synoptic climatology of the Arizona summer precipitation singularity. Bulletin of the American Meteorological Society 36: 329–339.

Bufkin, D. 1981. From mud village to modern metropolis: The urbanization of Tucson. Journal of Arizona History 22: 63–98.

Bull, W. B. 1964. History and causes of channel trenching in western Fresno County. American Journal of Science 262: 249–258.

Bull, W. B., and K. M. Scott. 1974. Impact of mining gravel from stream beds in the southwestern US. Geology 2: 171–174.

Burkham, D. E. 1970. Depletion of Streamflow by Infiltration in the Main Channels of the Tucson Basin, Southeastern Arizona. US Geological Survey Water-Supply Paper 1939-B.

Burkham, D. E. 1972. Channel Changes of the Gila River in Safford Valley, Arizona, 1846–1970. US Geological Survey Water-Supply Paper 655G.

Burkham, D. E. 1981. Uncertainties resulting from changes in river form. American Society of Civil Engineers, Journal of the Hydraulics Division 107: 593–610.

Burt, W. H., and R. P. Grossenheider. 1964. A Field Guide to the Mammals. Boston, MA: Houghton Mifflin.

Butler, E., and R. E. Marsell. 1972. Cloudburst Floods in Utah, 1939–69. US Geological Survey–Utah Division of Natural Resources Cooperative Investigations Report 11. Salt Lake City.

Butler, E., and J. C. Mundorff. 1970. Floods of December 1966 in Southwestern Utah. US Geological Survey Water-Supply Paper 1870-A.

Cable, D. R. 1975. Influence of precipitation on perennial grass production in the semidesert southwest. Ecology 56: 981–986.

Calkins, H. G. 1941. Man and gullies. New Mexico Quarterly Review 11: 69–78.

Cannon, W. A. 1911. The root habits of desert plants. Carnegie Institution of Washington Publication 131: 1–96.

Carleton, A. M. 1986. Synoptic-dynamic character of "bursts" and "breaks" in the south-west U.S. summer precipitation singularity. Journal of Climatology 6: 605–623.

Carothers, S. W., and R. R. Johnson. 1971. A summary of the Verde Valley breeding bird survey, 1970. Completion Report FW 16-10: 46–64. Phoenix: Arizona Game and Fish Department.

Carothers, S. W., R. R. Johnson, and S. W. Aitchison. 1974. Population structures and social organization of southwestern riparian birds. American Zoologist 14: 97–108.

Carothers, S. W., S. G. Mills, and R. R. Johnson. 1989. The creation and restoration of riparian habitat in southwestern arid and semi-arid regions. Pages 359–376 *in* J. A. Kusler and M. E. Kentula, editors, Wetland Creation and Restoration: The Status of the Science, vol. 1. EPA600/3-89/038a. Corvallis, OR: US Environmental Protection Agency Research Laboratory.

Carothers, S. W., S. G. Mills, and R. R. Johnson. 1990. The creation and restoration of riparian habitat in southwestern arid and semi-arid regions. Pages 351–366 *in* J. A. Kusler and M. E. Kentula, editors, Wetland Creation and Restoration: The Status of the Science. Washington, DC: Island Press.

Carpenter, M. C. 1999. South-central Arizona. Pages 65–78 *in* D. Galloway, D. R. Jones, and S. E. Ingebritsen, editors, Land Subsidence in the United States. US Geological Survey Circular 1182.

Carruth, R. L., D. R. Pool, and C. E. Anderson. 2007. Land Subsidence and Aquifer-System Compaction in the Tucson Active Management Area, South-Central Arizona, 1987–2005. US Geological Survey Scientific Investigations Report 5190.

Cartron, J.-L. E., and D. M. Finch, editors. 2000. Ecology and Conservation of the Cactus Ferruginous Pygmy-Owl in Arizona. US Forest Service General Technical Report RMRS-GTR-43. Ogden, UT: Rocky Mountain Research Station.

Castetter, E. F., and W. H. Bell. 1942. Pima and Papago Indian Agriculture. Albuquerque: University of New Mexico Press.

Cayan, D. R., and R. H. Webb. 1992. El Niño/Southern Oscillation and streamflow in the western United States. Pages 29–68 *in* H. F. Diaz and V. Markgraf, editors, El Niño: Historical and Paleoclimatic Aspects of the Southern Oscillation. Cambridge, UK: Cambridge University Press.

Christenson, G. E. 1985. Quaternary geology of the Montezuma Creek–lower Recapture Creek area, San Juan County, Utah. Pages 3–32 *in* G. E. Christenson, C. G. Oviatt, J. F. Shroder, and R. E. Sewell, editors, Contributions to Quaternary Geology of the Colorado Plateau. Special Studies 64. Salt Lake City: Utah Geological and Mineral Survey.

City of Tucson. 1981. Construction plans for Santa Cruz River flood control channelization. Tucson, AZ: City of Tucson Rio Nuevo Redevelopment Project, unpublished manuscript.

Clarke, A. B. 1852. Travels in Mexico and California. Boston, MA: Wright and Hasty Steam Printers.

Coale, H. K. 1894. Ornithological notes on a flying trip through Kansas, New Mexico, Arizona and Texas. Auk 11: 215–222.

Cockrum, E. L. 1960. The Recent Mammals of Arizona. Tucson: University of Arizona Press.

Condes de la Torre, A. 1970. Streamflow in the Upper Santa Cruz River Basin, Santa Cruz and Pima Counties, Arizona. US Geological Survey Water-Supply Paper 1939-A.

Cooke, P. St. George. 1878. The Conquest of New Mexico and California: An Historical Atlas and Personal Narrative. New York: G. P. Putnam's Sons.

Cooke, R. U., and R. W. Reeves. 1976. Arroyos and Environmental Change in the American South-west. Oxford, UK: Clarendon Press.

Cooley, M. E. 1962. Late Pleistocene and Recent Erosion and Alluviation in Parts of the Colorado River System, Arizona and Utah. US Geological Survey Professional Paper 450-B.

Cooper, J. G. 1861. New California animals. Proceedings of the California Academy of Sciences, 1st series, 2: 118–123.

Corbosiero, K. L., M. J. Dickinson, and L. F. Bosart. 2009. The contribution of eastern North Pacific tropical cyclones to the rainfall climatology of the Southwest United States. Monthly Weather Review 137: 2415–2435.

Corle, E. 1951. The Gila: River of the Southwest. Lincoln: University of Nebraska Press.

Corman, T. E., and C. Wise-Gervais. 2005. Arizona Breeding Bird Atlas. Albuquerque: University of New Mexico Press.

Cottam, C., and J. B. Trefethen, editors. 1968. Whitewings: The Life History, Status and Management of the White-winged Dove. With contributions by G. B. Saunders, P. B. Uzzell, S. Gallizioli, W. H. Kiel Jr., J. A. Neff, and J. Stair. Princeton, NJ: D. Van Nostrand.

Coues, E. 1866. From Arizona to the Pacific. Ibis, 2nd series, 2 (7): 259–275.

Coues, E. 1872a. A new bird to the United States. American Naturalist 6: 370.

Coues, E. 1872b. Key to North American Birds. New York: Dodd and Mead.

Coues, E. 1872c. Nest and eggs of *Helminthophaga luciae*. American Naturalist 6: 493.

Coues, E. 1873. Some United States birds, new to science, and other things ornithological. American Naturalist 7: 321–331.

Coues, E. 1878. Birds of the Colorado Valley. US Geological Survey of the Territories, Miscellaneous Publications 11: 1–807. Washington, DC.

Couts, C. J. 1961. Hepah, California: The Journal of Cave Johnson Couts. Tucson: Arizona Historical Society.

Cross, R. L. 1988. Eastern North Pacific hurricanes, 1987. Mariners Weather Log 32: 12–16.

Crossin, R. S. 1965. The history and breeding status of the Song Sparrow near Tucson, Arizona. Auk 82: 287–288.

Crother, B. I., editor. 2012. Scientific and Standard English Names of Amphibians and Reptiles of North America North of Mexico. SSAR Herpetological Circular 37. Shoreview, MN: Society for the Study of Amphibians and Reptiles.

Davidson, E. S. 1970. Geohydrology and Water Resources of the Tucson Basin, Arizona. US Geological Survey Open-File Report 70–95.

Davidson, E. S. 1973. Geohydrology and Water Resources of the Tucson Basin, Arizona. US Geological Survey Water-Supply Paper 1939-E.

Davis, G. P., Jr. 1982. Man and Wildlife in Arizona: The American Exploration Period, 1824–1865. Edited by N. B. Carmony and D. E. Brown. Phoenix: Arizona Game and Fish Department.

Davis, W. M. 1902. Base level, grade, and peneplain. Journal of Geology 10: 77–111.

Davis, W. M. 1903. An excursion to the Plateau province of Utah and Arizona. Bulletin of the Museum of Comparative Zoology, Harvard College, Geological Series 6: 1–50.

Dawson, W. L. 1921. The season of 1917. Journal of the Museum of Comparative Oology 2: 27–36.

Dawson, W. L. 1923. The Birds of California. 4 vols. San Diego, CA: South Moulton Company.

Dean, J. S. 1988. Dendrochronology and environmental reconstruction. Pages 119–167 *in* G. J. Gumerman, editor, The Anasazi in a Changing Environment. Cambridge: Cambridge University Press.

DeLoach, C. J., P. A. Lewis, J. C. Herr, R. I. Carruthers, J. L. Tracy, and J. Johnson. 2003. Host specificity of the leaf beetle *Diorhabda elongata deserticola* (Coleoptera: Chrysomelidae) from Asia, a biological control agent for salt-cedars (Tamarix: Tamaricaceae) in the western United States. Biological Control 27: 117–147.

Denevan, W. M. 1967. Livestock numbers in nineteenth-century New Mexico, and the problem of gullying in the Southwest. Annals of the Association of American Geographers 37: 681–783.

Diamond, J. M. 1986. The environmentalist myth. Nature 324: 19–20.

Dickerman, R. W., editor. 1997. The Era of Allan R. Phillips: A Festschrift. Albuquerque, NM: Horizon Communications.

Dickinson, W. R. 1991. Tectonic Setting of Faulted Tertiary Strata Associated with the Catalina Core Complex in Southern Arizona. Geological Society of America Special Paper 264. Boulder, CO.

Di Tomaso, J. M. 1998. Impact, biology, and ecology of saltcedar (*Tamarix* spp.) in the southwestern United States. Weed Technology 12: 326–336.

Dobyns, H. F. 1981. From Fire to Flood: Historic Human Destruction of Sonoran Desert River Oases. Socorro, NM: Ballena Press.

Dodge, R. E. 1902. Arroyo formation. Science 15: 746.

Doolittle, W. E. 2000. Cultivated Landscapes of Native North America. New York: Oxford University Press.

Douglas, A. V. 1976. Past air-sea interactions over the eastern North Pacific Ocean as revealed by tree-ring data. Tucson: University of Arizona, unpublished PhD dissertation.

Douglas, A. V. 1983. The Mexican Summer Monsoon of 1982. Proceedings of the Seventh Annual Climatic Diagnostics Workshop, 70–79. Boulder, CO: National Oceanic and Atmospheric Administration (NOAA).

Douglas, A. V., and P. J. Englehart. 1984. Factors Leading to the Heavy Precipitation Regimes of 1982–1983 in the United States. Proceedings of the Eighth Annual Climate Diagnostics Workshop, 42–54. Washington, DC: National Oceanic and Atmospheric Administration (NOAA).

Douglas, M. W., R. A. Maddox, and K. Howard. 1993. The Mexican monsoon. Journal of Climate 6: 1665–1677.

Drake, C. R. 1885. Papers. Transcript of court proceedings in the trial of *Dalton et al. versus Carrillo et al.* Tucson: Arizona Historical Society Library.

Dubois, S. M., and A. W. Smith. 1980. The 1887 Earthquake in San Bernardino Valley, Sonora: Historic Accounts and Intensity Patterns in Arizona. Special Paper 3. Tucson: University of Arizona, Bureau of Geology and Mineral Technology.

Dunbar, R. B. 1983. Stable isotope record of upwelling and climate from the Santa Barbara Basin, California. Pages 217–246 *in* B. J. Thiede and E. Suess, editors, Coastal Upwelling: Its Sediment Record. New York: Plenum Press.

Durivage, J. E. 1937. Letters and journal of John E. Durivage. Pages 159–255 *in* R. P. Bieber, editor, Southern Trails to California in 1849. Glendale, CA: Arthur H. Clark.

Durrenberger, R. W., and R. S. Ingram. 1978. Major Storms and Floods in Arizona, 1862–1977. Climatological Publications Precipitation Series no. 4. Tempe, AZ: Office of the State Climatologist.

Dutton, C. E. 1882. Tertiary History of the Grand Canyon District. US Geological Survey Monograph 2. Washington, DC: Government Printing Office.

Dwight, J., Jr., and L. Griscom. 1927. A revision of the geographical races of the Blue Grosbeak (*Guiraca caerulea*). American Museum Novitates 257: 1–5.

Dzerdzeevskii, B. L. 1969. Climatic epochs in the twentieth century and some comments on the analysis of past climate. Pages 49–60 *in* H. E. Wright Jr., editor, Quaternary Geology and Climate. Washington, DC: National Academy of Sciences.

Eccleston, R. 1950. Overland to California on the Southwestern Trail, 1849. Edited by G. P. Hammond and E. H. Howes. Berkeley: University of California Press.

Eden, S., J. Gelt, and M. Lamberton. 2008. River restoration: Arizona's oft neglected waterways get overdue attention. Arroyo. Tucson: Water Resources Research Center. https://wrrc.arizona.edu/publications/arroyo-newsletter/arroyo-2008-river-restoration (accessed 3 October 2013).

Eidemiller, D. I. 1978. The Frequency of Tropical Cyclones in the Southwestern United States and Northwestern Mexico. Climatological Publications, Scientific Papers 1. Phoenix: State of Arizona, Office of the State Climatologist.

Eidenbach, P., and M. Wimberly. 1980. Archeological Reconnaissance in White Sands National Monument, New Mexico. Tularosa, NM: Human Systems Research.

Elliott, J. G., A. C. Gellis, and S. B. Aby. 1999. Evolution of arroyos: Incised channels of the southwestern United States. Pages 153–185 *in* S. E. Darby and A. Simon, editors, Incised River Channels: Processes, Forms, Engineering and Management. New York: John Wiley and Sons.

Elliott, W. P., and J. K. Angell. 1988. Evidence for changes in Southern Oscillation relationships during the past 100 years. Journal of Climate 1: 729–737.

Ellis, L. 1992. River bank erosion and different hydrologic regimes. Nottingham, UK: University of Nottingham, Department of Geography, unpublished MS thesis.

Ely, L. E. 1997. Response of extreme floods in the southwestern United States to climatic variations in the late Holocene. Geomorphology 19: 175–201.

Ely, L. E., and V. R. Baker, compilers. 1985. Geomorphic surfaces in the Tucson Basin, Arizona. Tucson: University of Arizona, Department of Geosciences, unpublished field guidebook.

Ely, L. E., Y. Enzel, V. R. Baker, and D. R. Cayan. 1993. A 5000-year record of extreme floods and climate change in the southwestern United States. Science 262: 410–412.

Emery, W. J., and K. Hamilton. 1985. Atmospheric forcing of interannual variability in the northeast Pacific Ocean: Connections with El Niño. Journal of Geophysical Research 90: 857–868.

Emlen, J. M. 1977. Ecology: An Evolutionary Approach. Reading, MA: Addison-Wesley Publishing.

Emory, W. H. 1857. Report on the United States and Mexico Boundary Survey. Washington, DC: 34th Congress, 1st Session, Senate Executive Document 108.

Engel-Wilson, R. W., and R. D. Ohmart. 1979. Floral and attendant faunal changes on the lower Rio Grande between Fort Quitman and Presidio, Texas. Pages 139–147 *in* Proceedings of the 1978 National Symposium on Strategies for Protection and Management of Floodplain Wetlands and Other Riparian Ecosystems. US Forest Service General Technical Report WO-12. Washington, DC: US Department of Agriculture, Forest Service.

Engstrom, W. N. 1996. The California storm of 1862. Quaternary Research 46: 141–148.

Euler, R. C., G. J. Gumerman, T. N. V. Karlstrom, J. S. Dean, and R. H. Hevly. 1979. The Colorado Plateau: Cultural dynamics and paleoenvironment. Science 250: 1089–1101.

Evans, D. W., and D. R. Pool. 2000. Aquifer Compaction and Ground-Water Levels in South-Central Arizona. US Geological Survey Water-Resources Investigations Report 99-4249.

Evans, G. W. B. 1945. Mexican Gold Trail (The Journal of GWB Evans). Edited by G. S. Dumke. San Marino, CA: Huntington Library.

Everitt, B. L. 1979. The cutting of Bull Creek arroyo. Utah Geology 6: 39–44.

Eychaner, J. H. 1984. Estimation of Magnitude and Frequency of Floods in Pima County, Arizona, with Comparisons of Alternative Methods. US Geological Survey Water-Resources Investigations Report 84-4142.

Fabre, J., and C. Cayla. 2009. Riparian restoration efforts in the Santa Cruz River Basin. Tucson: Water Resources Center, University of Arizona. https://wrrc.arizona.edu/sites/wrrc.arizona.edu/files/Riparian_Restoration_Efforts_0.pdf.

Federal Emergency Management Agency (FEMA). 1982. Flood Insurance Study, Pima County, AZ. Washington, DC: Federal Emergency Management Agency Federal Insurance and Mitigation Administration.

Federal Emergency Management Agency (FEMA). 1986. A Unified National Program for Floodplain Management. Washington, DC: Federal Emergency Management Agency Federal Insurance and Mitigation Administration.

Federal Emergency Management Agency (FEMA). 2002. National Flood Insurance Program: Program Description. Washington, DC: Federal Emergency Management Agency Federal Insurance and Mitigation Administration.

Fergusson, D. 1863. Report of Major D. Fergusson on the Country, Its Resources, and the Route Between Tucson and Lobos Bay. Washington, DC: 37th Congress, Special Session, Senate Executive Document 1.

Fischer, D. L. 2001. Early Southwest Ornithologists, 1528–1900. Tucson: University of Arizona Press.

Fish, S. K., and W. B. Gillespie. 1987. Prehistoric use of riparian resources at the San Xavier Bridge Site. Pages 71–80 *in* J. C. Ravesloot, editor, The Archeology of the San Xavier Bridge Site (AZ BB:13:14), Tucson Basin, Southern Arizona. Archeological Series 7. Tucson: Arizona State Museum.

Fleischner, T. L. 1994. Ecological costs of livestock grazing in western North America. Conservation Biology 8: 629–644.

Fontana, B. 1971. Calabazas of the Rio Rico. Smoke Signal 24: 66–68.

Fontana, B. L. 1996. Biography of a Desert Church: The Story of Mission San Xavier del Bac. Revised edition. Tucson, AZ: Tucson Corral of the Westerners.

Force, E. R. 1997. Geology and Mineral Resources of the Santa Catalina Mountains, Southeastern Arizona. Monographs in Mineral Resource Science no. 1. Tucson, AZ: Center for Mineral Resources.

Foreman, S. W. 1872. Official Map of the City of Tucson, Situated in Pima County, Arizona Territory: Occupying Sections 12 & 13, Township 14 S., Range 13E., Gila and Salt River Meridian. From survey made by Foreman, surveyor. Theo. F. White, draughtsman. Tucson: University of Arizona Special Collections, 1 sheet.

Frederick, P. C., and D. Siegel-Causey. 2000. Anhinga. *In* The Birds of North America, no. 522. http://bna.birds.cornell.edu/bna (accessed 7 March 2011).

Friedman, J. M., K. R. Vincent, and P. B. Shafroth. 2005. Dating floodplain sediments using tree-ring response to burial. Earth Surface Processes and Landforms 30: 1077–1091.

Froebel, J. 1859. Seven Year Travel in Central America, Northern Mexico and the Far West of the United States. London: Richard Bentley.

Fuller, P. E. 1913. Report upon the Santa Cruz Reservoir Project. Tucson: Arizona Historical Society Library, unpublished manuscript.

Galloway, D. L., D. R. Jones, and S. E. Ingebritsen. 1999. Land Subsidence in the United States. US Geological Survey Circular 1182.

Ganey, J. L., J. A. Johnson, R. P. Balda, and R. W. Skaggs. 1988. Mexican Spotted Owl. Pages 145–150 *in* R. L. Glinski, B. G. Pendleton, M. B. Moss, M. N. LeFranc Jr., B. A. Millsap, and S. W. Hoffman, editors, Proceedings of the Southwest Raptor Management Symposium and Workshop. Washington, DC: National Wildlife Federation.

García, J. A., S. S. Castelan, and A. E. Betancourt. Ca. 1976. Trayectorias de Ciclones Tropicales. Mexico City: UNAM, Centro de Ciencias de la Atmosfera.

Gardner, A. L. 1962. Notes on the behavior of the Vermilion Flycatcher (*Pyrocephalus rubinus flammeus*) in southern Arizona. Washington, DC: Patuxent Wildlife Research Center, National Museum of Natural History. Unpublished manuscript.

Gavin, T. A., and L. K. Sowls. 1975. Avian fauna of a San Pedro Valley mesquite forest. Journal of the Arizona Academy of Science 10: 33–41.

Gaylean, K. 1996. Infiltration of Wastewater Effluent in the Santa Cruz River Channel, Pima County, Arizona. USGS Water Resources Investigations Report 96-4021.

Gelt, J., J. Henderson, K. Seasholes, B. Tellman, and G. Woodard. 1999. Water in the Tucson Area: Seeking Sustainability. Tucson, AZ: Water Resources Research Institute. http://cals.arizona.edu/AZWATER/publications /sustainability/report_html/cover.html (accessed 24 July 2009).

George, W. G. 1958. The hyoid apparatus of several nine-primaried Oscinine birds. Tucson: University of Arizona, unpublished MS thesis.

George, W. G. 1961. The taxonomic significance of the tongue musculature of Passerine birds. Tucson: University of Arizona, unpublished PhD dissertation.

George, W. G. 1962. The classification of the Olive Warbler, *Peucedramus taeniatus*. American Museum Novitates 2103: 1–41.

George, W. G. 1963. Phylogenetic riddle. Natural History 72: 45–47.

Gergis, J., and A. Fowler. 2005. Classification of synchronous oceanic and atmospheric El Niño–Southern Oscillation (ENSO) events for palaeoclimate reconstruction. International Journal of Climatology 25: 1541–1565.

Gergis, J., and A. Fowler. 2009. A history of El Niño–Southern Oscillation (ENSO) events since A.D. 1525: Implications for future climate change. Climatic Change 92: 343–387.

Gerrard, A. J. 1984. Multiple working hypotheses and equifinality in geomorphology: Comments on the recent article by Haines-Young and Petch. Transactions of the Institute of British Geographers, new series, 9: 364–366.

Gilman, M. F. 1909a. Among the thrashers in Arizona. Condor 11: 49–54.

Gilman, M. F. 1909b. Some owls along the Gila River in Arizona. Condor 11: 145–150.

Gilman, M. F. 1911. Doves on the Pima Reservation. Condor 13: 51–56.

Gilman, M. F. 1914. Breeding of the Bronzed Cowbird in Arizona. Condor 17: 151–163.

Gilman, M. F. 1915. Woodpeckers of the Arizona lowlands. Condor 16: 255–259.

Glennon, R. 2002. Water Follies, Groundwater Pumping and the Fate of America's Fresh Waters. Washington, DC: Island Press.

Goodfellow, G. E. 1888. The Sonora earthquake. Science 11: 162–167.

Gormally, J. F. 2002. Changes in riparian vegetation following release of reclaimed effluent water into the Santa Cruz River: As a corollary, the effects of channelization on vegetation in the Santa Cruz River. Tucson: University of Arizona, Department of Landscape Architecture, unpublished MS thesis.

Gormally, J. 2011. Plants found along the effluent dominated stretch of the middle Santa Cruz River. Desert Plants 27: 19–31.

Gould, P. J. 1960. Territorial relationships between Cardinals and Pyrrhuloxias. Tucson: University of Arizona, unpublished MS thesis.

Gould, P. J. 1961. Territorial relationships between Cardinals and Pyrrhuloxias. Condor 63: 246–256.

Graf, W. L. 1979. Mining and channel response. Annals of the Association of American Geographers 69: 262–275.

Graf, W. L. 1983. The arroyo problem—paleohydrology and paleohydraulics in the short term. Pages 279–302 *in* K. J. Gregory, editor, Background to Paleohydrology. New York: John Wiley and Sons.

Graf, W. L. 1984. A probabilistic approach to the spatial assessment of river channel instability. Water Resources Research 20: 953–962.

Graf, W. L. 1987. Late Holocene sediment storage in canyons of the Colorado Plateau. Geological Society of America Bulletin 99: 261–271.

Green, C. R., and W. D. Sellers, 1964. Arizona Climate. Tucson: University of Arizona Press.

Gregory, H. E. 1917. Geology of the Navajo Country: A Reconnaissance of Part of Arizona and Utah. US Geological Survey Professional Paper 93.

Gregory, H. E. 1945. Population of southern Utah. Economic Geography 21: 29–57.

Gregory, H. E. 1950. Geology and Geography of the Zion Park Region, Utah and Arizona. US Geological Survey Professional Paper 220.

Gregory, H. E., and R. C. Moore. 1931. The Kaiparowits Region: A Geographic and Geologic Reconnaissance of Parts of Utah and Arizona. US Geological Survey Professional Paper 164.

Griffiths, P. G., C. S. Magirl, R. H. Webb, E. Pytlak, P. A. Troch, and S. W. Lyon. 2009. Spatial distribution and frequency of precipitation during an extreme event: The July 2006 mesoscale convective complexes and floods in southeastern Arizona. Water Resources Research 45, W07419.

Grove, J. M. 1988. The Little Ice Age. London: Methuen.

Guber, A. L. 1988. Channel changes of the San Xavier reach of the Santa Cruz River, Tucson, Arizona, 1971–1988. Tucson: University of Arizona, Department of Geography and Regional Development, unpublished MS thesis.

Gutierrez, R. J., A. B. Franklin, and W. S. LaHaye. 1995. Spotted Owl (*Strix occidentalis*). *In* A. Poole and F. Gill, editors, The Birds of North America, no. 179. Philadelphia, PA: Birds of North America, Inc.

Guttman, N. B., J. J. Lee, and J. R. Wallis. 1993. Heavy winter precipitation in southwest Arizona. EOS 74: 482, 485.

Hack, J. T. 1939. The late Quaternary history of several valleys of northern Arizona: A preliminary announcement.

Museum of Northern Arizona Notes 11: 67–73. Flagstaff.

Hack, J. T. 1942. The changing physical environment of the Hopi Indians of Arizona. Papers of the Peabody Museum, Harvard University 35. Boston, MA.

Hadley, D., and T. E. Sheridan. 1995. Land Use History of the San Rafael Valley, Arizona (1540–1960). US Forest Service General Technical Report RM-GTR-269. Fort Collins, CO: Rocky Mountain Forest and Range Experiment Station.

Hales, J. E., Jr. 1974. Southwestern United States summer monsoon source—Gulf of Mexico or Pacific Ocean. Weatherwise 24: 148–155.

Hall, E. R., and K. R. Kelson. 1959. The Mammals of North America. New York: Ronald Press.

Hall, S. A. 1977. Late Quaternary sedimentation and paleoecologic history of Chaco Canyon, New Mexico. Geological Society of America Bulletin 88: 1593–1618.

Halpenny, L. C., and P. C. Halpenny. 1988. Review of the hydrogeology of the Santa Cruz basin in the vicinity of the Santa Cruz–Pima County line. Tucson, AZ: Water Development Corporation, unpublished report.

Hamlin, P. G., editor. 1966. Documents of Arizona history: An Arizona letter of R. S. Ewell. Journal of Arizona History 7 (Spring): 23–26.

Haney, J. A., D. S. Turner, A. E. Springer, J. C. Stromberg, L. E. Stevens, P. A. Pearthree, and V. Supplee. 2008. Ecological implications of Verde River flows: A report by the Arizona Water Institute, the Nature Conservancy, and the Verde River Basin Partnership. http://www.vwa.org/documents/verde_river_ecological_flows_final.pdf (accessed 12 July 2011).

Hansen, E. M., and F. K. Shwarz. 1981. Meteorology of Important Rainstorms in the Colorado River and Great Basin Drainages. Hydrometeorological Report 50. Silver Spring, MD: National Oceanic and Atmospheric Administration (NOAA).

Hansen, E. M., F. K. Shwarz, and J. Reidel. 1977. Probable Maximum Precipitation Estimates, Colorado River and Great Basin Drainages. Hydrometeorological Report 39. Silver Spring, MD: National Oceanic and Atmospheric Administration (NOAA).

Hansen, K. D., D. L. Richards, and M. E. Krebs. 2011. Performance of flood-tested soil-cement protected levees. In 21st Century Dam Design—Advances and Adaptations. 31st Annual USSD Conference, San Diego, California, April 11–15, 2011. http://pacewater.com/wp-content/uploads/2011/11/Performance-of-Flood-Tested-Soil-Cement-Protected-Levees-10-12-2010revised.pdf (accessed 30 October 2013).

Hanson, R. B. 2001. The San Pedro River: A Discovery Guide. Tucson: University of Arizona Press.

Hanson, R. T., M. D. Dettinger, and M. W. Newhouse. 2006. Relations between climatic variability and hydrologic time series from four alluvial basins across the south-

western United States. Hydrogeology Journal 14: 1122–1146.

Harden, T., M. G. Macklin, and V. R. Baker. 2010. Holocene flood histories in south-western USA. Earth Surface Processes and Landforms 35: 707–716.

Hardy, P. C., D. J. Griffin, A. J. Kuenzi, and M. L. Morrison. 2004. Occurrence and habitat use of passage neotropical migrants in the Sonoran Desert. Western North American Naturalist 64: 59–71.

Harris, B. B. 1960. The Gila Trail, the Texas Argonauts and the California Gold Rush. Edited and annotated by R. H. Dillon. Norman: University of Oklahoma Press. Originally published in 1849.

Hastings, J. R. 1959. Vegetation change and arroyo cutting in southeastern Arizona. Journal of the Arizona Academy of Science 1: 60–67.

Hastings, J. R., and R. M. Turner. 1965. The Changing Mile: An Ecological Study of Vegetation Change with Time in the Lower Mile of an Arid and Semiarid Region. Tucson: University of Arizona Press.

Haury, E. W. 1976. The Hohokam: Desert Farmers and Craftsmen; Excavations at Snaketown, 1964–1965. Tucson: University of Arizona Press.

Hayes, B. 1850. Diary of Judge Benjamin Hayes. Unpublished manuscript in Bancroft Library, University of California, Berkeley, and Arizona Historical Society, Tucson. .

Haynes, C. V. 1968. Geochronology of late Quaternary alluvium. Pages 591–631 in R. B. Morrison and H. E. Wright Jr., editors, Means of Correlation of Quaternary Successions. Salt Lake City: University of Utah Press.

Haynes, C. V., Jr. 2007. Quaternary geology of the Murray Springs Clovis site. Pages 16–56 in C. V. Haynes Jr. and B. B. Huckell, editors, Murray Springs: A Clovis Site with Multiple Activity Areas in the San Pedro Valley, Arizona. Anthropological Papers of the University of Arizona no. 71. Tucson: University of Arizona Press.

Haynes, C. V., Jr., and B. B. Huckell. 1986. Sedimentary Successions of the Prehistoric Santa Cruz River. Open-File Report 86-15. Tucson: Arizona Bureau of Geology and Mineral Technology.

Haynes, C. V., Jr., and B. B. Huckell, editors. 2007. Murray Springs: A Clovis Site with Multiple Activity Areas in the San Pedro Valley, Arizona. Anthropological Papers of the University of Arizona no. 71. Tucson: University of Arizona Press.

Hays, M. E. 1984. Analysis of historic channel change as a method for evaluating flood hazard in the semi-arid Southwest. Tucson: University of Arizona, unpublished MS thesis.

Helmick, W. R. 1986. The Santa Cruz River terraces near Tubac, Santa Cruz County, Arizona. Tucson: University of Arizona, Department of Geosciences, unpublished MS thesis.

Hendrickson, D. A., and W. L. Minckley. 1985. Ciénegas: Vanishing communities of the American Southwest. Desert Plants 6: 131–175.

Henshaw, H. W. 1875a. Annotated list of the birds of Arizona. Pages 153–166 (appendix I.2) *in* G. M. Wheeler, editor, Geographical and Geological Explorations and Surveys West of the 100th Meridian. Washington, DC: US Government Printing Office.

Henshaw, H. W. 1875b. Report upon the ornithological collections made in portions of Nevada, Utah, California, Colorado, New Mexico, and Arizona during the years 1871, 1872, 1873, and 1874. Pages 133–507 *in* G. M. Wheeler, editor, Geographical and Geological Explorations and Surveys West of the 100th Meridian, vol. 5. Washington, DC: US Government Printing Office.

Hereford, R. 1984. Climate and ephemeral-stream processes: Twentieth-century geomorphology and alluvial stratigraphy of the Little Colorado River, Arizona. Geological Society of America Bulletin 95: 654–668.

Hereford, R. 1986. Modern alluvial history of the Paria River drainage basin. Quaternary Research 25: 293–311.

Hereford, R. 1987. The short term: Fluvial processes since 1940. Pages 276–288 *in* W. L. Graf, editor, Geomorphic Systems of North America. Geological Society of America Centennial Special Paper 2. Boulder, CO.

Hereford, R. 1993. Entrenchment and Widening of the Upper San Pedro River, Arizona. Geological Society of America Special Paper 282. Boulder, CO.

Hereford, R. 2002. Valley-fill alluviation (ca. AD 1400–1800) during the Little Ice Age, Paria River basin and southern Colorado Plateau, United States. Geological Society of America Bulletin 114: 1550–1563.

Hereford, R. 2004. Map showing Quaternary geology and geomorphology of the Lonely Dell Reach of the Paria River, Lees Ferry, Arizona, with accompanying pamphlet by R. H. Webb and R. Hereford, titled Comparative Landscape Photographs of the Lonely Dell Area and the Mouth of the Paria River. US Geological Survey Geologic Investigations Series Map I-2771 (scale 1:5,000).

Hereford, R., and R. H. Webb. 1992. Historic variation of warm-season rainfall, southern Colorado Plateau, southwestern USA. Climatic Change 22: 235–256.

Hereford, R., G. C. Jacoby, and V. A. S. McCord. 1996. Late Holocene alluvial geomorphology of the Virgin River in the Zion National Park area, southwest Utah. Geological Society of America Special Paper 310. Boulder, CO.

Hereford, R., R. H. Webb, and S. Graham. 2002. Precipitation History of the Colorado Plateau Region, 1900–2000. US Geological Survey Fact Sheet 119-02.

Hereford, R., R. H. Webb, and C. I. Longpré. 2006. Precipitation history of the Mojave Desert region, 1893–2001. Journal of Arid Environments 67: 13–34.

Hinderlider, M. C. 1913. Irrigation of Santa Cruz valley— parts 1 and 2. Engineering Record 68: 200–201, 242–244.

Hirschboeck, K. K. 1985. Hydroclimatology of flow events in the Gila River basin, central and southern Arizona. Tucson: University of Arizona, unpublished PhD dissertation.

Hirschboeck, K. K. 1987. Catastrophic flooding and atmospheric circulation anomalies. Pages 23–56 *in* L. Mayer and D. Nash, editors, Catastrophic Flooding. Boston, MA: Allen and Unwin.

Hirschboeck, K. K. 2009. Flood flows of the San Pedro River. Pages 300–312 *in* J. C. Stromberg and B. Tellman, editors, Ecology and Conservation of the San Pedro River. Tucson: University of Arizona Press.

Hjalmarson, H. W. 1987. Discussion—large basin deterministic hydrology—a case study. American Society of Civil Engineers, Journal of the Hydraulics Division 113: 1461–1463.

Hoffmann, J. P., D. R. Pool, A. D. Konieczki, and M. C. Carpenter. 1998. Causes of sinks near Tucson, Arizona, USA. Hydrogeology Journal 6: 349–364.

Hoffmeister, D. R. 1986. Mammals of Arizona. Tucson: University of Arizona Press.

Horel, J. D., and J. M. Wallace. 1981. Planetary scale atmospheric phenomena associations with the Southern Oscillation. Monthly Weather Review 109: 813–829.

Horton, J. S., and C. J. Campbell. 1974. Management of Phreatophyte and Riparian Vegetation for Maximum Multiple Use Values. US Forest Service Research Paper RM-117. Fort Collins, CO: Rocky Mountain Forest and Range Experiment Station.

Hosmer, J., and the Ninth and Tenth Grade Classes of Green Fields Country Day School and University High School, Tucson. 1991. From the Santa Cruz to the Gila in 1850: An excerpt from the overland journal of William P. Huff. Journal of Arizona History 32: 41–110.

House, P. K., and K. K. Hirschboeck. 1997. Hydroclimatological and paleohydrological context of extreme winter flooding in Arizona, 1993. Pages 1–24 *in* R. A. Larson and J. E. Slosson, editors, Storm-Induced Geologic Hazards: Case Histories from the 1992–1993 Winter in Southern California and Arizona. Reviews in Engineering Geology 11. Boulder, CO: Geological Society of America.

Houser, B. B., L. Peters, R. P. Esser, and M. E. Gettings. 2004. Stratigraphy and Tectonic History of the Tucson Basin, Pima County, Arizona, Based on the Exxon State (32)-1 Well. US Geological Survey Scientific Investigations Report 2004-5076.

Hovingh, P. 2004. Intermountain freshwater mollusks, USA (*Margaritifera, Anodonta, Gonidea, Valvata, Ferrissia*): Geography, conservation, and fish management implications. Monographs of the Western North American Naturalist 2: 109–135.

Howard, O. W. 1906. The English Sparrow in the Southwest. Condor 8: 67–68.

Howell, A. B. 1916. Some results of a winter's observations in Arizona. Condor 18: 209–214.

Howell, A. B. 1918a. Arizona: Continental, Pima Co. Field notes, July 8 to August 15, 1918. Washington, DC: Smithsonian Institution Archives.

Howell, A. B. 1918b. Arizona: Continental, Pima Co. Field notes, August 15 to 19, 1918. Washington, DC: Smithsonian Institution Archives.

Hubbard, J. P. 1977. The Biological and Taxonomic Status of the Mexican Duck. Santa Fe: New Mexico Department of Game and Fish.

Hubbard, J. P. 1978. Revised Check-List of the Birds of New Mexico. New Mexico Ornithological Society Publication 6: 1–110.

Huckleberry, G. 1995. Archaeological implications of late-Holocene channel changes on the middle Gila River, Arizona. Geoarchaeology 10: 159–182.

Huckleberry, G. 1999. Assessing Hohokam canal stability through stratigraphy. Journal of Field Archaeology 26: 1–18.

Huels, T. R., D. E. Brown, and R. R. Johnson. 2013. Field notes of Herbert Brown: Arizona's pioneer ornithologist. Journal of the Arizona-Nevada Academy of Science 44: 79–142.

Hughes, S. 1885. Samuel Hughes reminiscences, 1838–1885. Tucson: Arizona Historical Society Library, unpublished manuscript in Donald Page Collection.

Hume, E. E. 1978. Ornithologists of the United States Army Medical Corps. New York: Arno Press.

Hunter, W. C., R. D. Ohmart, and B. W. Anderson. 1987. Status of riparian-obligate birds in southwestern riverine systems. Western Birds 18: 10–18.

Hunter, W. C., R. D. Ohmart, and B. W. Anderson. 1988. Use of exotic saltcedar (*Tamarix chinensis*) by birds in arid riparian systems. Condor 90: 113–123.

Huntington, E. 1914. The Climatic Factor as Illustrated in Arid America. Carnegie Institution of Washington Publication 192. Washington, DC.

Hyndman, D. W., R. A. Roberts, and R. H. Webb. 1991. Modeling sediment transport in an ephemeral river. Pages 8-101–8-108 *in* S. S. Fan and Y. H. Kuo, editors, Proceedings of the Fifth Federal Interagency Sedimentation Conference, vol. 2. Washington, DC: Federal Energy Regulatory Commission.

Jackson, E. 1951. Tumacacori's Yesterdays. Popular Series no. 6. Santa Fe, NM: Southwestern Monuments Association.

James, G. W. 1917. Arizona, the Wonderland. Boston, MA: Page Company.

James, J. D., and J. E. Thompson. 2001. Black-bellied Whistling-Duck. *In* A. Poole, editor, The Birds of North America Online, no. 578. Ithaca, NY: Cornell Lab of Ornithology. http://bna.birds.cornell.edu/bna (accessed 7 March 2011).

Johnson, R. R. 1971. Tree removal along southwestern rivers and effects on associated organisms. American Philosophical Society Yearbook 1970: 321–322.

Johnson, R. R. 1979. The lower Colorado River: A western system. Pages 41–55 *in* Proceedings of the 1978 National Symposium on Strategies for Protection and Management of Floodplain Wetlands and Other Riparian Ecosystems. US Forest Service General Technical Report WO-12. Washington, DC: US Department of Agriculture, Forest Service.

Johnson, R. R. 1991. Recent vegetational changes along the Colorado River in the Grand Canyon. Pages 178–206 *in* Colorado River Ecology and Dam Management: Proceedings of a Symposium, May 24–25, 1990, Santa Fe, New Mexico. Washington, DC: National Academy Press.

Johnson, R. R., and J. C. Barlow. 1971. Notes on the nesting of the Black-bellied Tree Duck near Phoenix, Arizona. Southwestern Naturalist 15: 394–395.

Johnson, R. R., and S. W. Carothers. 1982. Riparian habitat and recreation: Interrelationships and impacts in the Southwest and Rocky Mountain Region. Eisenhower Consortium Bulletin 12: 1–31.

Johnson, R. R., and S. W. Carothers. 1987. External threats: The dilemma of resource management on the Colorado River in Grand Canyon National Park, USA. Environmental Management 11: 99–107.

Johnson, R. R., and L. T. Haight. 1996. Canyon Towhee (*Pipilo fuscus*). *In* A. Poole and F. Gill, editors, The Birds of North America, no. 264. Philadelphia, PA: Academy of Natural Sciences; Washington, DC: American Ornithologists' Union.

Johnson, R. R., and J. M. Simpson. 1971. Important birds from Blue Point Cottonwoods, Maricopa County, Arizona. Condor 73: 379–380.

Johnson, R. R., and J. M. Simpson. 1988. Desertification of wet riparian ecosystems in arid regions of the North American Southwest. Pages 1383–1394 *in* E. E. Whitehead et al., editors, Arid Lands: Today and Tomorrow. Boulder, CO: Westview Press.

Johnson, R. R., and J. M. Simpson. No date. Manuscript a (MS a). Birds of the Salt River Valley, Maricopa County, Arizona. Tucson, AZ: Johnson-Simpson Archives, unpublished manuscript.

Johnson, R. R., and J. M. Simpson. No date. Manuscript b (MS b). Birds of Blue Point Cottonwoods, Maricopa County, Arizona. Tucson, AZ: Johnson-Simpson Archives, unpublished manuscript.

Johnson, R. R., L. T. Haight, and J. M. Simpson. 1977. Endangered species versus endangered habitats: A concept. Pages 68–79 *in* R. R. Johnson and D. A. Jones, technical coordinators, Importance, Preservation, and Management of Riparian Habitats. US Forest Service General Technical Report RM-43. Fort Collins, CO: Rocky Mountain Forest and Range Experiment Station.

Johnson, R. R., S. W. Carothers, and J. M. Simpson. 1984. A riparian classification system. Pages 375–382 *in* R. E. Warner and K. M. Hendrix, editors, California Riparian Systems. Berkeley: University of California Press.

Johnson, R. R., C. D. Ziebell, D. R. Patton, P. F. Ffolliott, and R. H. Hamre, technical coordinators. 1985. Riparian Ecosystems and Their Management: Reconciling Conflicting Uses. First North American Riparian Conference, April 16–18, 1985, Tucson, AZ. US Forest Service General Technical Report RM-120. Fort Collins, CO: Rocky Mountain Forest and Range Experiment Station.

Johnson, R. R., L. T. Haight, and J. M. Simpson. 1987. Endangered habitats versus endangered species: A management challenge. Western Birds 18: 89–96.

Johnson, R. R., J. M. Simpson, and L. T. Haight. 1997a. Allan R. Phillips and the development of ornithology as a science in Arizona. Pages 61–68 *in* R. W. Dickerman, editor, The Era of Allan R. Phillips: A Festschrift. Albuquerque, NM: Horizon Communications.

Johnson, R. R., H. K. Yard, and B. T. Brown. 1997b. Lucy's Warbler (*Vermivora luciae*). *In* A. Poole and F. Gill, editors, The Birds of North America, no. 318. Philadelphia, PA: Academy of Natural Sciences; Washington, DC: American Ornithologists' Union.

Johnson, R. R., J.-L. E. Cartron, L. T. Haight, R. B. Duncan, and K. J. Kingsley. 2000a. A historical perspective on the population decline of the Cactus Ferruginous Pygmy-Owl in Arizona. Pages 17–26 *in* J. L. E. Cartron and D. M. Finch, editors, Ecology and Conservation of the Cactus Ferruginous Pygmy-Owl in Arizona. US Forest Service General Technical Report RMRS-GTR-43. Fort Collins, CO: Rocky Mountain Forest and Range Experiment Station.

Johnson, R. R., R. L. Glinski, and S. W. Matteson. 2000b. Zone-tailed Hawk (*Buteo albonotatus*). *In* A. Poole and F. Gill, editors, The Birds of North America, no. 529. Philadelphia, PA: Birds of North America, Inc.

Johnson, R. R., J.-L. E. Cartron, L. T. Haight, R. B. Duncan, and K. J. Kingsley. 2003. The Cactus Ferruginous Pygmy-Owl in Arizona, 1872–1971. Southwestern Naturalist 48: 389–401.

Jones, L. C., and G. R. Lovich, editors. 2009. Lizards of the American Southwest. Tucson, AZ: Rio Nuevo Publishers.

Jones, P. 1973. Interview concerning the Tucson Farms Company. Tucson: University of Arizona Special Collections Library.

Jones, P. D., and T. M. L. Wigley. 1980. Northern hemisphere temperatures, 1881–1979. Climate Monitor 9: 43–47.

Jones, S. L., and J. S. Dieni. 1995. Canyon Wren (*Catherpes mexicanus*). *In* A. Poole and F. Gill, editors, The Birds of North America, no. 197. Philadelphia, PA: Academy of Natural Sciences; Washington, DC: American Ornithologists' Union.

Judson, S. 1952. Arroyos. Scientific American 187: 71–76.

Juillet-Le Clerc, A., and H. Schrader. 1987. Variations of upwelling intensity recorded in varved sediment from the Gulf of California during the past 3000 years. Nature 329: 146–149.

Kalmbach, E. R. 1958. Obituaries: Lee Weight Arnold. Journal of Wildlife Management 22: 105–107.

Kalnicky, R. A. 1974. Climatic change since 1950. Annals of the Association of American Geographers 64: 100–112.

Karlstrom, T. N. V. 1988. Alluvial chronology and hydrologic change of Black Mesa and nearby regions. Pages 45–91 *in* G. J. Gumerman, editor, The Anasazi in a Changing Environment. Cambridge: Cambridge University Press.

Kendeigh, S. C. 1944. Measurements of bird populations. Ecological Monographs 14: 67–106.

Kerlinger, P. 1993. Birding economics and birder demographics studies conservation tools. Pages 32–38 *in* D. M. Finch and D. W. Stangel, editors, Status and Management of Neotropical Birds: September 21–25, Estes Park, CO. US Forest Service General Technical Report RM-78. Fort Collins, CO: Rocky Mountain Forest and Range Experiment Station.

Kimball, H. H. 1923. Bird notes from Arizona and California. Condor 25: 109.

Kimball, S., A. L. Angert, T. E. Huxman, and D. L. Venable. 2010. Contemporary climate change in the Sonoran Desert favors cold-adapted species. Global Change Biology 16: 1555–1565, doi: 10.1111/j.1365-2486.2009.02106.x.

Klauber, L. M. 1945. The geckos of the genus *Coleonyx* with descriptions of new subspecies. Transactions of the San Diego Society of Natural History 10 (11): 133–216.

Knapp, F. C. 1937. Report on Santa Cruz watershed. Unpublished manuscript in possession of authors.

Knox, J. C. 1978. Arroyos and environmental change in the American South-west (book review). Annals of the Association of American Geographers 68: 137–139.

Knox, J. C. 1983. Response of river systems to Holocene climates. Pages 26–41 *in* H. E. J. Wright, editor, Late Quaternary Environments of the United States. Minneapolis: University of Minnesota Press.

Koch, P. L., and A. D. Barnosky. 2006. Late Quaternary extinctions: State of the debate. Annual Review of Ecology Evolution and Systematics 37: 215–250.

Konieczki, A. D., and J. A. Heilman. 2004. Water-Use Trends in the Desert Southwest—1950–2000. US Geological Survey Scientific Investigations Report 2004-5148.

Kresan, P. L. 1988. The Tucson, Arizona, flood of October 1983—Implications for land management along alluvial river channels. Pages 465–489 *in* V. R. Baker, R. C. Kochel, and P. D. Patton, editors, Flood Geomorphology. New York: John Wiley.

Krieger, D. J. 2001. Economic Value of Forest Ecosystem Services: A Review. The Wilderness Society. http://wilderness.org/files/Economic-Value-of-Forest-Ecosystem-Services.pdf (accessed 16 July 2011).

Kunzmann, M. R., R. R. Johnson, and P. S. Bennett. 1991. Birds of Chiricahua Mountains: An Annotated Checklist. Tucson: Cooperative National Park Resources Studies Unit, School of Renewable Natural Resources, University of Arizona.

Kupel, D. 2003. Fuel for Growth: Water and Arizona's Environment. Tucson: University of Arizona Press.

Kupel, D. E. 2003. Fuel for Growth: Water and Arizona's Urban Environment. Tucson: University of Arizona Press.

Lamberton, K. 2011. Dry River. Tucson: University of Arizona Press.

Lance, J. F. 1963. Alluvial stratigraphy in Lake and Moqui Canyons. University of Utah Anthropological Papers 63: 349–376.

Lehr, J. H. 1978. A Catalogue of the Flora of Arizona. Phoenix, AZ: Desert Botanical Garden.

León, C. S. No date. Personal reminiscences. Tucson: Arizona Historical Society Library, unpublished manuscript.

Leopold, L. B. 1951. Rainfall frequency: An aspect of climatic variation. Transactions of the American Geophysical Union 32: 347–357.

Leopold, L. B. 1976. Reversion of erosion cycle and climatic change. Quaternary Research 6: 557–562.

Leopold, L. B., and J. P. Miller. 1954. A Postglacial Chronology for Some Alluvial Valleys in Wyoming. US Geological Survey Water-Supply Paper 1261.

Leopold, L. B., W. M. Emmett, and R. M. Myrick. 1966. Channel and hillslope processes in a semiarid area, New Mexico. US Geological Survey Professional Paper 352G: 193–253.

Lesica, P., and T. H. DeLuca. 2004. Is tamarisk allelopathic? Plant and Soil 267: 357–365.

Lewis, D. D. 1963. Desert Floods: A Report on Southern Arizona Floods of September, 1962. Water Resources Report 13. Tucson: Arizona State Land Department.

Lewis, P. A., C. J. DeLoach, A. E. Knutson, J. L. Tracy, and T. O. Robbins. 2003. Biology of *Diorhabda elongata deserticola* (Coleoptera: Chrysomelidae), an Asian leaf beetle for biological control of saltcedars (Tamarix spp.) in the United States. Biological Control 27: 101–116.

Livingston, M. F., and S. D. Schemnitz. 1996. Summer birds/vegetation associations in tamarisk and native habitat along the Pecos River, southeastern New Mexico. Pages 171–180 *in* D. W. Shaw and D. M. Finch, technical coordinators, Bringing Interests and Concerns Together. US Forest Service General Technical Report RM-GTR-272. Fort Collins, CO: Rocky Mountain Forest and Range Experiment Station.

Logan, M. F. 2002. The Lessening Stream: An Environmental History of the Santa Cruz River. Tucson: University of Arizona Press.

Love, D. 1983. Quaternary facies in Chaco Canyon and their implication for geomorphic-sedimentologic models. Pages 79–90 *in* S. G. Wells, D. W. Love, and T. W. Gardner, editors, Chaco Canyon Country. American Geomorphological Field Group, 1983 Field Trip Guidebook.

Lowe, C. H. 1985. Amphibians and reptiles in Southwest riparian ecosystems. Pages 339–341 *in* R. R. Johnson and J. F. McCormick, editors, Strategies for Protection and Management of Floodplain Wetlands and Other Riparian Ecosystems. US Forest Service General Technical Report WO-12. Washington, DC: US Department of Agriculture, Forest Service.

Lowther, P. E., K. D. Groschupf, and S. M. Russell. 1999. Rufous-winged Sparrow (*Aimophila carpalis*). *In* A. Poole and F. Gill, editors, The Birds of North America, no. 422. Philadelphia, PA: Birds of North America, Inc.

Lowther, P. E., D. E. Kroodsma, and G. H. Farley. 2000. Rock Wren (*Salpinctes obsoletus*). *In* A. Poole and F. Gill, editors, The Birds of North America, no. 486. Philadelphia, PA: Birds of North America, Inc.

Lucero, C. 1928. Personal reminiscences. Tucson: Arizona Historical Society Library, unpublished manuscript.

Lumholtz, C. 1912. New Trails in Mexico: An Account of One Year's Exploration in North-western Sonora, Mexico, and South-western Arizona, 1909–1910. New York: Charles Scribner's Sons.

Lusby, G. C., V. H. Reid, and O. D. Knipe. 1971. Effects of Grazing on the Hydrology and Biology of the Badger Wash Basin in Western Colorado, 1953–66. US Geological Survey Water-Supply Paper 1532-D.

Mabry, J. B. 2002. Changing knowledge and ideas about the first farmers in southeastern Arizona. Pages 41–83 *in* B. J. Vierra, editor, Current Research on the Late Archaic Along the US-Mexican Borderlands. Austin: University of Texas Press.

Mabry, J. B. 2006a. Radiocarbon dating of the early occupations. Pages 19.1–19.5 *in* J. H. Thiel and J. B. Mabry, editors, Rio Nuevo Archaeology, 2000–2003: Investigations at the San Agustín Mission and Mission Gardens, Tucson Presidio, Tucson Pressed Brick Company, and Clearwater Site. Technical Report 2004-11. Tucson, AZ: Desert Archaeology, Inc. http://www.cdarc.org/pages/library/rio_nuevo/.

Mabry, J. B. 2006b. Geomorphology and stratigraphy. Pages 20.1–20.24 *in* J. H. Thiel and J. B. Mabry, editors, Rio Nuevo Archaeology, 2000–2003: Investigations at the San Agustín Mission and Mission Gardens, Tucson Presidio, Tucson Pressed Brick Company, and Clearwater Site. Technical Report 2004-11. Tucson, AZ: Desert Archaeology, Inc. http://www.cdarc.org/pages/library/rio_nuevo/.

MacArthur, R. H., and J. W. MacArthur. 1961. On bird species diversity. Ecology 42: 594–598.

MacArthur, R. H., H. Recher, and M. Cody. 1966. On the relation between habitat selection and species diversity. American Naturalist 100: 319–332.

Macklin, M. G., E. Johnstone, and J. Lewin. 2005. Pervasive and long-term forcing of Holocene river instability and flooding in Great Britain by centennial-scale climate change. The Holocene 15: 937–943.

Maddox, R. A., F. Canova, and L. R. Hoxit. 1980. Meteorological characteristics of flash flood events over the western United States. Monthly Weather Review 108: 1866–1877.

Maddox, R. A., D. M. McCollum, and K. W. Howard. 1995. Large-scale patterns associated with severe summertime thunderstorms over central Arizona. Weather Forecasting 10: 763–778.

Magirl, C. S., R. H. Webb, M. Schaffner, S. W. Lyon, P. G. Griffiths, C. Shoemaker, C. L. Unkrich, S. Yatheendradas, P. A. Troch, E. Pytlak, D. C. Goodrich, S. L. E. Desilets, A. Youberg, and P. A. Pearthree. 2007. Impact of recent extreme Arizona storms. Eos Transactions, American Geophysical Union 88 (17): 191, 193.

Malvick, A. J. 1980. A Magnitude-Frequency-Area Relation for Floods in Arizona. Research Report 2. Tucson: University of Arizona, Engineering Experimental Station.

Manje, J. M. 1954. Unknown Arizona and Sonora, 1693–1721, part 2. Translated by H. J. Karns. Originally titled Luz de Tierra Incógnita. Tucson: Arizona Silhouettes.

Mantua, N. J., S. R. Hare, Y. Zhang, J. M. Wallace, and R. C. Francis. 1997. A Pacific decadal climate oscillation with impacts on salmon. Bulletin of the American Meteorological Society 78: 1069–1079.

Margolin, A. S., compiler, et al. 1955. Fifty-fifth Christmas bird count at Salt River, Arizona, December 30, [1954]. Audubon Field Notes 9: 215.

Marshall, J. T. 1957. Birds of pine-oak woodland in southern Arizona and adjacent Mexico. Pacific Coast Avifauna 32: 1–125.

Marshall, J. T. 1960. Interrelations of Abert and brown towhees. Condor 62: 49–64.

Marshall, J. T. 1964. Voice in communication and relationships among brown towhees. Condor 66: 345–356.

Marshall, J. T. 1967. Parallel variation in North and Middle American screech-owls. Pages 1–72 *in* Western Foundation of Vertebrate Zoology Monograph 1. Los Angeles, CA.

Marshall, J. T., and R. R. Johnson. 1968. *Pipilo fuscus mesoleucus*, Canyon Brown Towhee. Pages 622–630 *in* O. L. Austin Jr., editor, Life Histories of North American Cardinals, Grosbeaks, Buntings, Towhees, Finches, Sparrows, and Allies. US National Museum Bulletin 237 (part 2).

Martin, P. S. 1963. The Last 10,000 Years: A Fossil Pollen Record of the American Southwest. Tucson: University of Arizona Press.

Martin, W. E., and R. L. Gum. 1978. Economic value of hunting, fishing, and general rural outdoor recreation. Wildlife Society Bulletin 6: 3–7.

Martin, W. E., R. L. Gum, and A. H. Smith. 1974. Demand for Hunting, Fishing and General Rural Outdoor Recreation in Arizona. Technical Bulletin 211. Tucson: University of Arizona, Agriculture Experiment Station.

Mattison, R. H. 1946. Early Spanish and Mexican settlements in Arizona. New Mexico Historical Review 21: 273–327.

Mauz, K. 2002. Plants of the Santa Cruz valley at Tucson. Desert Plants 18: 3–36.

McCarty, K. 1976. Desert Documentary: The Spanish Years, 1767–1821. Historical Monograph 4. Tucson: Arizona Historical Society.

McCollum, D. M., R. A. Maddox, and K. W. Howard. 1995. Case study of a severe mesoscale convective system in central Arizona. Weather Forecasting 10: 643–665.

McDonald, J. E. 1956. Variability of precipitation in an arid region: A survey of characteristics for Arizona. University of Arizona, Institute of Atmospheric Physics, Technical Report 1: 1–88.

McEwen, G. F. 1925. Ocean temperatures and seasonal rainfall in southern California. Monthly Weather Review 53: 483–494.

McFadden, L. D. 1978. Soils of the Cañada del Oro valley, southern Arizona. Tucson: University of Arizona, unpublished MS thesis.

McGrath, L. J., and C. van Riper III. 2005. Influences of Riparian Tree Phenology on Lower Colorado River Spring Migrating Birds: Implications of Flower Cueing. US Geological Survey Open-File Report 2005-1140.

McGuire, R. H. 1979. Rancho Punta de Agua: Excavations at a Historic Ranch near Tucson, Arizona. Contribution to Highway Salvage Archaeology in Arizona no. 57. Tucson: Arizona State Museum.

McHugh, C. P. 1995. Preparing public safety organizations for disaster response: A study of Tucson, Arizona's response to flooding. Disaster Prevention and Management 4: 25–36.

Mearns, E. A. 1886. Some birds of Arizona: Zone-tailed Hawk, *Buteo abbreviatus* Caban.; Mexican Black Hawk, *Urubitinga anthracina* (Licht.) (Lair.). Auk 3: 60–73.

Mearns, E. A. 1907. Mammals of the Mexican Boundary of the United States. US National Museum Bulletin 56 (part 1): 1–530.

Mehlman, D. W. 1997. Change in avian abundance across the geographic range in response to environmental change. Ecological Applications 7: 614–624.

Meinzer, O. E. 1923. Outline of Ground-Water Hydrology, with Definitions. US Geological Survey Water-Supply Paper 494.

Melton, M. A. 1965. The geomorphic and paleoclimatic significance of alluvial deposits in southern Arizona. Journal of Geology 73: 1–38.

Miller, A. M., D. R. Cayan, T. P. Barnett, N. E. Graham, and J. M. Oberhuber. 1994. The 1976–77 climate shift of the Pacific Ocean. Oceanography 7 (1): 21–26. http://dx.doi.org/10.5670/oceanog.1994.11.

Miller, J. P., and F. Wendorf. 1958. Alluvial chronology of the Tesuque Valley, New Mexico. Journal of Geology 66: 177–194.

Millsap, B. A., and R. R. Johnson. 1988. Ferruginous Pygmy-Owl. Pages 137–139 *in* R. L. Glinski, B. G. Pendleton, M. B. Moss, M. N. LeFranc Jr., B. A. Millsap, and S. W. Hoffman, editors, Southwest Raptor Management Symposium and Workshop Proceedings. Washington, DC: National Wildlife Federation.

Minckley, W. L. 1973. Fishes of Arizona. Phoenix: Arizona Game and Fish Department.

Minckley, W. L. 1999. Frederic Morton Chamberlain's 1904 survey of Arizona fishes with annotations. Journal of the Southwest 41: 177–237.

Minckley, W. L., J. N. Rinne, and J. E. Johnson. 1977. Status of the Gila Topminnow and Its Co-occurrence with Mosquitofish. USDA Forest Service Research Paper RM-198. Fort Collins, CO: Rocky Mountain Forest and Range Experiment Station.

Minckley, W. L., R. R. Miller, S. M. Norris, and S. A. Schaefer. 2002. Three new pupfish species, *Cyprinodon* (Teleostei, Cyprinodontidae), from Chihuahua, México, and Arizona, USA. Copeia (2002): 687–705.

Minobe, S. 1997. A 50–70 year climatic oscillation over the North Pacific and North America. Geophysical Research Letters 24: 683–686.

Minobe, S. 1999. Resonance in bidecadal and pentadecadal climate oscillations over the North Pacific: Role in climatic regime shifts. Geophysical Research Letters 26: 855–858.

Monson, G. 1942. Notes on some birds of southeastern Arizona. Condor 44: 222–225.

Monson, G. 1979. Lucy's Warbler: *Vermivora luciae*. Pages 66–67 *in* L. Griscom and A. Sprunt Jr., editors, The Warblers of America. 2nd edition. Garden City, NY: Doubleday.

Monson, G. 1998. Ferruginous Pygmy-Owl. Pages 159–161 *in* R. L. Glinski, editor, The Raptors of Arizona. Tucson: University of Arizona Press.

Monson, G., and A. R. Phillips. 1981. Annotated Checklist of the Birds of Arizona. 2nd edition. Tucson: University of Arizona Press.

Mooney, J. 1928. The Aboriginal Population of America North of Mexico. Edited by J. Swanton. Smithsonian Miscellaneous Collections 80 (7). Washington, DC: US Government Printing Office.

Muro, M. 1998. New finds explode old views of the American Southwest. Science 279: 653–654.

Murphy, R. W., K. H. Berry, T. Edwards, A. E. Leviton, A. Lathrop, and J. D. Riedle. 2011. The dazed and confused identity of Agassiz's land tortoise, *Gopherus agassizii* (Testudines, Testudinidae) with the description of a new species, and its consequences for conservation. ZooKeys 113: 39–71, doi: 10.3897/ZooKeys.113.1353.

Namias, J. 1960. The meteorological picture, 1957–58. Pages 31–41 *in* Symposium on the Changing Pacific Ocean in 1957–58. CalCOFI Report 7. Sacramento: California Marine Research Commission.

Namias, J. 1986. Persistence of flow patterns over North America and adjacent ocean sectors. Monthly Weather Review 114: 1368–1383.

Namias, J., X. Yuan, and D. R. Cayan. 1988. Persistence of North Pacific sea surface temperature and atmospheric flow patterns. Journal of Climate 1: 682–703.

National Research Council. 2002. Riparian Areas: Functions and Strategies for Management. Washington, DC: National Academy Press.

Nations, D., and E. Stump. 1981. Geology of Arizona. 1st edition. Dubuque, IA: Kendall-Hunt.

Nations, D., and E. Stump. 1996. Geology of Arizona. 2nd edition. Dubuque, IA: Kendall/Hunt Publishing.

Neff, J. A. 1940a. Range, population and game status of the White-winged Dove in Arizona. Journal of Wildlife Management 4: 117–127.

Neff, J. A. 1940b. Notes on nesting and other habits of the White-winged Dove in Arizona. Journal of Wildlife Management 4: 279–290.

Nelson, E. W. 1913a. Herbert Brown—a biographical note. Condor 15: 186–187.

Nelson, E. W. 1913b. Notes and news—Herbert Brown [obituary]. Auk 30: 472.

Nicholls, N. 1988. El Nino–Southern Oscillation and rainfall variability. Journal of Climate 1: 418–421.

Nicolson, J., editor. 1974. The Arizona of Joseph Pratt Allyn: Letters from a Pioneer Judge; Observations and Travels, 1863–1866. Tucson: University of Arizona Press.

Nigam, S., M. Barlow, and E. H. Berbery. 1999. Analysis links Pacific decadal variability to drought and streamflow in the United States. EOS Transactions, American Geophysical Union 80 (51): 621–625.

Oberholser, H. C. 1974. The Bird Life of Texas, vol. 2. Edited by E. B. Kincaid Jr. Austin: University of Texas Press.

Odum, E. P. 1953. Fundamentals of Ecology. Philadelphia, PA: W. B. Saunders.

Ohmart, R. D. 1996. Historical and present impacts of livestock grazing on fish and wildlife resources in western riparian habitats. Pages 245–279 *in* P. R. Krauseman, editor, Rangeland Wildlife. Denver, CO: Society of Range Management.

Ohmart, R. D., B. W. Anderson, and W. C. Hunter. 1988. The ecology of the lower Colorado River from Davis Dam to the Mexico–United States International Boundary: A community profile. US Fish and Wildlife Service Biological Report 85 (7.19): 1–296.

Olberg, C. R., and F. R. Schanck. 1913. Special Report on Irrigation and Flood Protection, Papago Indian Reservation, Arizona. Washington, DC: 62nd Congress, 3rd Session, Senate Executive Document 436.

Olson, C. E. 1940. Forests in the Arizona desert. Journal of Forestry 38: 956–959.

Palacios-Fest, M. R., J. B. Mabry, F. Nials, J. P. Holmlund, E. Miksa, and O. K. Davis. 2001. Early irrigation systems in southeastern Arizona: The ostracode perspective. Journal of South American Earth Sciences 14: 541–555.

Palmer, T. S., et al. 1954. Biographies of Members of the American Ornithologists' Union. Washington, DC: American Ornithologists' Union.

Parke, J. G. 1855. Report of explorations for that portion of a railroad route, near the 32nd parallel of north latitude, lying between Dona Ana, on the Rio Grande, and Pimas villages, on the Gila. Pages 1–28 *in* Reports of Explorations and Surveys . . . for a Railroad . . . to the Pacific Ocean, vol. 2. Washington, DC: 33rd Congress, 2nd Session, Senate Executive Document 78.

Parker, J. T. C. 1995a. Channel Change and Sediment Transport in Two Desert Streams in Central Arizona, 1991–92. US Geological Survey Water-Resources Investigations Report 95-4059.

Parker, J. T. C. 1995b. Channel Change on the Santa Cruz River, Pima County, Arizona, 1936–86. US Geological Survey Water-Supply Paper 2429.

Parkes, K. C. 1966. The Birds of Arizona *in* Reviews. Auk 83: 484–487.

Pashley, E. F., Jr. 1966. Structure and stratigraphy of the central, northern, and eastern parts of the Tucson Basin, Arizona. Tucson: University of Arizona, Department of Geology, unpublished PhD dissertation.

Patton, P. C., and S. A. Schumm. 1975. Gully erosion, northwestern Colorado: A threshold phenomenon. Geology 3: 88–90.

Patton, P. C., and S. A. Schumm. 1981. Ephemeral stream processes: Implications for studies of Quaternary valley fills. Quaternary Research 15: 24–43.

Paxton, E. H., T. C. Theimer, and M. K. Sogge. 2011. Tamarisk biocontrol using tamarisk beetles: Potential consequences for riparian birds in the southwestern United States. Condor 113: 255–265.

Pearthree, M. S., and V. R. Baker. 1987. Channel Change Along the Rillito Creek System of Southeastern Arizona, 1941 Through 1983. Special Paper 6. Tucson: Arizona Bureau of Geology and Mineral Technology, Geological Survey Branch.

Perla, B. S., and L. E. Stevens. 2008. Biodiversity and productivity at an undisturbed spring in comparison with adjacent grazed and upland habitats. Pages 230–243 *in* L. E. Stevens and V. J. Meretsky, editors, Aridland Springs in North America: Ecology and Conservation. Tucson: University of Arizona Press.

Peters, J. L. 1936. Records of two species new to Arizona. Condor 38: 218.

Philander, S. G. H. 1983. El Niño Southern Oscillation phenomena. Nature 302: 295–301.

Philander, S. G. H. 1985. El Niño and La Niña. Journal of Atmospheric Sciences 42: 2652–2662.

Phillips, A. R. 1939. The faunal areas of Arizona, based on bird distribution. Tucson: University of Arizona, unpublished MS thesis.

Phillips, A. R. 1940. Two new breeding birds for the United States. Auk 57: 117–118.

Phillips, A. R. 1946. The birds of Arizona. Ithaca, NY: Cornell University, unpublished PhD dissertation.

Phillips, A. R. 1949. Nesting of the Rose-throated Becard in Arizona. Condor 51: 137–139.

Phillips, A. R. 1950. Charles Taylor Vorhies [obituary]. Auk 67: 141.

Phillips, A. R. 1962. Notas sistemáticas sobre aves mexicanas, 2. Anales del Instituto de Biología, Universidad Nacional Autónoma de México 33 (1–2): 331–372.

Phillips, A. R. 1968. The instability of the distribution of land birds in the Southwest. Pages 129–162 *in* Collected Papers in Honor of Lyndon Lane Hargrave. Papers of the Archaeological Society of New Mexico Museum. Santa Fe: New Mexico Press.

Phillips, A. R. 1994. A tentative key to the species of kingbirds, with distributional notes. Journal of Field Ornithology 65: 295–306.

Phillips, A. R., and G. Monson. 1964. Annotated Checklist of the Birds of Arizona. Tucson: University of Arizona Press.

Phillips, A. R., and L. D. Yaeger. 1951. Fifty-first Christmas bird count at Salt River (Maricopa County), Arizona, December 26, [1950]. Audubon Field Notes 5: 170–171.

Phillips, A. R., J. T. Marshall Jr., and G. Monson. 1964. The Birds of Arizona. Tucson: University of Arizona Press.

Phillips, J. V., and T. L. Ingersoll. 1998. Verification of Roughness Coefficients for Selected and Natural Channels in Arizona. US Geological Survey Professional Paper 1584.

Phillips, J. V., D. McDoniel, J. P. Capesius, and W. Asquith. 1998. Method to Estimate Effects of Flow-Induced Vegetation Changes on Channel Conveyances of Streams in Central Arizona. US Geological Survey Water-Resources Investigations Report 98-4040.

Pielke, R. A., Jr., and M. W. Downton. 2000. Precipitation and damaging floods: Trends in the United States, 1932–97. Journal of Climate 13: 3625–3637.

Ponce, V. M., Z. Osmolski, and D. Smutzer. 1985. Large basin deterministic hydrology: A case study. Journal of Hydraulic Engineering 111: 1227–1245.

Pope, G. L., P. D. Rigas, and C. F. Smith. 1998. Statistical Summaries of Streamflow Data and Characteristics of Drainage Basins for Selected Streamflow-Gaging Stations in Arizona Through Water Year 1996. US Geological Survey Water-Resources Investigations Report 98-4225.

Poston, C. D. 1951. Poston's narrative. Pages 235–254 *in* J. R. Browne, A Tour Through Arizona 1864; or, Adventures in the Apache Country. Tucson: Arizona Silhouettes. Originally published in 1869 as Adventures in the Apache Country by Harper and Brothers, New York.

Powell, H. M. T. 1931. The Santa Fe Trail to California, 1849, 1852. San Francisco, CA: Grabhorn Press.

Powell, J. R. 1878. Report on the Lands of the Arid Regions of the United States. Washington, DC: US Government Printing Office.

Proudfoot, G. A., and R. R. Johnson. 2000. Ferruginous Pygmy-Owl (*Glaucidium brasilianum*). *In* A. Poole and F. Gill, editors, The Birds of North America, no. 498. Philadelphia, PA: Birds of North America, Inc.

Proudfoot, G. A., D. A. Sherry, and S. Johnson. 2000. Cactus Wren (*Campylorhynchus brunneicapillus*). *In* A. Poole and F. Gill, editors, The Birds of North America, no. 558. Philadelphia, PA: Birds of North America, Inc.

Pyke, C. B. 1972. Some Meteorological Aspects of the Seasonal Distribution of Precipitation in the Western United States and Baja California. Contribution 139. Los Angeles: University of California, Water Resources Center.

Quinn, W. H., V. T. Neal, and S. E. Antunez de Mayolo. 1987. El Niño occurrences over the past four and a half centuries. Journal of Geophysical Research 92: 14449–14461.

Ramage, C. S. 1983. Teleconnections and the siege of time. Journal of Climatology 3: 223–231.

Rasmusson, E. M. 1967. Atmospheric water vapor transport and the water balance of North America: part 1, Characteristics of the water vapor flux field. Monthly Weather Review 95: 403–426.

Rasmusson, E. M. 1984. El Niño: The ocean/atmosphere connection. Oceanus 27: 5–12.

Rasmusson, E. M. 1985. El Niño and variations in climate. American Scientist 73: 168–177.

Ravesloot, J. C., and S. M. Whittlesey. 1987. The Archaeology of the San Xavier Bridge Site (AZ BB:13:14), Tucson Basin, Southern Arizona. Tucson: Arizona State Museum.

Rea, A. M. 1983. Once a River: Bird Life and Habitat Changes on the Middle Gila. Tucson: University of Arizona Press.

Rea, A. M. 2007. Wings in the Desert: A Folk Ornithology of the Northern Pimans. Tucson: University of Arizona Press.

Reich, B. F. 1987. Estimating the regulatory flood on a degrading river. Pages 197–212 *in* V. Singh, editor, Regional Flood-Frequency Analysis. Dordrect, The Netherlands: D. Reidel Publishers.

Reich, B. F., and D. R. Davis. 1986a. The 1983 Santa Cruz flood: How should highway engineers respond? Transportation Research Record 1017: 1–8.

Reich, B. F., and D. R. Davis. 1986b. Reluctance to increase the regulatory flood on a degrading river. Pages 99–102 *in* Improving the Effectiveness of Floodplain Management in Arid and Semi-arid Regions. Las Vegas, NV: Association of State Floodplain Managers.

Reich, B. M. 1984. Recent changes in a flood series. Journal of the Arizona-Nevada Academy of Science 14: 231–238.

Reid, J. C. 1858. Reid's Tramp: A Journal of the Incidents of Ten Months Travel Throughout Texas, New Mexico, Arizona, Sonora, and California, Including Topogra-phy, Climate, Soil, Mineral, etc. Selma, AL: John Hardy and Company.

Reisner, M. 1986. Cadillac Desert: The American West and Its Disappearing Water. New York: Penguin Books.

Reitan, C. H. 1957. The role of precipitable water vapor in Arizona's summer rains. Technical Reports on the Meteorology and Climatology of Arid Regions 2. Tucson: University of Arizona Institute of Atmospheric Physics.

Reitan, C. H. 1960. Distribution of precipitable water vapor over the continental United States. Bulletin of the American Meteorological Society 41: 79–87.

Reyes, S., and D. L. Cadet. 1988. The Southwest branch of the North American monsoon during 1979. Monthly Weather Review 116: 1175–1187.

Rhoads, B. L. 1991. Impact of agricultural development on regional drainage in the lower Santa Cruz valley, Arizona, USA. Environmental Geology Water Science 18: 119–135.

Rhoads, S. N. 1892. The birds of southeastern Texas and southern Arizona observed during May, June, and July 1891. Proceedings of the Academy of Natural Sciences, Philadelphia 1892: 99–126.

Rich, J. L. 1911. Recent stream trenching in the semiarid region of southwestern New Mexico, a result of removal of vegetative cover. American Journal of Science, 4th series, 32: 237–245.

Ridgway, R. 1884. Descriptions of some new North American birds. Proceedings of the Biological Society of Washington 2: 89–95.

Ridgway, R. 1885a. Descriptions of a new Cardinal Grosbeak from Arizona. Auk 2: 343–345.

Ridgway, R. 1885b. *Icterus cucullatus*, Swainson and its geographical variations. Proceedings of the United States National Museum 8: 18–19.

Ridgway, R. 1898. Descriptions of supposed new genera, species, and subspecies of American birds: I. Fringillidae. Auk 15: 223–230.

Ridgway, R. 1903. Descriptions of new genera, species and subspecies of American birds. Proceedings of the Biological Society of Washington 16: 105–111.

Ritchie, E. A., K. M. Wood, D. S. Gutzler, and S. R. White. 2011. The influence of Eastern Pacific tropical cyclone remnants on the southwestern United States. Monthly Weather Review 139: 192–210.

Roberts, R. A. 1990. Analysis of sediment transport in an ephemeral stream. Tucson: University of Arizona, Department of Civil Engineering and Engineering Mechanics, unpublished MS thesis.

Robinson, T. W. 1965. Introduction, Spread, and Areal Extent of Salt Cedar (*Tamarix*) in the Western United States. US Geological Survey Professional Paper 491-A.

Roeske, R. H. 1978. Methods for Estimating the Magnitude and Frequency of Floods in Arizona. US Geological Survey Open File Report 78-711.

Roeske, R. H., M. E. Cooley, and B. N. Aldridge. 1978. Floods of September 1970 in Arizona, Utah, Colorado, and New Mexico. US Geological Survey Water-Supply Paper 2052.

Roeske, R. H., J. M. Garrett, and J. H. Eychaner. 1989. Floods of October 1983 in Southeastern Arizona. US Geological Survey Water-Resources Investigations Report 85-4225-C.

Rogers, G. F., H. E. Malde, and R. M. Turner. 1984. Bibliography of Repeat Photography for Evaluating Landscape Change. Salt Lake City: University of Utah Press.

Ropelewski, C. F., and M. S. Halpert. 1986. North American precipitation and temperature patterns associated with El Niño/Southern Oscillation (ENSO). Monthly Weather Review 114: 2352–2362.

Ropelewski, C. F., and P. D. Jones. 1987. An extension of the Tahiti-Darwin Southern Oscillation Index. Monthly Weather Review 115: 2161–2165.

Rosen, P. C. 2001. Biological values of the West Branch of the Santa Cruz River with an outline for a potential park or reserve. Tucson, AZ: Report to Pima County Board of Supervisors, 15 October 2001.

Rosen, P. C. 2003. Herpetology of the West Branch of the Santa Cruz River, Tucson. Sonoran Herpetologist 16 (4): 38–42.

Rosen, P. C. 2008. Urban amphibian and reptile biodiversity: Report to the technical advisory committee (TAC) for City of Tucson Habitat Conservation Plan (HCP). Tucson, AZ: City of Tucson, unpublished report.

Rosenberg, K. V., R. D. Ohmart, and B. W. Anderson. 1982. Community organization of riparian breeding birds: Response to an annual resource peak. Auk 99: 260–274.

Rosenberg, K. V., R. D. Ohmart, W. C. Hunter, and B. W. Anderson. 1991. Birds of the Lower Colorado River Valley. Tucson: University of Arizona Press.

Saarinen, T. F., V. R. Baker, R. Durrenberger, and T. Maddock Jr. 1984. The Tucson, Arizona, Flood of October 1983. Washington, DC: National Academy Press.

Schrader, F. C., R. W. Stone, and S. Sanford. 1917. Useful Minerals of the United States. US Geological Survey Bulletin 624.

Schumm, S. A. 1979. Geomorphic thresholds: The concept and its applications. Institute of British Geographers Transactions 4: 485–515.

Schumm, S. A., and R. F. Hadley. 1957. Arroyos and the semi-arid cycle of erosion. American Journal of Science 255: 161–174.

Schumm, S. A., and R. S. Parker. 1973. Implications of complex response of drainage systems for Quaternary alluvial stratigraphy. Nature, Physical Science 243: 99–100.

Schumm, S. A., M. D. Harvey, and C. C. Watson. 1984. Incised Channels: Morphology, Dynamics and Control. Littleton, CO: Water Resources Publications.

Schwalen, H. C. 1942. Rainfall and Runoff in the Upper Santa Cruz River Drainage Basin. Agricultural Experiment Station Technical Bulletin 1. Tucson: University of Arizona.

Schwalen, H. C., and R. J. Shaw. 1957. Ground Water Supplies of Santa Cruz Valley of Southern Arizona Between Rillito Station and the International Boundary. Agricultural Experiment Station Bulletin 288. Tucson: University of Arizona.

Schwalen, H. C., and R. J. Shaw. 1961. Water in the Santa Cruz Valley, Arizona. Agricultural Experiment Station Report 205. Tucson: University of Arizona.

Scott, V. E., and D. R. Patton. 1975. Cavity-Nesting Birds of Arizona and New Mexico Forests. US Forest Service General Technical Report RM-10. Fort Collins, CO: Rocky Mountain Forest and Range Experiment Station.

Scott, V. E., K. Evans, D. R. Patton, and C. P. Stone. 1975. Cavity-Nesting Birds of North American Forests. Agriculture Handbook 511. Washington, DC: US Department of Agriculture, Forest Service.

Scott, W. E. D. 1886. On the avifauna of Pinal County with remarks on some birds of Pima and Gila Counties, Arizona. Auk 3: 249–258, 383–389, 421–432.

Scott, W. E. D. 1887. On the avifauna of Pinal County with remarks on some birds of Pima and Gila Counties, Arizona. Auk 4: 16–24, 196–205.

Scott, W. E. D. 1888. On the avifauna of Pinal County with remarks on some birds of Pima and Gila Counties, Arizona. Auk 5: 19–36, 159–168.

Sellers, W. D. 1960. Precipitation trends in Arizona and western New Mexico. Proceedings of the 28th Annual Western Snow Conference. Santa Fe, NM.

Sellers, W. D., and R. H. Hill, editors. 1974. Arizona Climate, 1971–1972. Tucson: University of Arizona Press.

Seymour, D. J. 1989. The dynamics of Sobaipuri settlement in the eastern Pimería. Journal of the Southwest 31: 205–222.

Seymour, D. J. 2012. Santa Cruz River: The origin of a place name. Journal of Arizona History 53: 81–88.

Shafroth, P. B., J. R. Cleverly, T. L. Dudley, J. P. Taylor, C. Van Riper, E. P. Weeks, and J. N. Stuart. 2005. Control of *Tamarix* in the western United States: Implications for water salvage, wildlife use, and riparian restoration. Environmental Management 35: 231–246.

Shamir, E., D. M. Meko, N. E. Graham, and K. P. Georgakakos. 2007. Hydrologic model framework for water resources planning in the Santa Cruz River, southern Arizona. Journal of the American Water Resources Association 43: 1155–1170.

Sheridan, T. E. 1986. Los Tucsonenses: The Mexican Community in Tucson, 1854–1941. Tucson: University of Arizona Press.

Sheridan, T. E. 1995. Arizona: A History. Tucson: University of Arizona Press.

Siegel, S. 1956. Nonparametric Statistics for the Behavioral Sciences. New York: McGraw-Hill.

Simpson, J. M., and J. R. Werner. 1958. Some recent records from the Salt River valley, central Arizona. Condor 60: 68–70.

Skagen, S. K., C. P. Melcher, W. H. Howe, and F. L. Knopf. 1998. Comparative use of riparian corridors and oases by migrating birds in southeast Arizona. Conservation Biology 12: 896–909.

Slezak-Pearthree, M., and V. R. Baker. 1987. Channel Change Along the Rillito Creek System of Southeastern Arizona, 1941 through 1983. Special Paper 6. Tucson: Arizona Bureau of Geology and Mineral Technology Geological Survey Branch.

Smith, C. F., K. M. Sherman, G. L. Pope, and P. D. Rigas. 1998. January and February 1993, in Arizona. *In* C. A. Perry and L. J. Combs, editors, Summary of Floods in the United States, January 1992 Through September 1993. US Geological Survey Water-Supply Paper 2499.

Smith, G. E. P. 1910. Groundwater supply and irrigation in the Rillito Valley. University of Arizona Agricultural Experiment Station Bulletin 64: 18–244.

Smith, G. E. P. 1936. Water supplies. Pages 37–46 *in* H. L. Shantz, B. S. Butler, and E. D. Wilson, editors, Arizona and Its Heritage. General Bulletin 3. Tucson: University of Arizona.

Smith, G. E. P. 1938. The Physiography of Arizona Valleys and the Occurrence of Ground Water. Agricultural Experiment Station Bulletin 77. Tucson: University of Arizona.

Smith, G. E. P. 1940. The Ground-Water Supply of the Eloy District in Pinal County, Arizona. Agricultural Experiment Station Technical Bulletin 87. Tucson: University of Arizona.

Smith, H. N. 1975. Virgin Land: The American West as Symbol and Myth. Cambridge, MA: Harvard University Press.

Smith, W. 1986. The Effects of Eastern North Pacific Tropical Cyclones on the Southwestern United States. Technical Memorandum NWS, WS-197. Salt Lake City, UT: National Oceanic and Atmospheric Association.

Snyder, N. F. R., E. C. Enkerlin-Hoeflich, and M. A. Cruz-Nieto. 1999. Thick-billed Parrot (*Rhynchopsitta pachyrhyncha*). *In* A. Poole and F. Gill, editors, The Birds of North America, no. 406. Philadelphia, PA: Birds of North America, Inc.

Sogge, M. K., S. J. Sferra, and E. H. Paxton. 2008. *Tamarix* as habitat for birds: Implications for riparian restoration in the southwestern United States. Restoration Ecology 16: 146–154.

Sonoran Institute. 2008. Santa Cruz River Riparian Vegetation Mapping Project: Santa Cruz County, Arizona. Tucson: Sonoran Institute and University of Arizona Office of Arid Lands Studies.

Soutar, A., and P. A. Crill. 1977. Sedimentation and climatic patterns in the Santa Barbara basin during the 19th and 20th centuries. Geological Society of America Bulletin 88: 1161–1172

Soutar, A., and J. D. Isaacs. 1974. Abundance of pelagic fish during the 19th and 20th centuries as recorded in anaerobic sediments off the Californias. Fisheries Bulletin 72: 257–273.

Spalding, V. M. 1909. Distribution and movement of desert plants. Carnegie Institution of Washington Publication 113: 1–144.

Spring, J. 1966. John Spring's Arizona. Tucson: University of Arizona Press.

Stamp, N. E. 1978. Breeding birds of riparian woodland in south-central Arizona. Condor 80: 64–71.

Stebbins, R. C. 1985. A Field Guide to Western Reptiles and Amphibians. 2nd edition, revised. The Peterson Field Guide Series. Boston, MA: Houghton Mifflin.

Stebbins, R. C. 2003. A Field Guide to Western Reptiles and Amphibians. 3rd edition, revised. The Peterson Field Guide Series. Boston: Houghton Mifflin.

Stejskal, D., and G. H. Rosenberg. 2011. Finding Birds in Southeastern Arizona. 8th edition. Tucson, AZ: Tucson Audubon Society.

Stevens, J. E. 1988. Hoover Dam: An American Adventure. Norman: University of Oklahoma Press.

Stevens, L. E. 1987. The status of ecological research on tamarisk (Tamaricaceae: *Tamarix ramosissima*) in Arizona. Pages 99–105 *in* Tamarisk Control in Southwestern United States: Proceedings of Tamarisk Conference, University of Arizona, Tucson, Arizona September 2 and 3, 1987. Cooperative National Park Resources Studies Unit Special Report 9. Tucson, AZ.

Stevens, L. E., K. A. Buck, B. T. Brown, and N. Kline. 1997. Dam and geomorphic influences on Colorado River waterbird distribution, Grand Canyon, Arizona. Regulated Rivers: Research and Management 13: 151–169.

Stevenson, M., editor. 2007. Finding Birds in Southeastern Arizona. 7th edition. Tucson, AZ: Tucson Audubon Society.

Stone, W. 1928. Notes and news—William Leon Dawson [obituary]. Auk 45: 417.

Stouffer, P. C., and R. T. Chesser. 1998. Tropical Kingbird (*Tyrannus melancholis*). *In* A. Poole and F. Gill, editors, The Birds of North America, no. 358. Philadelphia, PA: Birds of North America, Inc.

Stromberg, J. C. 1993a. Fremont cottonwood–Goodding willow riparian forests: A review of their ecology, threats, and recovery potential. Journal of the Arizona-Nevada Academy of Sciences 27: 97–110.

Stromberg, J. C. 1993b. Riparian mesquite forests: A review of their ecology, threats, and recovery potential. Journal of the Arizona-Nevada Academy of Sciences 27: 111–124.

Stromberg, J. C., and B. Tellman, editors. 2009. Ecology and Conservation of the San Pedro River. Tucson: University of Arizona Press.

Swarth, H. S. 1904. Birds of the Huachuca Mountains, Arizona. Pacific Coast Avifauna 4: 1–70.

Swarth, H. S. 1905. Summer birds of the Papago Indian Reservation and of the Santa Rita Mountains, Arizona. Condor 7: 22–27, 47–50.

Swarth, H. S. 1914. A distributional list of the birds of Arizona. Pacific Coast Avifauna 10: 1–133.

Swarth, H. S. 1929. The faunal areas of southern Arizona: A study in animal distribution. Proceedings of the California Academy of Sciences, 4th series, 18: 267–383.

SWCA Environmental Consultants (SWCA). 2000. Avian surveys along the lower Santa Cruz River. Tucson, AZ: Unpublished report prepared for US Bureau of Reclamation.

Swift, T. T. 1926. Date of channel trenching in the Southwest. Science 63: 70–71.

Sykes, G. G. 1967. Scraps from the past. Tucson, AZ: Unpublished manuscript in the collection of the Sykes family.

Taber, W. 1955. In memoriam: Arthur Cleveland Bent. Auk 72: 332–339.

Tang, M., and E. R. Reiter. 1984. Plateau monsoons of the northern hemisphere: A comparison between North America and Tibet. Monthly Weather Review 112: 617–637.

Tellman, B., R. Yarde, and M. G. Wallace. 1997. Arizona's Changing Rivers: How People Have Affected the Rivers. Tucson: University of Arizona, Water Resources Center.

Terres, J. K. 1980. The Audubon Encyclopedia of North American Birds. New York: Knopf.

Tevis, J. H. 1954. Arizona in the 50's. Albuquerque: University of New Mexico Press.

Thiel, J. H., and J. B. Mabry. 2006. Summary: The archaeology of a changing community. Pages 22.1–22.14 *in* J. H. Thiel and J. B. Mabry, editors, Rio Nuevo Archaeology, 2000–2003: Investigations at the San Agustín Mission and Mission Gardens, Tucson Presidio, Tucson Pressed Brick Company, and Clearwater Site. Technical Report 2004-11. Tucson, AZ: Desert Archaeology, Inc. http://www.cdarc.org/pages/library/rio_nuevo/.

Thomas, B. E., and D. R. Pool. 2006. Trends in Streamflow in the San Pedro River, Southeastern Arizona, and Regional Trends in Precipitation and Streamflow in Southeastern Arizona and Southwestern New Mexico. US Geological Survey Professional Paper 1712.

Thomas, W. O. 1985. A uniform technique for flood frequency analysis. Journal of Water Resources Planning and Management 111: 321–337.

Thompson, J. D., editor. 2008. New Mexico Territory During the Civil War: Wallen and Evans Inspection Reports, 1862–1863. Albuquerque: University of New Mexico Press.

Thompson, L. G., E. Mosley-Thompson, W. Dansgaard, and P. M. Grootes. 1986. The Little Ice Age as recorded in the stratigraphy of the tropical Quelccaya ice cap. Science 234: 361–364.

Thornber, J. J. 1909. Relation of plant growth and vegetation forms to climatic conditions. Plant World 12: 1–7.

Thornthwaite, C. W., C. F. S. Sharpe, and E. F. Dosch. 1942. Climate and Accelerated Erosion in the Arid and Semiarid Southwest, with Special Reference to the Polacco Wash Drainage Basin. Technical Bulletin 808. Washington, DC: US Department of Agriculture.

Treese, S., T. Meixner, and J. F. Hogan. 2009. Clogging of an effluent dominated semiarid river: A conceptual model of stream-aquifer interactions. Journal of the American Water Resources Association 45: 1047–1062.

Tuan, Y. 1966. New Mexican gullies: A critical review and some recent observations. Annals of the Association of American Geographers 56: 573–597.

Tucson Audubon Society. 2012. Sweetwater Wetlands checklist. http://www.tucsonaudubon.org/component/content/article/48-visit/212-sweetwater.html (accessed 30 October 2013).

Turner, R. M. 1974. Quantitative and Historical Evidence of Vegetation Changes Along the Upper Gila River, Arizona. US Geological Survey Professional Paper 655-H.

Turner, R. M., and M. M. Karpiscak. 1980. Recent Vegetation Changes Along the Colorado River Between Glen Canyon Dam and Lake Mead, Arizona. US Geological Survey Professional Paper 1132.

Turner, R. M., J. E. Bowers, and T. L. Burgess. 1995. Sonoran Desert Plants. Tucson: University of Arizona Press.

Turner, R. M., R. H. Webb, J. E. Bowers, and J. R. Hastings. 2003. The Changing Mile Revisited. Tucson: University of Arizona Press.

Turner, S. F., G. A. Waring, J. D. Hem, M. J. Scott, H. M. Babcock, W. T. Stuart, E. M. Cushing, and H. R. McDonald. 1943. Ground-Water Resources of the Santa Cruz Basin, Arizona. US Geological Survey Open-File Report, unnumbered.

Turner, S. F., et al. 1947. Further Investigations of the Ground-Water Resources of the Santa Cruz Basin, Arizona. US Geological Survey Open-File Report 46–80.

Tweit, R. C., and C. W. Thompson. 2000. Pyrrhuloxia (*Cardinalis sinuatus*). *In* A. Poole and F. Gill, editors, The Birds of North America, no. 391. Philadelphia, PA: Birds of North America, Inc.

US Army Corps of Engineers. 1972. Interim Report on a Survey for Flood Control, Santa Cruz River Basin. Los Angeles: US Army Corps of Engineers.

US Bureau of the Census. 1982. 1980 Census of the Population. Vol. 1, Characteristics of the Population. Chapter A, Number of Inhabitants. Part 4, Arizona. PC80-1-A4. Washington, DC: US Department of Commerce, Bureau of the Census.

US Court of Private Land Claims. 1881. Review of San Ignacio de la Canoa petition. *In* "Records relating to cases decided by the United States Court of Private Land Claims, Arizona District, 1800–1904," University of Arizona Special Collections Library, Tucson. Original papers at the National Archives, Special Collections, MS 312, Reel 3, Washington, DC.

US Court of Private Land Claims. 1882. Review of Rancho de Martinez petition. *In* "Records relating to cases decided by the United States Court of Private Land

Claims, Arizona District, 1800–1904," University of Arizona Special Collections Library, Tucson. Original papers at the National Archives, Special Collections, MS 312, Reel 19, Washington, DC.

US Department of Agriculture. 1939. Range Management and Agronomic Practices on the San Xavier Indian Reservation, Arizona, and Land Classification of San Xavier Indian Reservation. Denver, CO: US Bureau of Indian Affairs.

US Fish and Wildlife Service. 1997. Endangered and threatened wildlife and plants: Determination of endangered status for the Cactus Ferruginous Pygmy-Owl; Final rule. Federal Register 62: 10730–10747.

US Fish and Wildlife Service. 2011a. Endangered species profile for Masked Bobwhite. http://ecos.fws.gov /speciesProfile/profile/speciesProfile.action ?spcode=B00Z (accessed 15 January 2011).

US Fish and Wildlife Service. 2011b. Endangered species profile for Willow Flycatcher. http://www.fws.gov /species/#endangered (accessed 5 February 2011).

US Fish and Wildlife Service. 2011c. Endangered species profile for Santa Cruz pupfish (extirpated) and Gila topminnow (endangered). http://www.fws.gov/species /#endangered (accessed December 2011).

US Geological Survey. 1985. Summary of the Technical Symposium on Flood Frequency of Large Rivers in Southern Arizona, November 13–14, Tucson. Tucson: US Geological Survey, Water Resources Division, Arizona District.

US Signal Service. 1890. Characteristics of the weather for July and August, 1890. Monthly Weather Review 18: chart 2.

US Surgeon General's Office. 1870. Report on the Barracks and Hospitals, with Descriptions of Military Posts. War Department Circular 4. Washington, DC: US Government Printing Office.

US Water Resources Council. 1981. Guidelines for Determining Flood Flow Frequency. Bulletin 17B. Washington, DC: US Water Resources Council.

Van Rossem, A. J. 1934. Notes on some types of North American birds. Transactions of the San Diego Society of Natural History 7: 347–362.

Vinnikov, K., G. V. Gruza, V. K. Zakharov, A. A. Kirilov, N. P. Kovyneva, and E. Ya. Ran'kova. 1980. Current climate changes in the northern hemisphere. Soviet Meteorology and Hydrology 6: 1–10.

Visher, S. S. 1909. The capture of the Red-eyed Cowbird in Arizona. Auk 26–307.

Vorhies, C. T. 1926. A new Ophidian record for Arizona. Copeia 157: 156–157.

Vorhies, C. T., R. Jenks, and A. R. Phillips. 1935. Bird records from the Tucson region, Arizona. Condor 37: 243–247.

Walker, G. T. 1924. Correlation in seasonal variation of weather: IX, A further study of world weather. Memoirs of the Indian Meteorological Department 3: 1830–1837.

Walker, G. T., and E. W. Bliss. 1932. World weather V. Memoirs of the Royal Meteorological Society 4: 57–84.

Walker, M., S. Johnsen, S. O. Rasmussen, T. Popp, J.-P. Steffensen, P. Gibbard, W. Hoek, J. Lowe, J. Andrews, S. Björck, L. C. Cwynar, K. Hughen, P. Kershaw, B. Kromer, T. Litt, D. J. Lowe, T. Nakagawa, R. Newnham, and J. Schwander. 2009. Formal definition and dating of the GSSP (Global Stratotype Section and Point) for the base of the Holocene using the Greenland NGRIP ice core, and selected auxiliary records. Journal of Quaternary Science 24 (1): 3–17.

Wallis, J. R., and E. F. Wood. 1985. Relative accuracy of log Pearson III procedures. Journal of Hydraulic Engineering 111: 1043–1056.

Warner, S. 1884. Papers. Tucson: Arizona Historical Society Library.

Waters, M. R. 1985. Late Quaternary alluvial stratigraphy of White Water Draw, Arizona: Implications for regional correlation of fluvial deposits in the American Southwest. Geology 13: 705–708.

Waters, M. R. 1988. Holocene alluvial geology and geoarcheology of the San Xavier reach of the Santa Cruz River, Arizona. Geological Society of America Bulletin 100: 479–491.

Waters, M. R. 1992. Principles of Geoarchaeology: A North American Perspective. Tucson: University of Arizona Press.

Waters, M. R., and C. V. Haynes. 2001. Late Quaternary arroyo formation and climate change in the American Southwest. Geology 29: 399–402.

Waters, M. R., and J. C. Ravesloot. 2001. Landscape change and the cultural evolution of the Hohokam along the middle Gila River and other river valleys in south-central Arizona. American Antiquity 66: 285–299.

Way, P. R. 1960. Overland via Jackass Mail in 1858: The diary of Phocian R. Way. Arizona and the West 2: 147–164, 279–292, 335–353, 353–370.

Webb, G. 1959. A Pima Remembers. Tucson: University of Arizona Press.

Webb, R. H. 1985. Late Holocene flooding on the Escalante River, south-central Utah. Tucson: University of Arizona, unpublished PhD dissertation.

Webb, R. H. 1996. Grand Canyon: A Century of Change. Tucson: University of Arizona Press.

Webb, R. H., and V. R. Baker. 1987. Changes in hydrologic conditions related to large floods on the Escalante River, south-central Utah. Pages 306–320 in V. Singh, editor, Regional Flood-Frequency Analysis. Dordrect, The Netherlands: D. Reidel Publishers.

Webb, R. H., and J. L. Betancourt. 1992. Climatic Variability and Flood Frequency of the Santa Cruz River, Pima County, Arizona. US Geological Survey Water-Supply Paper 2379.

Webb, R. H., and J. Hasbargen. 1998. Floods, groundwater levels, and arroyo formation on the Escalante River, south-central Utah. Pages 335–357 *in* Learning from the Land, Grand Staircase–Escalante National Monument, Science Symposium Proceedings. Report BLM/UT /GI-98/006+1220. Salt Lake City: Utah Bureau of Land Management.

Webb, R. H., and R. Hereford. 2010. Historical arroyo formation: Documentation of magnitude and timing of historical changes using repeat photography. Pages 89–104 *in* R. H. Webb, D. E. Boyer, and R. M. Turner, editors, Repeat Photography: Methods and Applications in the Natural Sciences. Washington, DC: Island Press.

Webb, R. H., and S. A. Leake. 2006. Ground-water surface-water interactions and long-term change in riverine riparian vegetation in the southwestern United States. Journal of Hydrology 320: 302–323.

Webb, R.H., and S. L. Rathburn. 1989. Paleoflood hydrologic research in the southwestern United States. Transportation Research Board Record 1201: 9–21.

Webb, R. H., and R. M. Turner. 2010. A debt to the past: Long-term and current plant research at Tumamoc Hill (the Desert Laboratory) in Tucson, Arizona. Desert Plants 26: 3–18.

Webb, R. H., J. E. O'Connor, and V. R. Baker. 1988. Paleohydrologic reconstruction of flood frequency on the Escalante River, south-central Utah. Pages 403–418 *in* V. R. Baker, R. C. Kochel, and P. C. Patton, editors, Flood Geomorphology. New York: John Wiley and Sons.

Webb, R. H., S. S. Smith, and V. A. S. McCord. 1991. Historic Channel Change of Kanab Creek, Southern Utah and Northern Arizona. Monograph 9. Grand Canyon, AZ: Grand Canyon Natural History Association.

Webb, R. H., T. S. Melis, and R. A. Valdez. 2002. Observations of Environmental Change in Grand Canyon. US Geological Survey Water Resources Investigations Report 02-4080.

Webb, R. H., R. Hereford, and G. J. McCabe. 2005. Climatic fluctuations, drought, and flow in the Colorado River. Pages 59–69 *in* S. P. Gloss, J. E. Lovich, and T. S. Melis, editors, The State of the Colorado River Ecosystem in Grand Canyon. US Geological Survey Circular 1282.

Webb, R. H., D. E. Boyer, and R. M. Turner. 2007a. The Desert Laboratory Repeat Photography Collection—An Invaluable Archive Documenting Landscape Change. US Geological Survey Fact Sheet 2007-3046.

Webb, R. H., S. A. Leake, and R. M. Turner. 2007b. The Ribbon of Green: Change in Riparian Vegetation in the Southwestern United States. Tucson: University of Arizona Press.

Webb, R. H., C. S. Magirl, P. G. Griffiths, and D. E. Boyer. 2008. Debris Flows and Floods in Southeastern Arizona from Extreme Precipitation in Late July 2006: Magnitude, Frequency, and Sediment Delivery. US Geological Survey Open-File Report 2008-1274.

Wells, S. G., R. S. Balling Jr., and R. S. Cerveny. 1988. Arroyo-cutting and flood events in the American Southwest: Role of historic climatic patterns, atmospheric circulation patterns, and volcanic activity. *In* Abstracts with Programs, Cordilleran Section Meeting. Las Vegas, NV, March 2008. Denver, CO: Geological Society of America.

West, R. L., and G. K. Hess. 2002. Purple Gallinule. *In* A. Poole and F. Gills, editors, The Birds of North America, no. 626. http://bna.birds.cornell.edu/bna (accessed 7 March 2011).

Wilkinson, R. E. 1966. Seasonal development of anatomical structures of saltcedar foliage. Botanical Gazette 127: 231–234.

Willard, F. C. 1912. A week afield in southern Arizona. Condor 14: 53–63.

Williams, G. P. 1978. The Case of the Shrinking Channels: The North Platte and Platte Rivers in Nebraska. US Geological Survey Circular 781.

Willis, E. L. 1939. Plant associations of the Rillito floodplain in Pima County, Arizona. Tucson: University of Arizona, unpublished MS thesis.

Wilson, R. P., and W. B. Garrett. 1989. Water-Resources Data for Arizona, Water Year 1987. US Geological Survey Water-Data Report AZ-87-1.

Wolman, M. G. 1977. Changing needs and opportunities in the sediment field. Water Resources Research 13: 50–54.

Wood, M. L., P. K. House, and P. A. Pearthree. 1999. Historical Geomorphology and Hydrology of the Santa Cruz River. Open-File Report 99-13. Tucson: Arizona Geological Survey.

Woodhouse, C. A., K. E. Kunkel, D. R. Easterling, and E. R. Cook. 2005. The 20th century pluvial in the western United States. Geophysical Research Letters 32. doi: 1029/2005GL022413.

Woolley, R. R. 1946. Cloudburst Floods in Utah, 1850–1938. US Geological Survey Water-Supply Paper 994.

Wright, P. B. 1989. Homogenized long-period Southern Oscillation indices. International Journal of Climatology 9: 33–54.

Yarnal, B., and H. F. Diaz. 1986. Relationships between extremes of the Southern Oscillation and the winter climate of the Anglo-American Pacific coast. Journal of Climatology 6: 197–219.

Youngs, F. O. 1931. Soil Survey of the Tucson Area, Arizona. Washington, DC: US Department of Agriculture, Bureau of Chemistry and Soils.

Zeller, M. E. 1984. Estimating the magnitude and frequency of a 100-year (design) flood on the Santa Cruz River in the vicinity of Tucson, Arizona from the storm and flood of September 28 through October 3, 1983—an alternative approach to flood-frequency analysis. Tucson, AZ: Pima County Department of Transportation and Flood Control District, unpublished report.

Index

Acequia de Punta de Agua, 41
   (fig. 4.2)
Acequia Madre, 57 (fig. 5.3)
Active Management Areas, 146
aerial photography, 124–25 (fig. 8.8),
   128, 138–40 (fig. 9.4), 142 (fig. 9.6), 173
   (fig. 11.3)
aggradation, periods of, 14, 30
agricultural clearing, 172
agricultural water demands, in 1920s
   and 1930s, 92–94
agriculture: in prehistoric times,
   29–30; in 19th century, 49, 59; in
   mid-20th century, 112–19; in late 20th
   century, 146
Agua de la Misión, 14, 50, 123
Aguirre Wash, 82
Aldrich, Lorenzo D., 40
Aldrich, Mark, 45
Aleutian Low, 17
algal growth, 163–64
Allison, Frank and Warren, 76–77
Allyn, Joseph Pratt, 48
Altithermal, 13, 31
Amado, AZ, 9
American Ornithologists' Union, 127
A Mountain. See Sentinel Peak
Ancestral Puebloans, 29
anecdotal evidence, 3–4, 29, 32, 35, 52
annual flow volume, 20–21
annual peak discharge, compared to
   climate and hurricane frequency, 169
   (fig. 11.1)
Anza, Juan Bautista de, 38
Apache, 38, 40, 47, 52–53, 63
aquifer compaction, 119
aquifer depletion, 123
aquifer recharge, 141, 164–65
archival evidence, 4, 129, 247n71;
   photographs as, 7–8
Argonauts, 40–42
Arizona ash, 108, 128
Arizona Daily Star, 69, 130
Arizona Department of Environmental
   Quality, 161–63

Arizona legislature, and ratification of
   Colorado River compact, 146
Arizona Mining Index, 62
Arizona National Guard, 146
Arizona state legislation: for creation
   of AMAs, 146; for creation of
   Tucson AMA (1980), 248n21; House
   Bill 2010, 132
Arizona sycamore, 108
Arizona walnut, 108, 128
army, Mexican, 45
army, US, 53–54; arrival of, in 1856, i, 45,
   47–49
Arnold, L. W., 102, 106, 110
arrowweed, 23, 93 (fig. 7.1), 128
arroyo cutting-and-filling, regional
   generalizations of, 145–46
arroyo development, 166–67;
   generalized stages of, 34 (fig. 3.1)
arroyo downcutting, 3–4, 13, 35, 123,
   146, 170–72; causes of, 27–35; cultural
   impacts of, 14; and groundwater, 80,
   121–23; and heavy *vs.* light rains,
   168–70; historical data (1889–1915),
   68–91; initiated by floods, 71; and
   intercept ditch, 68–71; regional
   synchroneity, 6–7. *See also*
   paleoarroyo downcutting
arroyo filling, 31, 33–35
arroyos, 3; continuous, 166–67;
   discontinuous, 36–51; historical
   data (1691–1872), 36–51; historical
   data (1889–1915), 68–91; historical data
   (1943–1975), 112–29; historical data
   (1976–1995), 130–48; Holocene
   development of, 12–14; management
   of, in time of floods (1976–1995),
   130–48; reactions to creation of, 73–78;
   widening, 68–91
artesian wells (flowing wells), 58,
   242n49
Athel tamarisk, 46–47 (fig. 4.5), 51, 61
   (fig. 5.4), 113, 116 (fig. 8.3), 120–21
   (fig. 8.6). *See also* saltcedar
Atkinson, Henry, 7

avifauna diversity, 25–26, 174–77;
   extirpation of species, 128, 175–76.
   See also bird species
Avra Valley aquifer, 148

bank protection, 141–42, 145, 152–53
   (fig. 10.1), 172
Bartlett, John R., 40–42
Bashford, Levi, 52
beaver, 29, 50, 241n73
bed clogging, 163–64, 181
Bell, William Alexander, 48
Bendire, Charles, 53, 100
Bent, Arthur C., 79, 102, 106, 109
Berger, J. M., 67
biodiversity, 4, 24, 54, 108
biogeography, of Santa Cruz and San
   Pedro rivers, 25–26
birding, 177. *See also* bird species;
   ornithology
bird migration patterns, 23–25
bird populations, quantifying, 174–77
bird species: Anhinga, 54, 79; Bell's
   Vireo, 106; Bendire's Thrasher, 53;
   Black-bellied Whistling-Duck, 54, 79,
   174, 249n70; Black-necked Stilt, 164;
   Black Phoebe, 175; Blue Grosbeak,
   106; Brown Towhee, 127; cavity
   nesting, 107–8, 107 (table 7.1), 179;
   Common Black-Hawk, 103, 175–76,
   250n98; Common Gallinule, 164;
   Common Yellowthroat, 164, 175;
   Crested Caracara, 53, 106, 128, 175–76;
   European Starling, 127; Ferruginous
   Pygmy-Owl, 53, 102, 175–76, 249n88;
   Flammulated Owl, 103, 245n61; Gray
   Hawk, 53, 102–3, 106, 128, 175–76,
   246n56, 250n98; Great-blue Heron,
   164; Great-tailed Grackle, 127; Green
   Heron, 101, 164, 175, 244n20, 249n73;
   Inca Dove, 127; Killdeer, 164, 175;
   Ladder-backed Woodpecker, 53,
   242n19; Lucy's Warbler, 106; Mallard,
   164; Marsh Wren, 164; Masked
   Bobwhite, 242n36; Mexican Spotted

bird species (*continued*)
Owl, 53–54, 176; Mourning Dove, 180; Northern Beardless-Tyrannulet, 103; Northern (Masked) Bobwhite, 54; Northern Cardinal, 127; Pacific Loon, 78–79; Painted Redstart, 53; Pied-billed Grebe, 101, 164, 175, 244n20, 249n73; Purple Gallinule, 54, 79; Pyrrhuloxia, 127; Rose-throated Becard, 103, 106, 128, 176, 250n98; Ruddy Duck, 164; Rufous-winged Sparrow, 53, 127; Scarlet Ibis, 54; Scarlet Tanager, 54; Song Sparrow, 53, 127, 175; Spotted Owl, 103, 175, 242n21, 245n61; Streak-backed Oriole, 54; Tropical Kingbird, 103; Vermilion Flycatcher, 106; Western Screech-Owl, 127; Whiskered Screech-Owl, 103, 245n61; White-winged Dove, 79–80, 104–5 (fig. 7.6), 106–7, 180; Wild Turkey, 42, 175; Willow Flycatcher, 106, 175; Yellow-billed Cuckoo, 175; Zone-tailed Hawk, 54, 175–76
bison, 28
Black Mountain, 49–50
black (Goodding) willow, 23, 103, 128, 151, 162–63 (fig. 10.8), 163
blue paloverde, 23, 162–63 (fig. 10.8), 239n123
bobcat, 108
bosque, 23–25, 96–97 (fig. 7.2), 172–77; early observations of, 42–45. *See also* Great Mesquite Forest; Komatke Thicket
Bourke, Capt. John G., 49
Brandt, Herbert, 101–3, 109–10, 128, 245n28–245n29, 249n72
bridges: and channelization, 127; construction of, 73, 152–53 (fig. 10.1), 154–55 (fig. 10.2); failure of, 90, 96–97 (fig. 7.2), 124–25 (fig. 8.8), 133, 138–39
Brown, Herbert, 54, 78, 127–28
Brown, J. K., 76
Brown, Rollin C., 73–75
Browne, J. Ross, 48
Bryan, Kirk, 30–31
Buehman, Henry, 6 (fig. 1.2)
Bulletin 17B, 131, 136, 141, 247n8
burrobrush, 23, 162–63 (fig. 10.8)

cadastral surveys, 52
California, 31, 146
California Endangered Species Act (CESA), 106
California floater (clam), 54, 176
Cameron, Alice F., 69
Camp Lowell, 48, 54, 170. *See also* Fort Lowell
Cañada del Oro, 10, 12, 24, 45, 51
canal systems: ancient, 14, 33; and CAP, 146–48

CAP water: chemistry of, 147–48; direct *vs.* recovered, 147 (fig. 9.9); recharge into Avra Valley aquifer, 148; and river restoration, 160
Carnegie Desert Botanical Laboratory (Tumamoc Hill), 79; observations, 90–91
Carothers, Steven W., 129
Carrillo, Leopoldo, 55–57, 56 (fig. 5.2), 59, 62
Carson, Mose, 45
Casa Grande ruins, 36
cattle, 28. *See also* livestock grazing
cavity nesting birds, 107–8, 107 (table 7.1), 179
census data, 113 (fig. 8.1)
Central Arizona Project (CAP), 146–48. *See also* CAP water
Central Arizona Project Association, 146–47
Central Arizona Water Conservation District, 147
centrifugal pump, 123
Chaco Canyon, 3
change, irreversible, 182
change-in-rainfall-intensity hypothesis, and arroyo downcutting, 31–32
channel aggradation, 151–53 (fig. 10.1), 159
channel change, 152–53 (fig. 10.1), 158 (fig. 10.4)–158 (fig. 10.5), 167, 171–72; and large floods, 34–35, 170–71; lateral, 133–36, 137 (fig. 9.3), 138, 145; and riparian ecosystems, 23–26; role of climate in, 8
channel downcutting, 14; caused by effluent discharge, 164. *See also* arroyo downcutting
channel erosion, 14, 33
channel filling (1996–2012), 149–65
channel incision, 32
channelization, 92, 95–100, 127, 145; and riparian vegetation, 150–51
channel management, 178–79 (fig. 11.4)
channel manipulation, in Great Mesquite Forest, 82–86
channel stabilization, 133–36
channel topography, changes in, 144–45
channel within a channel, 151
Charleston, AZ, 63
Chihuahuan biogeographic region, 26
Chillson, L. D., 65
cholera epidemic (1850–1851), 42
chubascos, 17–18
cienegas, 13, 23–24, 42, 166; and arroyo downcutting, 79; development of, 13–14; draining of, 112; at Spring Branch, 51
Civil War, 47–49
Clark, John, 52
Clarke, Asa Bennett, 40, 50

climate: and arroyo downcutting, 27, 30–31; historical data, 52–67; role in channel change, 8; variation, 14–17, 167–68
climate periods, 15–17, 167–68; late 19th century pluvial, 16, 168; early 20th century drought, 16, 76; early 20th century pluvial, 16, 168; midcentury drought (1950s drought), 16, 112–19, 126, 130, 134 (fig. 9.1), 137 (fig. 9.3), 138–40 (fig. 9.4), 146; late 20th century pluvial, 130–48; early 21st century drought, 16–17, 146, 149–50, 165
Clovis culture, 14
Colorado Plateau: and changes in arroyo morphology, 34; and drought-causes-erosion hypothesis, 31
Colorado River, as water source, 146–48
Colorado River compact, 146–48
Compact Point, AZ, 146
Congress Street, 145; from Powderhouse Hill, 114–15 (fig. 8.2)
Congress Street Bridge, 143 (fig. 9.7), 158 (fig. 10.5)
Congress Street Bridge gaging station (Santa Cruz River at Tucson), 10, 20–22, 80, 86, 95, 133, 136, 142, 145, 159
Continental, AZ, 10
Contzen, Fritz, 45, 47
Cooke, Capt. Phillip St. George, 39
Cordonnes surface, 12
Cortaro Farms, 94
Cortaro Road Bridge gaging station, 20, 150, 160
cotton, 94, 123
cottonwood, 108, 156–57 (fig. 10.3)
cottonwood-willow forests, 177
Couts, Lt. Cave J., 39
coyote willow, 23, 108, 128
creosotebush, 23
critical valley oversteepening, 29, 32
Crosscut Canal, 81–82, 82 (fig. 6.6), 83 (fig. 6.7), 84 (fig. 6.8)–84 (fig. 6.9), 94, 100
crosscut system, 81–82, 82 (fig. 6.6), 83 (fig. 6.7), 84 (fig. 6.8)–84 (fig. 6.9)
Crossin, Richard S., 127
cutoff low-pressure systems, 17
Cuver, J. P., 67
cyclones, 17–19, 126, 133, 150, 168; recurvature of, 19, 136, 168

Dalton, W. A., 62
dams and damming, 45–47 (fig. 4.5), 60, 61 (fig. 5.4), 76, 171, 241n73; Indian Dam, 94–97 (fig. 7.2); Silver Lake Dam, 68–71; Spring Branch Dam, 83, 90
Davis, W. M., 3, 237n5
Dawson, William L., 101–2, 106, 109, 244n19

deforestation, and arroyo downcutting, 29
desert broom, 23
desert river characteristics, 9–26
desert streams, myths about, 49–51
design flood, establishment of, 141
Dobyns, Henry, 103
downcutting. *See* arroyo downcutting
Driscoll, Thomas, 58, 62, 65
drought-causes-erosion hypothesis, 31
droughts, 112; and arroyo downcutting, 27, 30–31; early 20th century, 16, 76; early 21st century, 16–17, 146, 149–50, 165; effect on river flow, 123–27; historical data, 58; midcentury (1950s drought), 16, 112–19, 126, 130, 134 (fig. 9.1), 137 (fig. 9.3), 138–40 (fig. 9.4), 146
Durivage, John E., 40

Eagle Steam Flour Mill, 55
earth fissures, 119
earthquakes, 52, 63, 119
East Side Canal, 77
Eccleston, Robert, 42
ecological sink, 175, 249n88
ecosystem services, 23
ecotourism, 110, 164, 177
effluent discharge, 161 (fig. 10.7), 163–64. *See also* wastewater
El Cordonazo de San Francisco, 17–18
Elías, Tomás, 55
El Niño, 8, 16–17, 19–20, 33, 112, 136, 142, 145, 167–68, 170. *See also* ENSO
El Ojito, 58
Eloy, AZ, 10
Eloy Basin, 119
El Rancho Viejo, 43
El Rio Medio Project, 164
Emory, Maj. William, 42
ENSO (El Niño–Southern Oscillation) phenomenon, 19–20, 167–68, 170–71
entrenchment, periods of, 30
equifinality, and arroyo downcutting, 27, 35
erosion, 99; drought-causes-erosion hypothesis, 31; historical, 35; Pleistocene, 49–50. *See also* channel erosion; headcut erosion; lateral erosion; soil erosion
erosion control, 83–86, 178–79
Esconolea (Apache chief), 63
Evans, George W. B., 42
explorers, Spanish, 36–38
extreme-flood hypothesis, and arroyo downcutting, 33

facultative riparian vegetation, 23
Farmer's Canal, 80
farmlands, 80, 134 (fig. 9.1)
Federal Emergency Management Agency (FEMA), 131, 133, 141

Federal Insurance Administration, 133
Fergusson, Maj. D., 47–48, 52
fires, 54–58
Fish, E. N., 49
fish breeding, 61 (fig. 5.4)
fish species, 128; Gila chub, 54, 108; Gila topminnow, 108; longfin dace, 108; Santa Cruz pupfish, 108, 176; Sonora sucker, 108
flash flooding, 21–22, 22 (fig. 2.8), 150
flood control, 4, 181–82; and arroyo creation, 76–78
flood damage, 136–39
Flood Disaster Protection Act (1973), 131
flood events: *1886*, 62–63; *1887*, 63–65, 78; *1890*, 68–71, 170; *1891*, 75–76; *1905*, 80; *1908*, 83; *1914–1915*, 86–90, 86 (fig. 6.11), 87 (fig. 6.12), 127, 131; *1926*, 244n1; *1929*, 92; *1951*, 126; *1961*, 126; *1962*, 126; *1977*, 124–25 (fig. 8.8), 127, 132–36; *1983*, 96–97 (fig. 7.2), 126, 130, 136–42, 141 (fig. 9.5), 142 (fig. 9.6); *1993*, 126, 142–46, 143 (fig. 9.7); *2005*, 159; *2006*, 150
flood frequency, 21, 149–50; analysis, 131, 144 (fig. 9.8), 145; and floodplain management, 130–33; nonstationarity in, 143–45
flood hazard: mitigation of, 145; planning for, 113, 170–71
floodplain, 137 (fig. 9.3); development of, 138–40 (fig. 9.4), 151–55, 178–79 (fig. 11.4); formation of, 151; human impacts on, 171–72; management of, 33, 130–33, 145–46; reestablishment of, 134 (fig. 9.1), 135 (fig. 9.2); urbanization of, 121
floodplain agriculture, 112–19
floodplain legislation, federal, 130–31
flood records (Tucson), 21 (fig. 2.7)
floods and flooding, 126 (fig. 8.9), 167; and arroyo downcutting, 33; and arroyo management, 130–48; and channel change, 34–35; climatic conditions for, 17; determination of regulatory flood, 131, 141; during drought, 30, 126; economic consequences, 33; floods of record, 142; historical data, 52–67; hundred-year flood estimates, 132 (table 9.1), 143–45, 144 (fig. 9.8); large regional floods, 33, 47; late 20th century floods, 170–71; monsoonal, 17, 19, 145; and monsoonal storms, 19; patterns of, 20; periods of, 16
flour mills, 45, 54–57, 55 (fig. 5.1), 57 (fig. 5.3), 61 (fig. 5.4)
Flowing Wells Irrigation District, 94
Font, Pedro, 38
foothill paloverde, 239n123
Foreman, S. W., 49–50, 53

Fort Lowell, 53–54, 78, 102–3, 108
Fort Lowell Formation, 12
Forty-Niners, 39–42
Franciscan order, 38
Frémont cottonwood, 23, 103, 108, 128, 163
Fremont River, 240n79
Froebel, Julius, 42
frontal systems, 17
fuelwood, demand for, 76, 95. *See also* woodcutting
Fuller, P. E., 82
Fulton, James C., 77

Gadsden Purchase, 36, 41–42, 45, 241n30
gaging stations, for streamflow, 20–23, 33, 80, 136, 142, 155. *See also* Congress Street Bridge gaging station; Cortaro Road Bridge gaging station
garbage disposal, 112, 119–21
geographic cycle, 3
geologic history, of Tucson Basin reach, 10–12
geomorphic surfaces, in Tucson Basin, mapping of, 12
geomorphic thresholds concept, 27
George, William G., 127–28
Gila River, 4, 10, 14, 25, 40–41, 54, 133, 160
Gila River Indian Reservation, 25
Gilbert, G. K., 3
gold panning, 45
Gold Rush, 40–42
Gonzales, Ignacio Elías, 39
Gould, Patrick J., 127
grade-control structures, 136, 145, 158 (fig. 10.5)
graded stream, 3
Grand Canyon, 180
Grant, William S., 45
graythorn, 103
Great Drought, 31
Great Mesquite Forest, 24–25, 38, 44 (fig. 4.4), 53–54, 78–80, 109 (fig. 7.7), 123, 145, 164–65, 172–75, 173 (fig. 11.3), 177, 241n31, 244n19, 244n21; aerial photography, 101 (fig. 7.5), 124–25 (fig. 8.8), 128; channel manipulation in, 82–86; decimation of, 76; decline and demise of, 108–11, 127–29; early observations of, 42–45; historical data (1916–1942), 92–111; non-avian fauna, 108; and ornithology, 100–111, 127–28; as outdoor laboratory, 180–81; vegetation of, 103
Greene, Col. William C., 82
Greene's Canal, 10, 82, 85 (fig. 6.10), 89–90, 100, 133, 140–41, 142 (fig. 9.6), 167, 172
Green Valley, 51
Gregg, Judge, 62

Gregory, Herbert, 3
Grossetta, S. W., 76
groundwater, 146; aquifer, 12; dependence on, 121–23; development of, 58–60, 77–78, 94; discharge of, 13; early extraction of, 76; reduction of usage, 146–48; and riparian vegetation, 23; withdrawal of, 111, 119, 174
groundwater levels, 31, 100 (fig. 7.4); decreasing, 67, 79–82, 119, 123, 129, 146, 172–75, 173 (fig. 11.2); rising, 148
groundwater overdraft, 96–97 (fig. 7.2)

Hack, John, 31
Hartt, W. A., 76
Hassayampa River, 23
Hastings, J. Rodney, 129
Hayes, Judge Benjamin, 42
headcut erosion, 69–71
headcut migration, 140–41, 142 (fig. 9.6), 171–72
Hereford, Richard, 33–34
highway construction (I-19), 127, 129
historical portrait, of Santa Cruz River basin (1691–1872), 36–51
historical sources, use of, 7
Hohokam, 14, 51; canal system, 14, 33; land-use practices, 29
Holocene, 7, 12–14, 13 (fig. 2.3), 31, 35, 182
Holocene alluvium, in Tucson Basin, 12
Howard, O. W., 79
Howell, A. Brazier, 102, 106
Huff, William P., 241n22
Hughes, Sam, 45, 47, 49, 65, 66 (fig. 5.7), 68–71, 73, 171. See also Sam Hughes's Ditch
human occupation, early evidence of, 14
Huntington, Ellsworth, 30, 88–89 (fig. 6.13)
hurricane frequency, 169 (fig. 11.1)
hurricanes, 18–19; Heather (1977), 133; Nora (1997), 150
hydroclimatology, 14–20
hydrograph, of flash flood, 21–22, 22 (fig. 2.8)
hydrologic data collection, 20–23
hydrologic effects, of 1887 earthquake, 63, 64 (fig. 5.5)
hydroriparian ecosystem, 23

ice house, 56 (fig. 5.2)
Ina Road Wastewater Reclamation Facility, 10, 160, 161 (fig. 10.7)
intercept ditch, 65, 65 (fig. 5.6), 66 (fig. 5.7), 68–71, 171
intrinsic geomorphic factors, and arroyo downcutting, 32–33
intrinsic responses concept, 27

irrigation, 52, 65–67, 76, 80–82, 99, 129, 171, 174; water sources, 21
irrigation canals, 55, 66 (fig. 5.7); and flood patterns, 50; and flow velocity, 29; historical evidence, 45. See also canal systems

Jackass Mail Route, 45
javelina, 108
Jaynes terrace, 12
Jeffords, Thomas J., 58–59
Jesuit order, 38
Johnson, President Lyndon B., 147
Johnson, R. Roy, 127, 129
Jones, Percy, 82

Kearny, Col. Stephen W., 39
Kino, Eusebio, 7, 9, 14, 24, 36, 38, 50, 172
Kitt, Edith Stratton, 75
Knapp, F. C., 132
Knapp, F. H., 99–100
Komatke, AZ, 10
Komatke Thicket (New York Thicket), 25, 106, 239n151, 245n82

Lake Havasu, AZ, 147
landfill, 70–71 (fig. 6.1), 118 (fig. 8.5), 119–21 (fig. 8.6), 122 (fig. 8.7)
land surveys, 52–53
land-tenure conflicts, historical, 42
land-use patterns, 42, 171–72; historical data, 52–67
land-use practices: and arroyo downcutting, 29–30; and flooding, 167
La Niña, 19–20, 149, 167, 170
lateral aggradation, 151
lateral channel change, 133–36, 137 (fig. 9.3), 138, 145
lateral erosion, 139–40
Laveen, AZ, 10
Leatherwood, Robert N., 59
Leedy, C. F., 50
León, Cirilio Solano, 75
León, Manuel de, 38–40
Leopold, Luna, 31
Levin, Alex "Boss," 69
litigation, water rights, 60–62
Little Ice Age, 170
livestock data, 113 (fig. 8.1)
livestock grazing, 4, 8, 29, 167; and arroyo downcutting, 28–29; historical patterns, 28–29, 58
Lochiel, AZ, 9, 69
log-Pearson type III distribution, used for flood-frequency analysis, 131, 133, 144 (fig. 9.8), 247n8
Lovell, Hereford, 60
lowland leopard frog, 176, 250n108
Lucero, Carmen, 50
Lumholtz, Carl, 110

MacDougal, D. T., 90–91
Maish, Fred, 58, 62, 65, 69
malaria, 48–49, 112
Manje, Captain Juan Mateo, 36, 38
Manning, Levi H., 77, 81 (fig. 6.5), 94
Manning Ditch, 80, 81 (fig. 6.5)
"many possible causes" argument, 27
Marshall, Joe T., 110, 127
Martinez, José María, 40, 50
Martinez Hill, 10, 36, 49–51, 124–25 (fig. 8.8)
McKay, Alex, 55
Mearns, Dr. Edgar A., 78, 128
mesophytic forest, 90
mesoriparian ecosystems, 23
mesquite, 23, 76, 103, 116 (fig. 8.3), 127, 162–63 (fig. 10.8), 177
mesquite mouse, 108
Mexican elder, 103, 128
Mexican garter snake, 108
Mexican Independence, 38
Mexican period, 38–39
Mexican War, 39–40
Mexico, 146
Midvale Farms, 94
military occupation, during and after Civil War, 47–49
milling, 45, 47, 54–57, 55 (fig. 5.1)
mining, of sand and gravel, 112, 119–21, 138–40 (fig. 9.4)
Mission San Xavier, 9, 42, 48–50, 63
mixed-population analysis, 145
Moeur, Gov. Benjamin B., 146
Mogollon Rim, 25
Monson, Gale, 110
monsoon, 63, 150, 167; and arroyo downcutting, 31
monsoonal frontal systems, 19
monsoonal precipitation, 17
monsoonal storms, 19, 168, 242n48
Mormons: McGee colony, 94; Mormon Battalion, 39
mosquito abatement, 165
mosquito habitat, enhancement of, 165
multipurpose channels, 181
muskrat, 50, 176

National Flood Insurance Act (1968), 130–31
National Flood Insurance Program, 141
natural history, early observations, 53–54
negative evidence, reliability of, 7
netleaf hackberry, 23, 103, 128
New Mexico Territory, and Civil War, 47–49
newspapers, as evidence, 7
Nine Mile Water Hole, 39, 49, 51
Noble Savage concept, 29–30
Nogales, 9

Nogales International Wastewater
    Treatment Plant, 9, 175
nonstationarity, 131, 143–45, 170–71
northern Mexican garter-snake, 176,
    250n108
North Pacific Ocean, storms in, 18–19,
    19 (fig. 2.6)

Oberholser, Harry C., 128
obligate riparian vegetation, 23–24, 151,
    172–75
ornate tree lizard, 176
ornithology, 4, 23–25, 53–54, 78–80,
    100–111, 127–28, 164, 174–76, 179–81,
    244n19–244n20, 249n70, 249n72–
    249n73, 249n88, 250n98–250n99.
    See also bird species
Oro Valley, 12
Ortíz, Tomás and Ignacio, 38–40, 58
overbank inundation, 131, 138
overgrazing, 27–29
Overland Stage, 47

Pacific Decadal Oscillation (PDO), 169
    (fig. 11.1), 171
Pacific Ocean, effects on climate of
    Arizona, 17–20
paleoarroyo downcutting, 14, 28
paleoarroyos, 50–51
palynological evidence, 31
Pantano Formation, 12
Pantano Wash, 150
Paria River, 23
Parke, Lt. John G., 42
Parker, AZ, 146
Parker, J. W., 59
partial duration series, 126
Paseo de las Iglesias, 164
perennial flow, 108, 160, 165; end of, 123,
    127, 174; historical data, 36–51;
    restoration of, 141
perennial vegetation, decimation of, 128
phenology, 24
Phillips, Allan R., 102, 110
Phoenix metropolitan area, 119; and
    CAP water, 147
photographs, as archival evidence, 7–8,
    29, 129
photography. *See* aerial photography;
    repeat photography
phreatophyte, 179, 250n124
Phy, Joseph, 58
Pima County, AZ, 12, 52, 110, 113,
    132–33, 141–42, 146, 148–49, 164, 167,
    177; Board of Supervisors, 53;
    floodplain management ordinance
    (1974), 132; Sheriff's Department, 143
Pima Farms Company, 94
Pima Indians, 25, 38, 166
Pineapple Express, 17
Pioneer Mill, 55

placer mining, and arroyo
    downcutting, 29
pluvials: late 19th century, 16, 168; early
    20th century, 16, 168; late 20th
    century, 130–48, 168
point bars, 151
Polk, President James K., 40
population growth, 52, 58–60, 67, 112–13,
    113 (fig. 8.1), 114–15 (fig. 8.2), 149, 160
Post, Edwin R., 94
Post Project, 94
Powell, H. M. T., 40
Powell, John Wesley, 3
precipitation patterns, 14–20; and
    arroyo downcutting, 31–32; and
    ENSO, 168; heavy *vs.* light rains,
    168–70
precipitation records, 14–17, 15 (fig. 2.4),
    48, 167–68
presettlement conditions, 8, 166
pseudoriparian ecosystems, 23
pump-well technology, 76
Punta de Agua, 48, 50, 123
Punta de los Llanos, 38

Quinton, J. H., 83

radiocarbon dating, of floods and
    channel change, 240n69
railroad, 62, 112
rainfall-runoff model, 141
Rancho Punta de Agua, 45, 47
reclaimed water, increasing value of,
    164–65
reclamation disclimax, 180
recreation, 110, 177. *See also* ecotourism
regional water conservation plan, 148
regulatory flood, establishment of,
    131, 141
repeat photography, 129, 174, 247n71;
    Tucson, 7–8
requiem, 182
resorts, development of, 59
revetment, 98–99 (fig. 7.3), 135 (fig. 9.2)
Rillito (town), 133
Rillito Peak, views from, 88–89 (fig. 6.13)
Rillito phase, 14
Rillito River, 4, 10, 23–24; and
    agriculture, 78; annual flow volume,
    20–21; avifauna, 25, 54, 102, 174–75,
    180; confluence with Santa Cruz
    River, 162–63 (fig. 10.8); earliest
    description of, 49; early natural
    history observations, 53; floods, 49,
    63, 68, 71, 77, 92, 136, 142, 150, 170,
    181; naming of, 237n8; riparian
    ecosystems, 108, 176; riparian
    vegetation, 25
Rincon Creek, 150
Rincon Mountains, 12
Rincon phase, 14

Rio Grande (upper), and historic
    livestock grazing, 28–29
Rio Nuevo redevelopment project,
    114–15 (fig. 8.2), 133–36
riparian ecosystems, 161–63 (fig. 10.8),
    166, 172–75; and channel change,
    23–26; and impact of nonnative
    species, 177–80; long-term change in,
    24–25; regional significance of, 25–26
riparian marshes. *See* cienegas
riparian vegetation, 10, 108, 113, 123–25
    (fig. 8.8), 138–40 (fig. 9.4), 148, 151,
    156–57 (fig. 10.3); decline of, 162–63
    (fig. 10.8); observations and
    descriptions, 23, 51–52, 90; return of,
    150–51, 160–65. *See also species names*
riparian woodlands. *See* bosque
riverine change, problem of, 3–8
river restoration efforts, 160–64, 166;
    future of, 181–82; historical data
    (1996–2012), 149–65
Roger Road Wastewater Reclamation
    Facility, 10, 160, 161 (fig. 10.7)
Romero, Juan, 45
Roskruge, George, 69–71
Rowlett, Alfred, 45–47 (fig. 4.5)
Rowlett, William, 45–47 (fig. 4.5)
rubber, 94

Sabino Canyon, 150
Sabino Canyon Ranch, 59
Salpointe, Bishop J. B., 55
saltcedar, 51, 98–99 (fig. 7.3), 156–57
    (fig. 10.3), 162–63 (fig. 10.8), 177–80,
    241n88; and bird habitat, 179–80
Salt River, 14, 23, 25, 51, 78, 128, 160,
    175, 179
Sam Hughes's Ditch, 74–75 (fig. 6.4), 171
San Agustín Mission, 57 (fig. 5.3)
Sanders, Adam, 58
San Ignacio de la Canoa Land Grant,
    38–39, 51, 58, 81, 94
San Pedro River, 4, 7, 25, 133
Santa Catalina Mountains, 12, 150
Santa Cruz arroyo, 73 (fig. 6.3)
Santa Cruz Flats, 10
Santa Cruz Reservoir Project, 80–82
Santa Cruz River, figures and tables:
    aerial photographs, 124–25 (fig. 8.8),
    138–40 (fig. 9.4), 173 (fig. 11.3); at
    confluence with Rillito River, 162–63
    (fig. 10.8); at Congress Street Bridge,
    120–21 (fig. 8.6), 138–40 (fig. 9.4), 152–53
    (fig. 10.1), 154–55 (fig. 10.2), 156–57 (fig.
    10.3); drainage basin, 5 (fig. 1.1); at
    Drexel Road ford, 137 (fig. 9.3);
    Holocene stratigraphy of, 13 (fig. 2.3);
    at Irvington Road Bridge, 178–79 (fig.
    11.4); map, 10 (fig. 2.1); from Martinez
    Hill, 96–97 (fig. 7.2), 104–5 (fig. 7.6),
    173 (fig. 11.3); from A Mountain,

Santa Cruz River, figures and tables
(*continued*)
134 (fig. 9.1); near 22nd Street, 98–99
(fig. 7.3); at Silverlake Road, 135 (fig.
9.2); at St. Mary's Road, 70–71 (fig. 6.1),
72 (fig. 6.2), 93 (fig. 7.1); at Valencia
Road Bridge, 178–79 (fig. 11.4); view
(north of 22nd St. in Tucson), 11 (fig.
2.2); West Branch confluence, 74–75
(fig. 6.4)
Santa Cruz River Park, 156–57 (fig. 10.3)
Santa Cruz River valley, 57 (fig. 5.3)
Santa Cruz Valley, 6 (fig. 1.2), 37
(fig. 4.1)
Santa Rita Experimental Range, 32
Santa Rita Mountains, 12
San Xavier del Bac, 36, 38–39
San Xavier Indian Reservation, 10,
144–45; historical maps of, 43 (fig. 4.3),
44 (fig. 4.4); as source of wood, 76–77;
water development, 94–95
San Xavier Indian Reservation Riparian
Restoration Project, 164
Schmutzdecke, 248n23
Schwalen, Harold C., 95, 123
screwbean mesquite, 128
sediment production, and livestock
grazing, 28–29
sediment transport, 22–23, 30
sediment trapping, and riparian
vegetation, 151
seep willow, 23, 108
semiarid watershed, 26
semiriparian ecosystems, 23
Sentinel Peak (A Mountain), 24, 50, 55
(fig. 5.1), 60, 116 (fig. 8.3), 117 (fig. 8.4),
118 (fig. 8.5), 122 (fig. 8.7)
setbacks, 247n16
settlers, Anglo-European, 4, 36–51
Shaw, Richard J., 123
Shultz, Mrs. T. L., 62
Sierrita Mountains, 12
Silver Lake, 45–47 (fig. 4.5), 59, 68–71
Silver Lake resort, 59, 69
sinkholes, 119
Sky Islands, 24, 53, 180
Smith, C. K., 95
Smith, George E. P., 123
Snaketown AZ, 14
soapberry, 128
Sobaipuri, 4, 9, 14, 29, 36
soil cement, 46–47 (fig. 4.5), 70–71 (fig.
6.1), 72 (fig. 6.2), 74–75 (fig. 6.4), 98–99
(fig. 7.3), 116 (fig. 8.3), 117 (fig. 8.4), 118
(fig. 8.5), 120–21 (fig. 8.6), 130, 133–36,
134 (fig. 9.1), 139–40, 145, 154–55
(fig. 10.2), 156–57 (fig. 10.3), 162–63
(fig. 10.8), 178–79 (fig. 11.4), 247n1
soil deposition, and cyclical
drought, 30
soil erosion, 28, 30

Sonoita Creek, 9
Sonora, Mexico, 9
Sonora mud turtle, 176, 250n108
Sonoran biogeographic region, 26
Sonoran spiny lizard, 176
Southern Oscillation. *See* ENSO
(El Niño–Southern Oscillation)
phenomenon
Southern Oscillation Index (SOI), 168
Spalding, Volney, 73, 90, 101
Spanish colonial period, 36–38
Spanish colonists, 4; and alteration of
watercourses, 24
Spring, John, 48
Spring Branch, 14, 40, 42, 50–51, 82–83,
96–97 (fig. 7.2)
stage, 155
stage-discharge relations, 155–59, 159
(fig. 10.6)
stage-rating curves, 160 (table 10.1)
stationarity, 130–31
statistical methods, for flood-frequency
analysis, 131
Stephens, Frank, 54, 79, 128
St. George, UT, 51
St. Mary's Road headcut, 66 (fig. 5.8)
stopover habitat, 24–25
streamside margin, 151
Struby's Ranch, 48
subsidence, 119, 146
surface water usage, 148
Swarth, Harry S., 78–80, 100–101, 106,
244n19
Sweetwater Wetlands, 161–65, 174–75,
180–81
sycamore, 23
Sycamore Creek, 25
Sykes, Glenton, 78
Sykes, Godfrey, 91

Taliefero, T. W., 45
tamarisk. *See* saltcedar
tamarisk beetle, 180
Tanque Verde Creek, 150
Tanque Verde phase, 14
teleconnections, 20, 167
temperature records, 15, 48
teraque, 241n16
thalweg, 65 (fig. 5.6), 151
thunderstorms, 17, 19
Tinaja beds, 12
Tohono O'odham, 4, 36, 40, 45, 47, 54;
and fuelwood, 95; land-use practices,
29; and water development, 94
Tohono O'odham Reservation, 119, 128;
San Xavier District, 45
Treaty of Guadalupe Hidalgo, 40–41
tree planting projects, 177
Tres Alamos, 65
trigger-pull analogy, for arroyo
downcutting, 30

tropical storms, 17–19; Claudia (1962),
126; Norma (1970), 126; Octave (1983),
136–42
Tubac, AZ, 9, 12, 38
Tucson, AZ, 4, 6 (fig. 1.2), 10, 117 (fig. 8.4),
118 (fig. 8.5); and CAP water, 147; and
cholera epidemic of 1850–1851, 42;
communication systems, 54;
development of water supply, 58–60;
downtown revitalization, 133–36;
earliest map of (1862), 52; early
business activities, 45; early surveys
of, 52–53; earthquake damage, 63;
emergency management agency, 143;
and Gadsden Purchase, 45; historical
data (1943–1975), 112–29; hundred-
year flood estimates, 132 (table 9.1);
and land subsidence, 119; and
land-tenure conflicts, 42; as
ornithological capital, 25, 102; and
subsidence problem, 146; temperature
records, 15; transportation systems,
62; and wastewater, 160–64; water-
distribution system, 123. *See also* Pima
County, AZ; population growth
Tucson Active Management Area,
248n21
Tucson Audubon Society, 164
Tucson Basin, 10–12
Tucson Farms Company, 80–83, 86, 89,
94; crosscut system, 81–82, 82 (fig. 6.6),
83 (fig. 6.7), 84 (fig. 6.8)–84 (fig. 6.9),
94, 100
Tucson gaging station. *See* Congress
Street Bridge gaging station
Tucson International Airport,
precipitation records, 14
Tucson metropolitan area, and
channelization, 145–46
Tucson Mountains, 12
Tucson Reclaimed Water Treatment
Plant (RWTP), 148, 160
Tucson University of Arizona,
precipitation records, 15 (fig. 2.4)
Tucson Water Company, 60, 67, 76–77,
80, 94, 165; water deliveries (1899–
2011), 147 (fig. 9.9)
Tumacácori, 36
Tumamoc Hill, 50
turbine pump, 123
Turner, Raymond M., 129

University of Arizona, 180; precipitation
records, 14
Upper Sonoran Desert, precipitation
records, 15, 16 (fig. 2.5)
US Army Corps of Engineers,
130, 164
US Bureau of Reclamation, 146–47
US General Land Office (GLO), 7, 52
US Geological Survey, 123, 136, 142

US Senate, and Tucson water suppy, 95
US Soil Conservation Service, 95, 99, 101
US War Department, flood-control
    hearings (1937), 100

Valencia Road headcut, 50–51
valley aggradation, 12–13
velvet mesquite, 23. 163
Verde River, 25
vertebrates, non-avian, 108, 176. *See also
    species names*
vertical aggradation, 151
Vorhies, C. T., 101, 244n20

wall of water, 21–22
Warner, Solomon, 45, 47, 50, 54–57, 55
    (fig. 5.1), 56 (fig. 5.2), 57 (fig. 5.3), 59–60,
    61 (fig. 5.4), 122 (fig. 8.7)
Warner's Lake, 60, 61 (fig. 5.4), 74–75
    (fig. 6.4)
Wasson, John, 53
wastewater, 88–89 (fig. 6.13); discharge,
    161 (fig. 10.7), 163–64; reclamation,

148, 160–63 (fig. 10.8), 164–65; and
    Sweetwater Wetlands, 161–64
wastewater treatment plants, 160, 175,
    248n17
water conservation, 148, 165
water development, 112–19; Colorado
    River as water source, 146–48;
    historical data (1916–1942), 92–111; on
    San Xavier Reservation, 94–95
water distribution system, 65–67, 123
water diversion, 174
water harvesting, 164
water rights, 60–62, 65–67, 242n62; case
    of 1885, 49
watershed, description of, 9
water supply, 54–60
water use: Arizona statewide, 146;
    restrictions, 76
Watts, J. R., 67, 77
Watts, Sylvester, 59
Way, Phocion, 45
well drilling, 112, 119, 121–23
well pumping, optimization of, 148

wells, 94; artesian (flowing wells), 58,
    242n49; expansion of, 77–78, 81–82;
    powered by jet-driven pumps, 99;
    use of, 76–78
West, Col. Joseph R., 47
West Branch, 60, 74–75 (fig. 6.4)
wetlands. *See* cienegas
Wetmore, E. L., 69
White, Theodore, 7, 50, 53
white-tailed deer, 108
Wilbur, R. A., 54
Willard, F. C., 78–80, 101–2, 109, 128
willow, 23
woodcutting, 108–9, 127, 172
Works Progress Administration (WPA),
    95, 98–99 (fig. 7.3), 127
World War I, 92–94
World War II, 123

xeroriparian vegetation, 23, 151, 172, 174,
    177, 182

Zuñiga, José de, 38

About the Authors

Robert H. Webb is a hydrologist and plant ecologist retired from the US Geological Survey, and he is currently an adjunct professor at the University of Arizona. He is the author or editor of fourteen books, including *Grand Canyon: A Century of Change*, *The Ribbon of Green*, and *Repeat Photography*, and more than two hundred scientific articles. He maintains the largest archive of repeat photography in the world.

Julio L. Betancourt is a geoscientist with the National Research Program, Water Mission Area, US Geological Survey, and is an adjunct professor at the University of Arizona. He is the editor of *Packrat Middens: The Last 40,000 Years of Biotic Change*, and has published more than 150 scientific articles. He co-founded the Southern Arizona Buffelgrass Coordination Center and the USA National Phenology Network.

R. Roy Johnson is an ornithologist who is retired both as a senior research scientist for the National Park Service and professor in the School of Natural Resources at the University of Arizona. He is a co-author of *Grand Canyon Birds*, and has published more than two hundred scientific papers.

Raymond M. Turner is a retired plant ecologist, US Geological Survey, and is an emeritus professor at the University of Arizona. He has been studying the plant ecology of the desert regions of Arizona and Mexico since 1954. He is the author or editor of several books, including *Kenya's Changing Landscape*, *The Changing Mile*, and *The Changing Mile Revisited*, as well as numerous scientific articles.